INSECTS

and Sustainability of Ecosystem Services

Social-Environmental Sustainability Series

Series Editor

Chris Maser

Published Titles

INSECTS

and Sustainability of Ecosystem Services

Timothy D. Schowalter

CRC Press
Taylor & Francis Group
Boca Raton London New York

CRC Press is an imprint of the
Taylor & Francis Group, an **informa** business

CRC Press
Taylor & Francis Group
6000 Broken Sound Parkway NW, Suite 300
Boca Raton, FL 33487-2742

First issued in paperback 2019

© 2013 by Taylor & Francis Group, LLC
CRC Press is an imprint of Taylor & Francis Group, an Informa business

No claim to original U.S. Government works

ISBN-13: 978-1-4665-5390-3 (hbk)
ISBN-13: 978-0-367-86728-7 (pbk)

Visit the Taylor & Francis Web site at
http://www.taylorandfrancis.com

and the CRC Press Web site at
http://www.crcpress.com

Contents

Series Preface

In this book about the role of insects as components of the biophysical services of nature that humans rely on, Dr. Schowalter examines not only the various ecosystem functions provided by insects but also our human perceptions of their respective values. In the context of general human perceptions, it needs to be understood that, since biblical times, most insects that interfere in one way or another with the plants humans value for our own uses have been considered to have only negative effects on the resource and so are thought of as "pests." However, insects are not considered pests—if they are noticed at all by the lay populous—when they feed on plants for which we find no social or economic value.

The term *pest* reflects this traditional bias and the perceived necessity of *always having to battle insects* for control of the resources that humans value as commodities or for the maintenance of our physical health. Only within the past three decades or so has evidence become sufficiently available to show that many of the so-called "insect pests"—like all other species—enrich the world and in the process provide largely unrecognized benefits. Dr. Schowalter has been a pioneer in raising the level of consciousness in science, forestry, and agriculture with respect to the beneficial contributions insects make to our overall socio-environmental well-being.

As Dr. Schowalter points out in this book, insects are critical pollinators of our food crops and medicinal plants and are essential in breaking down and recycling the nutrient resources in dead plants and animal waste, thereby allowing this organic matter to be reused in the ecosystem. In addition, insects are important sources of food in many cultures and are the primary food for numerous commercial fisheries and game animals. And this says nothing of their significance as cultural icons, such as Egyptian scarabs and oriental crickets, or their vital nature as regulatory instruments in ecosystems wherein plant production is nearing the environmental carrying capacity. Finally, some medicinal and industrial products benefit from the existence of certain insects as part of their ingredients—all of which are elucidated within the pages of the book you are holding.

Chris Maser, Series Editor

Preface

Humans depend on a variety of services provided by ecosystems, including food, pharmaceuticals, building materials and fuel, clean water, spiritual retreat, carbon sequestration, and climate moderation. Many, perhaps most, people, especially in North America, do not know where their food, wood products, or fresh water originate, beyond their market or faucet, and do not care as long as the supply is sufficient and inexpensive. Factors that affect the sustainable supply of these services have the capacity to cause famine and disease epidemics. Insects are primary factors that affect all of these ecosystem services in a variety of ways, positive and negative. How we deal with insects, as much as how we manage ecosystems, will determine the capacity of ecosystems to continue to supply those services.

This book challenges traditional perceptions of the value of insects in our lives. With few exceptions, insects are perceived in industrialized countries as undesirable pests. In reality, relatively few insects interfere with humans or our resources. Most have benign or positive effects on ecosystem services, and many represent useful resources in nonindustrialized countries. Some insect groups are critical to sustained supply of ecosystem services.

Although some insects will continue to be viewed as undesirable, believing that insects should, or can, be "managed" is unrealistic for reasons that are addressed in this book. Policies and decisions to suppress insect populations should be based on the multiple, often compensatory, effects of insects on various resources or ecosystem services and on the consequences of control tactics for those resources or services. This book reviews aspects of insect physiology, behavior, and ecology that affect their interactions with other ecosystem components and ecosystem services, emphasizes critical effects of insects on the sustainability of ecosystem processes and services, and integrates short- and long-term effects of insects on multiple ecosystem services, as well as the consequences of current management practices and environmental policies for insect populations and ecosystem services. Most importantly, this volume recommends changes in perspectives and policies regarding insects that will contribute to sustainability of ecosystem services. This book is intended for anyone interested in developing more sustainable approaches to "managing" insects and their effects on natural resources and ecosystem services, including amateur entomologists, conservation practitioners, environmentalists, as well as natural resource managers, land use planners, and environmental policymakers.

I bring a unique background and perspective to the preparation of this book. First, unlike most modern urbanites, I have worked on farms and ranches, driven tractors and combines, loaded wheat into silos and repaired equipment, and have worked in the urban offices of a farmers' cooperative.

I am familiar with the long hours and risks that characterize farming and am aware of the deep ecological knowledge of many family farmers and ranchers and the tradeoffs they make between good land stewardship as a legacy to their heirs and making a sufficient profit to maintain the farm or ranch. I have witnessed the shift from family farms to corporate farms, managed by local labor but directed from urban offices for maximum profit for owners and investors, with intensified agricultural use and increased insect problems.

Second, I have pursued my career as a professional entomologist at Land Grant universities. Land Grant entomologists are trained as biologists but, unlike most biologists, have an explicit mission to improve the efficiency and productivity of agricultural and silvicultural practices and the health of rural communities. Although Land Grant universities traditionally have served agricultural and rural constituents, insects have universal concern, and entomologists increasingly have addressed urban issues involving structural and household pests.

Third, I am trained in both insect ecology and ecosystem ecology, a combination shared by few ecologists. I began collecting insects and appreciating their diversity as a child. My early graduate work was part of the International Biological Programme, a pioneering multidisciplinary, multinational program to investigate ecosystem processes (particular energy fluxes) that drive ecosystem development. I subsequently addressed effects of insect herbivores and detritivores on nutrient cycling processes and how consequent changes in ecosystem conditions select for adaptation to environmental changes. This work has been synthesized in the textbook *Insect Ecology: An Ecosystem Approach*, which is now in its third edition.

During my lifetime, I have seen the effects of unsustainable use of ecosystem services. The coastal swamp forests of the Gulf Coast have been overharvested and channeled for oil exploration, allowing storm surge during hurricanes to move far inland without hindrance and become an increasingly serious threat to coastal communities. The fertile grasslands of midwestern North America, highly valued for agriculture, lost much topsoil during the great Dust Bowl years; its major aquifers were depleted by unsustainable agricultural practices. The arid grasslands of the Southwest have become shrub lands because of overgrazing. This is demonstrated by the high-quality remnant grasslands that have never been grazed by livestock and are separated only by a fence line from overgrazed desert shrub land. Since 1900, the Pacific Northwest has seen a shift from a landscape that was 75% old-growth forest to a landscape that now is 75% young plantation forests that are more vulnerable to a variety of insects and pathogens and less capable of reaching sufficient age to sustain production of decay-resistant heartwood.

Finally, as my career has progressed, I have had opportunities, rare among scientists, to become engaged in development of environmental policy. I have been a program director for ecosystem studies at the National Science

Foundation—a position in which I helped to direct science policy. I served on a National Research Council Committee for Environmental Issues in Pacific Northwest Forest Management (report published by the National Academy of Sciences in 2000). I was elected Ecological Society of America's Vice President for Public Affairs (1999–2002) and have provided invited testimony on insect and forest health legislation to congressional leaders, governors, and state and federal committees.

This broad background has instilled a perspective that combines empathy for people struggling with insect/pest problems with an appreciation for the amazing diversity, adaptive ability, and natural roles of insects. Environmental policy must be based on solid science in support of sustainable use of ecosystem resources and services, as well as on social and economic needs. Clearly, we need environmental policies that meet needs for pest control where warranted but do not undermine the important contributions of insects to sustaining ecosystem processes and services.

I am grateful to a large number of colleagues around the world with whom I have had the privilege to work and who have provided valuable data and perspective over several decades, including my daughter, Dr. Shannon Heuberger, also an entomologist and former Congressional Science Fellow. I have been a colleague of series editor Chris Maser for many years and share his enthusiasm for encouraging a broad audience to improve our environmental management strategies. I am delighted and honored to have the privilege of developing this book. In addition to Chris, I am indebted to Richard Pouyat for providing helpful comments on the manuscript, Taylor & Francis editors Irma Shagla-Britton and Jennifer Stair, Senior Project Coordinator Kari Budyk, Cenveo Publishing Services Senior Project Manager Marc Johnston, and my wife, Catherine Schowalter, for support, ideas, and encouragement in the preparation of this book.

Author

Timothy D. Schowalter has been a professor and head of the department of entomology at Louisiana State University Agricultural Center since 2003. He received his bachelor's degree (biology and anthropology) from Wichita State University (1974), master's degree (biology) from New Mexico State University (1976), and his doctoral degree (entomology) from the University of Georgia (1979). He was a postdoctoral fellow in the department of entomology at Texas A&M University before moving to Oregon State University as assistant professor of entomology in 1981. He was promoted to professor in 1993 and was interim head (2001–2003) prior to moving to LSU. His research and teaching expertise are in forest entomology and insect ecology, where he focuses on the effects of arthropod herbivores and detritivores on primary production, decomposition, and biogeochemical fluxes and on arthropod responses to natural and anthropogenic disturbances and environmental change. He has published a textbook, *Insect Ecology: An Ecosystem Approach*, now in its third edition, in addition to 70 peer-reviewed journal articles, 3 invited reviews, 14 books and book chapters, and additional symposia proceedings and other publications. He was the program director for ecosystem studies at the National Science Foundation (1992–1993), a panel member for the USDA NRICGP Forest/Range/Crop/Aquatic Ecosystems Program (1996), a member of the National Research Council Committee on Environmental Issues in Pacific Northwest Forest Management (1995–2000; report published in 2000), a member of the Oregon governor's task force on forest health (1995–2003), and the vice president for public affairs for the Ecological Society of America (1999–2002). He is currently a member of the editorial boards for *Frontiers in Ecology and the Environment* and *Population Ecology* and the Louisiana Structural Pest Control Commission. He was recently named a fellow of the Ecological Society of America.

1

Introduction

Does the flap of a butterfly's wings in Brazil set off a tornado in Texas?

Edward Lorenz (1993)

The year is 1973, and the fledgling U.S. Environmental Protection Agency (EPA), created in 1970 to enforce environmental protection and conservation, is under intense pressure to approve renewed use of dichlorodiphenyltrichloroethane (DDT) to control a Douglas-fir tussock moth (*Orgyia pseudotsugata*) outbreak in the Pacific Northwest (PNW). The EPA had cancelled use of DDT in the United States in 1972—a decade after Rachel Carson's popular book *Silent Spring* (1962) raised public awareness of its accumulation in food chains and its detrimental effects on bird populations. In 1974, appeals by PNW foresters were joined by crop producers who hoped to have the ban repealed, citing their belief that DDT was necessary to control the tussock moth and crop insects. However, the case against DDT was based as much on adaptation by target insects that rendered it less effective for control as on its nontarget environmental effects (e.g., Roussel and Clower 1957).

The EPA finally bowed to this pressure, issuing an emergency authorization in February 1974, based on the apparent lack of practical alternatives for control of an outbreak perceived as threatening to PNW forestry (Stark 1978). However, as part of its authorization, the EPA mandated that research on alternative methods of control be intensified and that replicated experimental plots be established to demonstrate the efficacy of, and need for, DDT (Brookes et al. 1978). Following application in 1974, tussock moth populations declined in all experimental plots, regardless of DDT treatment, leading to recognition that nuclear polyhedrosis virus (NPV), *Baculovirus* spp., naturally ends tussock moth outbreaks in three to four years. Scientists had known since 1965 that NPV caused high mortality in tussock moth populations, but its importance had been masked previously by chemical control programs that prevented development of epizootics (a sharp increase in incidence of a disease in an animal population) (Thompson 1978). As a result, aerial application of technical grade NPV became the preferred means of control for incipient outbreaks of this insect.

Modern humans, particularly in industrialized societies, think of themselves as separated from, and in control of, nature. Many, perhaps most, people in industrialized societies have no concept of where their food, wood products, or fresh water originate, beyond their local market or faucet, and do not care as long as the supply is adequate and inexpensive. Relatively few have experience working on farms. In fact, finding enough people willing to do such work has become a major challenge for food production. Few people are willing to work the long hours and take the high risks that are required for farming. Even if environmental conditions cooperate and crop yield is good, prices for produce may be low. Without subsidies, most farmers would not be able to continue to produce crops. It is small wonder that their children have moved to cities for jobs with better pay, better work hours, and better benefits. Rural communities are universally poorer than are the urban communities that they support (Rural Sociological Society Task Force on Persistent Rural Poverty 1993; Christensen et al. 2000).

Nevertheless, all humans still depend on natural or managed ecosystems for all of our food, fresh water, and organic building materials and for many of our pharmaceutical and industrial products (Daily 1997; Millenium Ecosystem Assessment 2005; Carpenter 2009; Maser 2009, 2010). Ecosystems also provide spiritual retreat, carbon sequestration, climate moderation, and buffers against wind and wave action, services that strike home when storm surge from hurricanes is exacerbated by loss of coastal marshes and swamps that historically abated this destructive force. Factors that affect the sustainable supply of ecosystem services have the capacity to cause famine and water shortages that, in turn, can cause social unrest, war, and epidemics of insect-vectored, as well as other, diseases (Riley 1878; Bray 1996; Smith 2007; Zhang et al. 2007; Bora et al. 2010; Hsiang et al. 2011).

Insects affect virtually all ecosystem services in complex, often complementary, ways that represent important feedback mechanisms that have maintained ecosystem structure and function for millennia without human intervention. In fact, human life on this planet probably would not persist more than a few months in the absence of insects and their contributions to ecosystem services (Wilson 1987). Unfortunately, insects also represent significant health threats and are our primary competitors for many ecosystem resources. Furthermore, they have proven extremely difficult to control, requiring stronger measures that undermine ecosystem processes that provide the services on which we depend.

Clearly, we need to ensure sustainability of the services on which our lives, and those of our children, depend. Sustainability requires that the rate and manner of utilization do not deplete or damage the resource or service (Ehrlich and Goulder 2007). In other words, sustainability requires a balance between rates of provision and use (Maser 2009, 2010). However, whereas we are technologically capable of feeding the world's population, our capacity to deliver sufficient food, fresh water, and other ecosystem services to local markets currently depends on unsustainable levels of fertilizer and pesticide application and fossil fuel consumption (primarily for transportation of services from rural to urban markets). Ecosystem services also will be affected by global climate change and sea level rise (Wall 2007; Hatfield et al. 2011; Izurralde et al. 2011; Soboll et al. 2011). Achieving sustainability in supply and delivery of ecosystem services requires changes in our perspectives and approaches to managing insects and ecosystems. As shown in the example that opened this chapter, a narrow focus on controlling pests can interfere with ecosystem processes without improving our ability to achieve desired goals. In fact, efforts to control pests can disrupt ecosystem processes that are critical to the sustainability of ecosystem services.

The following sections provide a basis for understanding ecosystem services, ecosystem processes that maintain these services, and why insects need to be a focus of sustainability policies. These topics will be described in greater detail in following chapters.

1.1 Ecosystem Services

Natural and managed ecosystems provide a variety of services on which humans and other organisms depend for survival (Costanza et al. 1997; Daily 1997; Dasgupta et al. 2000; Carpenter et al. 2009; Jørgensen 2010). Ecosystem services have been categorized as provisioning (harvestable products such as food, fiber, water, and other resources), cultural (spiritual and recreational values), supporting (primary production, pollination, and soil formation), and regulating (contributing to a consistent supply of other ecosystem services through density-dependent feedback) (Millenium Ecosystem Assessment 2005). These services are provided to us at no expense as a result of natural ecosystem processes, but we may incur significant expense if ecosystem deterioration forces us to employ labor or artificial means to replace these.

Insects, especially during outbreaks of herbivorous species, affect a variety of ecosystem services through changes in primary production and vegetation cover, pollination, removal of litter, contributions to soil fertility and aeration, and fluxes of energy and nutrients (Klock and Wickman 1978; Leuschner 1980; White and Schneeberger 1981). "Pest" management has been advocated when the cost of controlling an undesirable effect is less than the anticipated value of losses to insects. However, insect outbreaks contribute to some ecosystem services, and complex long-term effects on supporting and regulating services can compensate for short-term losses in provisioning services. Management decisions, including whether or not to let an outbreak run its course, should be based on predicted population growth and trade-offs among short- and long-term effects on multiple ecosystem services (see following sections), as described in Chapters 4 and 7.

1.1.1 Provisioning Services

Ecosystems are the source of food, fresh water, fiber, biofuels, and medical and industrial resources for humans. Many plants produce edible fruits, seeds or tubers, wood, fiber, or other tissues that have become the basis for their cultivation as crop plants (Figure 1.1). Insects can contribute to, or detract from, provisioning services depending on particular resources affected. Wildlife and fish are important food sources worldwide, and many of these animals feed primarily or exclusively on insects (Losey and Vaughan 2006). Woody materials are used for housing, furnishings, and fences but are also widely used for firewood. Phytochemicals provide important pharmaceutical compounds (Zenk and Juenger 2007) such as salicylic acid, morphine, quinine, epinephrine, and taxol. Synthesis of complex compounds often is difficult or expensive, and exploration continues for new medically important materials (Helson et al. 2009). Many plant and animal products, including insects (see following paragraph), are widely used in traditional remedies. Plant-derived

FIGURE 1.1 (SEE COLOR INSERT.)
Food production is a primary ecosystem service necessary to support human populations. (a) Sorghum is a major crop around the world. (b) Manual planting of sugarcane in South Africa.

tannins, resins, and other compounds are used in various industrial applications, as shellac, adhesives, and so forth. Insects feeding on plants can reduce provision of these resources.

Insects or their products are valuable ecosystem resources in many cultures (see Chapter 2). About 1,500 edible insect species are consumed by 3,000 ethnic groups in 113 countries (MacEvilly 2000). Silkworms, *Bombyx mori*, remain the primary source of silk. Silk production has been practiced at least since 2000–3000 B.C. in China and is among the most widely traded commercial products. Other insects have provided valuable medical and industrial products.

Natural ecosystems are valued sources of fresh water, and adequate water supply often is the primary management goal for municipal watersheds (Figure 1.2). Herbivorous insects reduce canopy cover and increase the volume of precipitation reaching the ground and flowing into streams. Terrestrial insects falling into nutrient-poor lakes and headwater streams contribute substantial amounts of carbon, nitrogen, and phosphorus to these nutrient-poor ecosystems (Carlton and Goldman 1984; Mehner et al. 2005;

FIGURE 1.2 (SEE COLOR INSERT.)
Fresh water is a primary ecosystem service necessary to support human populations. (a) Iguaçu Falls, South America. (b) Fishing and laundry from the banks of the Songhua River in China.

Nowlin et al. 2007; Menninger et al. 2008; Pray et al. 2009). Soil and litter insects affect soil porosity and decomposition rate, factors that affect the rate of water movement through the substrate (Eldridge 1994; Coleman et al. 2004; Schowalter 2011). Insects can affect provisioning services in positive as well as negative ways, as described in Chapters 7 and 8.

1.1.2 Cultural Services

Ecosystems provide various spiritual, recreational, and other cultural services, including hiking, backpacking, hunting, and fishing, and educational and scientific activities. For example, most remnant forests in Ghana are sacred sites set aside by religious leaders and protected by sanctions and taboos (Bossart et al. 2006). Many people use natural or seminatural ecosystems for bird-watching, hunting, or fishing. Such ecosystems also are prized by scientists for comparison with altered ecosystems to assess the effects of alteration. The global value of recreational services alone (which often can be calculated from usage fees) has been estimated at US$815 billion by Costanza et al. (1997). Insect effects on cultural services can be positive or negative, as described in Chapters 7 and 8.

1.1.3 Supporting Services

Supporting services include primary production, pollination, and soil formation. Primary production is the energy and matter accumulated by plants; it determines the production of provisioning services. Insect feeding on plants can reduce production of these resources in the short term but may maintain plant production within carrying capacity and contribute to sustainability over the long term (Schowalter 2011). Primary production also is key to carbon sequestration, an increasingly important supporting service that can offset accumulating atmospheric carbon dioxide that contributes to global warming. Pollination is essential for reproduction by many plants, including 35% of global fruit and seed production (Clausen 1954; Crane 1999; Losey and Vaughn 2006; Klein et al. 2007). A rich diversity of insects, including bees, flies, beetles, butterflies, and moths, is critical to reproduction of many plant species, especially in the tropics (Figure 1.3; Bawa 1990; Momose et al. 1998; Steffan-Dewenter and Tscharntke 1999; Ricketts et al. 2008). Soil formation provides resources necessary for primary production. Litter-feeding insects are instrumental in breakdown and removal of organic litter (Coleman et al. 2004). Accumulation of livestock dung in the absence of dung-feeding species can result in fouling of pastures and avoidance by livestock (Ferrar 1975; Tyndale-Biscoe and Vogt 1996). Burrowing species increase soil aeration and facilitate penetration of water during precipitation events (Eldridge 1994; Schowalter 2011). Disruption of these processes interferes with provisioning and cultural services, as well as with carbon sequestration and climate modification services that depend on primary production and vegetation cover (see following section).

FIGURE 1.3 (SEE COLOR INSERT.)
Pollination by insects is a critical ecosystem service that supports reproduction of a large proportion of plant species, especially in the tropics. (From Schowalter, T. D., *Insect Ecology: An Ecosystem Approach*, Elsevier, San Diego, 2011. With permission.)

With the exception of pollination services, insects traditionally have been viewed as affecting supporting services negatively. However, under some conditions, herbivorous insects can stimulate plant growth; they and soil/litter insects contribute to soil formation and long-term primary productivity (Pedigo et al. 1986; Trumble et al. 1993; Feeley and Terborgh 2005; Dungan et al. 2007). Long-term compensatory effects require consideration of trade-offs in evaluating insect effects on supporting services. Surprisingly, insect effects on supporting services are largely positive, at least in the sense of sustaining processes that contribute to ecosystem services, as described in Chapters 5, 7, and 8.

1.1.4 Regulating Services

In contrast to supporting services, regulating services control the supply of other services. For example, availability of water and nutrients controls rates of plant growth, which supports other provisioning services. Some interactions among species within ecosystems provide positive or negative feedback that maintains a more consistent (i.e., sustainable) supply of other services, much like the thermostat in a house or office controls temperature within a narrow range around the set point. For example, the combination of positive and negative feedback among organisms that maintains plant production near ecosystem carrying capacity enables more consistent provision of services than would occur in the absence of such cybernetic regulation (Foley et al. 2003a, b; Schowalter 2011). Insects are among the natural regulatory mechanisms that provide positive and negative feedback to maintain primary productivity near carrying capacity (Mattson and Addy 1975; Schowalter 2011), thereby stabilizing other ecosystem processes and conditions (see Chapters 5, 7, and 8). Therefore, effects of insect outbreaks on regulating services warrant consideration in policy decisions.

1.2 Ecosystem Characteristics

Ecosystem services are not provided primarily for our use but rather relect the integration of direct and indirect interactions among the multitude of species (the community) that control rates and directions of energy and matter fluxes among organisms (including humans) and abiotic pools (e.g., atmosphere, oceans, and sediments) that compose an ecosystem (see Chapter 5). Ecosystems generally are considered to represent the integration of a more or less discrete community of organisms and the abiotic conditions at a site (Figure 1.4). However, research and environmental policy decisions increasingly recognize the importance of scale in ecosystem studies, that is, extending research or extrapolating results to landscape, regional, and global scales (Turner 1989; Holling 1992).

Ecosystems are interconnected, just as the species within them are interconnected. Exports from one ecosystem become imports for others (Figure 1.5). Energy, water, organic matter, and nutrients from terrestrial ecosystems are major sources of these resources for many aquatic ecosystems. Atmospheric carbon dioxide concentration is governed by ocean temperature and salinity that affect precipitation of carbonates into sediments or evaporation into the atmosphere and by terrestrial vegetation that fixes carbon in biomass (through photosynthesis). Organic matter and nutrients eroded by wind from arid ecosystems are filtered from the airstream by ecosystems downwind. Some ecosystems within a landscape or watershed are the sources of colonists for recently disturbed ecosystems. Insect outbreaks can spread from one ecosystem to another. Toxic or exogenous materials introduced into one ecosystem can adversely affect remote ecosystems, for example, agricultural chemicals causing hypoxic (dead) zones in coastal waters (Krug 2007; Howarth et al. 2011). Our traditional management of individual ecosystems

FIGURE 1.4 (SEE COLOR INSERT.)
Patchwork of old-growth and harvested sites in western Oregon. Each patch can be treated as an ecosystem.

FIGURE 1.5 (SEE COLOR INSERT.)
Interconnected ecosystems. A swamp represents two integrated ecosystems, with terrestrial material providing input to the aquatic ecosystem and flooding providing input to the terrestrial ecosystem. Coastal swamps and marshes also represent important filters that slow water flow and capture sediments and nutrients that otherwise would be exported to estuaries and oceans.

and services as local resources has led to widespread degradation and loss of services. Therefore, our approach to managing ecosystems must address consequences globally because these affect the supply of resources and services on which we depend.

Ecosystems are subject to natural changes in environmental conditions that affect abundances of insects, as well as other organisms, and ecosystem services. Some changes are abrupt, such as those resulting from fire, storm, drought, or flood disturbances (Figure 1.6). Others occur more gradually, over time periods of decades to millennia. Insects have adapted to much of the natural environmental variation to which they have been exposed, and they often respond similarly to anthropogenic (human induced) changes.

Individuals better adapted to survive environmental changes are more likely to pass adaptive genes to their offspring, increasing the frequency of adaptive genes in the population. Populations characterizing particular ecosystems generally are adapted to regularly occurring environmental changes and disturbances to which they have been exposed. Diapause (a period of suspended development during adverse environmental conditions) during cold winters or dry summers is one obvious example. Plant adaptations to disturbances include fire- or wind-resistant structure, underground rhizomes, buried seed banks, and early successional species that facilitate rapid colonization or replacement of vegetation cover and protection of soil conditions. Insect adaptations include flight, buried or protected life stages, and ability to withstand hypoxia (see Chapter 3). Larger animals often can escape a disturbed area. Ecosystems with more stable conditions impose less selective pressure on species than do ecosystems with more variable conditions. Therefore, temperate species have wider tolerances to

FIGURE 1.6 (SEE COLOR INSERT.)
Examples of disturbances to illustrate variation in type, magnitude, extent, and contrast between disturbed and surrounding landscape patches. (a) Fire in oak savanna (note sites of high and low flame height and intensity); (b) hurricane effect on coastal deciduous forest; (c) landslide resulting from heavy rainfall; (d) volcanic eruption (note fresh lava flow on left and zones of burning and exposure to fumes in vegetation fragment). (From Schowalter, T. D., *Annual Review of Entomology*, 57, 1–20, 2012. With permission.)

abiotic conditions than do tropical species and are more likely to become invasive in new habitats.

Insect outbreaks traditionally have been viewed as disturbances (White and Pickett 1985; Willig and Walker 1999). If "disturbance" is defined as any relatively discrete event (such as fire, storm, or flood) that causes measurable change in population, community, or ecosystem structure or function (White and Picket 1985; Willig and Walker 1999), then insect outbreaks could be considered disturbances. Insects can defoliate or kill most host plants over large areas, up to 10^3 to 10^6 ha (Furniss and Carolin 1977). For example, 39% of a montane forest landscape in Colorado has been affected by insect outbreaks (spruce beetle, *Dendroctonus rufipennis*) since about 1633, compared with 59% by fire and 9% by snow avalanches (Veblen et al. 1994), with an average return interval of 117 years, compared with 202 years for fire.

However, unlike abiotic disturbances, insect outbreaks typically are triggered by environmental changes that stress, and/or increase the abundance of, their host plants (Chapter 4). Locust outbreaks, for example, often reflect drought conditions (Konishi and Itô 1973; Stige et al. 2007) that increase availability of stressed vegetation and suitable oviposition sites but also follow flooding that results in the same conditions as water recedes (Stige et al. 2007).

Recent outbreaks most commonly reflect anthropogenic redistribution of resources, especially increased density of commercially valuable (often exotic) plant species (especially varieties bred to express higher growth and fruit production at the expense of defenses) and exotic insect species. Deforestation and road building into previously undeveloped ecosystems have increased abundance of disease vectors and increased their contact with humans (Vittor et al. 2006).

Furthermore, interactions among species composing ecosystems tend to regulate abundances and fluxes of energy and matter through feedback mechanisms that stabilize ecosystem conditions. Herbivores (including insects) tend to respond to overabundant, or weakened, plants in ways that reduce abundance of those plants and favor better adapted plant species, thereby maintaining vegetation cover and primary production near the carrying capacity of the ecosystem, just as predators remove excess or weakened prey and maintain healthier prey populations near their carrying capacity (Schowalter 2011).

Currently, humans are altering earth's ecosystems at an unprecedented and accelerating rate (Vitousek et al. 1997; Thomas et al. 2004; Burney and Flannery 2005). Comparable change in temperature in the fossil record occurred over 10,000 years (Currano et al. 2008). Anthropogenic changes to the global environment affect insects in various ways. Combustion of fossil fuels has elevated atmospheric concentrations of CO_2 (Keeling et al. 1995; Beedlow et al. 2004), methane, ozone, nitrous oxides, and sulfur dioxide, leading to increasingly acidic precipitation and prospects of global warming. Altered distribution of vegetation cover and impoundment of rivers have resulted in changes in global albedo that affect temperature and precipitation patterns (Foley et al. 2003a, b; Juang et al. 2007; Janssen et al. 2008). Conversion of natural ecosystems is altering and isolating natural communities at an unprecedented rate, disconnecting interaction between herbivores and predators and leading to outbreaks of insect "pests" in crop monocultures and fragmented ecosystems (Roland 1993). Transport of invasive species by humans affects community and ecosystem structure and processes directly and indirectly (Kizlinski et al. 2002; Orwig 2002; Sanders et al. 2003).

More immediate anthropogenic disturbances introduce conditions to which organisms are not adapted, delaying or preventing ecosystem recovery. Spills from medical, industrial, and agricultural sources introduce novel compounds into the environment; these hazardous wastes can be lethal or teratogenic to many organisms. Land conversion clears vegetation and exposes soil over larger areas than can be colonized before soil desiccation and erosion make the site largely uninhabitable for most organisms (Amaranthus and Perry 1987). Furthermore, land clearing typically creates straight edges between converted areas and nonconverted remnants. Straight edges maximize the force of wind along the edge, often resulting in a domino effect of treefall and/or soil desiccation far into remaining fragments of natural habitats (Chen et al. 1995). Several studies have demonstrated that as the proportion of converted land area increases, remnant natural areas lose their capacity to modify environmental conditions and enter a positive feedback cycle of desertification (Schlesinger et al. 1990; Janssen et al. 2008; see Chapter 5). These human activities have seriously undermined the capacity of many ecosystems to maintain integrity of processes and services on which we depend.

1.3 Insects and Insect Control

Insects are among the oldest groups of terrestrial organisms, tracing their origin back 400 million years to the Devonian Period (Figure 1.7). They soon dominated terrestrial ecosystems and were the first organisms to fly. In the absence of faster vertebrate predators, these ancient insects reached large sizes, with foot-long (30 cm) cockroaches and dragonflies with 30-inch (75 cm) wingspans. The weight of exoskeleton required to support such large insects limited movement, and the later appearance of faster, and winged, vertebrate predators forced insects to adapt by becoming smaller and faster. Nevertheless, insects have survived all five major extinction events that eliminated many other plant and animal groups, and they continue to show remarkable ability to adapt to anthropogenic changes. Insects currently are the dominant group of organisms on earth in terms of taxonomic diversity (>50% of all described species) and ecological function (Wilson 1992) (Figure 1.8), and their biomass (total weight per unit area) in terrestrial and freshwater ecosystems frequently exceeds that of more conspicuous vertebrates, especially during outbreaks (Odum, 1957, 1970; Watts et al. 1982; Whitford 1986; Wilson 1987; see Chapter 5).

Insects represent the vast majority of species in terrestrial and freshwater ecosystems (Schowalter 2011), and are important components of nearshore

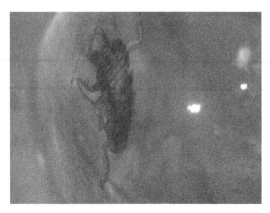

FIGURE 1.7 (SEE COLOR INSERT.)
Cockroach in Dominican amber estimated to be approximately 25 million years old.

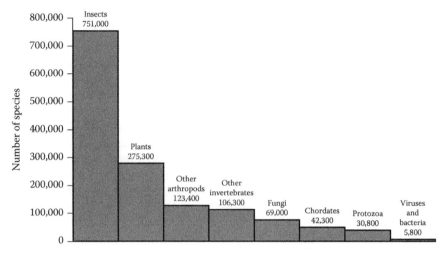

FIGURE 1.8
Distribution of described species within major taxonomic groups. Species numbers for insects, bacteria, and fungi likely will increase greatly as these groups become better known. Data from Wilson (1992). (From Schowalter, T. D., *Insect Ecology: An Ecosystem Approach*, Elsevier, San Diego, 2011. With permission.)

marine ecosystems as well. At present, more than one million insect species have been described, and estimates of the total number of insect species yet to be described range from four million to thirty million (Erwin 1982; Novotny et al. 2002). Most people have difficulty comprehending such diversity (and the inability of entomologists to identify many specimens). Some related species differ only in subtle ways, such as color or pattern of "hairs" on particular body parts, but differ enormously in how they respond to environmental changes.

Insects are able to adapt quickly, often in less than a decade, to environmental changes because of their small size, short generation times, and high reproductive capacity (see Chapter 3). Many insects, especially those characterizing the most variable ecosystems, have considerable capacity for long-distance dispersal, enabling them to find and colonize new habitats and resources as these appear. Other insects, especially those characterizing more stable forest ecosystems, are flightless and vulnerable to environmental change or habitat fragmentation. Abundances of some species can change several orders of magnitude within a year in response to environmental changes, making their management a challenge at best (see Chapter 4). Such changes are easily detectable and make insects more useful indicators of environmental changes than are larger or longer-lived organisms that respond more slowly.

Insects and how we respond to them affect ecosystem services. Three insect functional groups—herbivores (including pollinators), detritivores, and blood-feeders—are particularly important. Herbivores and detritivores can affect ecosystem services directly by altering rates of growth, survival, and reproduction of plants or decomposition of dead material (Schowalter 2011), whereas blood-feeders are most important when famine, conflict, or persistent poverty increase human crowding and vulnerability to insect-transmitted diseases (Bray 1996; Diamond 1999; Zhang et al. 2007).

It is critical for sustainable management of ecosystem services that we acknowledge that the term "pest" has no meaning outside of human perspective (Kogan and Jepson 2007). Humans have valued many insects and been threatened by others (see Chapter 2). Insects are adapted for the roles they play in integrated ecosystems, and their interference with our health or resource acquisition occurs because we share the same food webs. In fact, insects play critical roles in ecosystem development and function, and many plants would not survive without insects. They represent important food resources for many other organisms, including humans, and they have the capacity to alter primary production in ways that potentially affect global climate. These natural roles often bring insects into conflict with humans but also are fundamental to the sustainability of many ecosystem services.

Insects are not the only factors that affect the sustainability of ecosystem services, but their effects are pervasive, even under nonoutbreak conditions. Insects cause billions of dollars in crop losses, destroying 18% of global crop production annually (Oerke 2006) and representing a major impediment to our ability to feed a growing human population. Some detritivorous species fail to distinguish wood or other plant fibers in our houses from detritus on which they feed in natural ecosystems. Other insects transmit devastating diseases that sicken or kill millions of people each year. It is difficult to accept that such mortality prevented human populations from overexploiting their resources for millennia before technology disrupted this balance. As a result, our response to their presence is driven largely by entomophobia rather than careful consideration of the costs and benefits of insect activity and management options.

Insects are integral to virtually all ecosystem processes that provide the services on which we depend. Adaptation and explosive population growth in response to environmental changes, especially those resulting from anthropogenic activities, have the capacity to exacerbate or mitigate changes in ecosystem conditions and, perhaps, global processes. Efforts to control insects often have unintended and/or undesirable consequences for environmental quality and ecosystem services. At best, such efforts interfere with natural regulatory mechanisms, as in the example of the Douglas-fir tussock moth that opened this chapter, but also can undermine ecosystem processes that are necessary to sustain ecosystem services. Clearly, understanding how insects affect ecosystems is critical to our approaches to managing ecosystems in a way that promotes sustainability of resources and services. How we deal with insects, as much as how we manage ecosystems, will determine the capacity of ecosystems to supply those services into the future.

Sustainability of ecosystem services is possible only if we protect the ecosystem processes that control their supply. The goal of sustainability is a consistent rate of production or supply of ecosystem services, not a maximum rate that cannot be sustained under changing environmental conditions. Our actions and policies, including insect control, often undermine the ecological processes that sustain the supply of ecosystem services. We cannot anticipate all the consequences of our environmental policies and actions, but we know that our actions, like the flap of a butterfly's wings, can have unintended long-term and far-reaching effects on the sustainability of the ecosystem services on which we depend.

1.4 Scope of This Book

This book addresses the complex effects of insects on ecosystem services and the consequences of our attitudes and policies toward management of insect "pests." The "butterfly effect" is an underlying theme. As originally popularized by Ray Bradbury (1952), the concept of small actions having large, unintended consequences can describe the loss of pollination services as a result of agricultural intensification that removes pollinator habitat from the landscape, as well as contamination of fish tissues and demise of fisheries resulting from insecticide runoff. As used by Edward Lorenz (1993), who coined the term, the "butterfly effect" also underlies the difficulty of accurately predicting complex phenomena because small changes in model parameters can cause large changes in the result. This is particularly true for prediction of climate change, which reflects the integration of many subsystems, including the effects of species interactions on vegetation cover and albedo at regional and global scales and for effects of our "management" activities on ecosystem integrity and services.

In particular, our efforts to control insects often have unintended consequences for ecosystem services and may be more damaging to ecosystem conditions and services than are the insects themselves. For example, insecticide application introduces novel chemicals into the environment that affect nontarget organisms or processes within or among interconnected ecosystems that affect ecosystem services. Insecticides may be even more counterproductive in forests or other seminatural areas to the extent that insect outbreaks represent natural feedback mechanisms that maintain plant abundance within the carrying capacity of the environment. This calls to question the wisdom of believing that we are "managing" insects and ecosystem services, rather than simply tampering. We are manipulating our global environment without regard for our lack of knowledge about ecosystem responses that may produce consequences in opposition to our long-term benefit.

This book is organized to describe (1) the variety of ways in which humans and insects have interacted through time (Chapter 2), (2) aspects of insect biology that explain their ecological success and their often amazing ability to adapt quickly to environmental changes, including control tactics (Chapters 3–4), (3) how ecosystem processes produce services (Chapter 5), (4) how our management of insects and ecosystems affects ecosystem services (Chapters 6 and 8), (5) complex (and often complementary) effects of insects on ecosystem resources and services (Chapters 7 and 8), and (6) more effective ways of dealing with insects that can promote sustainability of ecosystem resources and services (Chapter 9). Spiders and mites typically are regarded as "bugs" in the same sense as insects and will be addressed in this book as appropriate. For simplicity, "policymaker" will refer to anyone with rule-making authority, including local, regional, and national legislators and land management and environmental quality directors. "Manage" and "management" commonly refer to human manipulation of insects, resources, and ecosystems, ostensibly for public benefit. However, as will be demonstrated throughout this book, manipulation without adequate understanding of consequences is not truly management and is likely to jeopardize the ecosystem services on which we depend.

References

Amaranthus, M. P. and D. A. Perry. 1987. Effect of soil transfer on ectomycorrhiza formation and the survival and growth of conifer seedlings on old, nonreforested clear-cuts. *Canadian Journal of Forest Research* 17: 944–950.

Bawa, K. S. 1990. Plant-pollinator interactions in tropical rain forests. *Annual Review of Ecology and Systematics* 21: 399–422.

Beedlow, P. A., D. T. Tingey, D. L Phillips, W. E. Hogset, and D. M. Olszyk. 2004. Rising atmospheric CO_2 and carbon sequestration in forests. *Frontiers in Ecology and the Environment* 2: 315–322.

Bora, S., I. Ceccacci, C. Delgado, and R. Townsend. 2010. World Development Report 2011: Food Security and Conflict. Washington, DC: Agriculture and Rural Development Department, World Bank.

Bossart, J. L., E. Opuni-Frimpong, S. Kuudaar, and E. Nkrumah. 2006. Richness, abundance, and complementarity of fruit-feeding butterfly species in relict sacred forests and forest reserves of Ghana. *Biodiversity and Conservation* 15: 333–359.

Bradbury, R. 1952. Sound of thunder. *Colliers Magazine* June 28: 20–21, 60–61.

Bray, R. S. 1996. *Armies of Pestilence: The Impact of Disease on History.* New York: Barnes and Noble.

Brookes, M. H., R. W. Stark, and R. W. Campbell, eds. 1978. *The Douglas-Fir Tussock Moth: A Synthesis.* USDA Forest Service Technical Bulletin 1585. Washington, DC: USDA.

Burney, D. A. and T. F. Flannery. 2005. Fifty millennia of catastrophic extinctions after human contact. *Trends in Ecology and Evolution* 20: 395–401.

Carlton R. G. and C. R. Goldman. 1984. Effects of a massive swarm of ants on ammonium concentrations in a subalpine lake. *Hydrobiologia* 111: 113–117.

Carpenter, S. R., H. A. Mooney, J. Agard, D. Capistrano, R. S. DeFries, S. Díaz, T. Dietz, A. K. Duraiappah, A. Oteng-Yeboah, H. M. Pereira, et al. 2009. Science for managing ecosystem services: Beyond the Millennium Ecosystem Assessment. *Proceedings of the National Academy of Sciences USA* 106: 1305–1312.

Carson, R. 1962. *Silent Spring.* New York: Houghton-Mifflin.

Chen, J., J. F. Franklin, and T. A. Spies. 1995. Growing-season microclimatic gradients from clearcut edges into old-growth Douglas-fir forests. *Ecological Applications* 5: 74–86.

Christensen, N. L., Jr., S. V. Gregory, P. R. Hagenstein, T. A. Heberlein, J. C. Hendee, J. T. Olson, J. M. Peek, D. A. Perry, T. D. Schowalter, K. Sullivan, et al. 2000. *Environmental Issues in Pacific Northwest Forest Management.* Washington, DC: National Academy Press.

Clausen, L. W. 1954. *Insect Fact and Folklore.* New York: MacMillan.

Coleman, D. C., D. A. Crossley, Jr., and P. F. Hendrix. 2004. *Fundamentals of Soil Ecology,* 2nd ed. Amsterdam: Elsevier.

Costanza, R., R. d'Arge, R. de Groot, S. Farger, M. Grasso, B. Hannon, K. Limburg, S. Naeem, R. V. O'Neill, J. Paruelo, R. G. Raskin, et al. 1997. The value of the world's ecosystem services and natural capital. *Nature* 387: 253–260.

Crane, E. 1999. *The World History of Beekeeping and Honey Hunting.* New York: Routledge.

Currano, E. D., P. Wilf, S. L. Wing, C. C. Labandeira, E. C. Lovelock, and D. L. Royer. 2008. Sharply increased insect herbivory during the Paleocene-Eocene Thermal Maximum. *Proceedings of the National Academy of Sciences USA* 105: 1960–1964.

Daily, G. C., ed.. 1997. *Nature's Services: Societal Dependence on Natural Ecosystems.* Washington, DC: Island Press.

Dasgupta, P., S. Levin, and J. Lubchenco. 2000. Economic pathways to ecological sustainability. *BioScience* 50: 339–345.

Diamond, J. 1999. *Guns, Germs, and Steel: The Fates of Human Societies.* New York: W.W. Norton.

Dungan, R. J., M. H. Turnbull, and D. Kelly. 2007. The carbon costs for host trees of a phloem-feeding herbivore. *Journal of Ecology* 95: 603–613.

Ehrlich, P. R. and L. H. Goulder. 2007. Is current consumption excessive? A general framework and some indications for the United States. *Conservation Biology* 21: 1145–1154.

Eldridge, D. J. 1994. Nests of ants and termites influence infiltration in a semi-arid woodland. *Pedobiologia* 38: 481–492.

Erwin, T. L. 1982. Tropical forests: Their richness in Coleoptera and other arthropod species. *Coleopterists Bulletin* 36: 74–75.

Feeley, K. J. and J. W. Terborgh. 2005. The effects of herbivore density on soil nutrients and tree growth in tropical forest fragments. *Ecology* 86: 116–124.

Ferrar, P. 1975. Disintegration of dung pads in north Queensland before the introduction of exotic dung beetles. *Australian Journal of Experimental Agriculture and Animal Husbandry* 15: 325–329.

Feyereisen, R. 1999. Insect P450 enzymes. *Annual Review of Entomology* 44: 507–533.

Foley, J. A., M. T. Coe, M. Scheffer, and G. Wang. 2003a. Regime shifts in the Sahara and Sahel: Interactions between ecological and climatic systems in northern Africa. *Ecosystems* 6: 524–539.

Foley, J. A., M. H. Costa, C. Delire, N. Ramankutty, and P. Snyder. 2003b. Green surprise? How terrestrial ecosystems could affect earth's climate. *Frontiers in Ecology and the Environment* 1: 38–44.

Furniss, R. L. and V. M. Carolin. 1977. *Western Forest Insects.* USDA Forest Service Misc. Publ. 1339. Washington, DC: USDA Forest Service.

Gahan, L. G., Y.-T. Ma, M. L. MacGregor Coble, F. Gould, W. J. Moar, and D. G. Heckel. 2005. Genetic basis of resistance to Cry1Ac and Cry2Aa in *Heliothis virescens* (Lepidoptera: Noctuidae). *Journal of Economic Entomology* 98: 1357–1368.

Hatfield, J. L., K. J. Boote, B. A. Kimball, L. H. Ziska, R. C. Izaurralde, D. Ort, A. M. Thomson, and D. Wolfe. 2011. Climate impacts on agriculture: Implications for crop production. *Agronomy Journal* 103: 351–370.

Helson, J. E., T. L. Capson, T. Johns, A. Aiello, and D. M. Windsor. 2009. Ecological and evolutionary bioprospecting: Using aposematic insects as guides to rainforest plants active against disease. *Frontiers in Ecology and the Environment* 7: 130–134.

Holling, C. S. 1992. Cross-scale morphology, geometry, and dynamics of ecosystems. *Ecological Monographs* 62: 447–502.

Howarth, R., F. Chan, D. J. Conley, J. Garnier, S. C. Doney, R. Marino, and G. Billen. 2011. Coupled biogeochemical cycles: Eutrophication and hypoxia in temperate estuaries and coastal marine ecosystems. *Frontiers in Ecology and the Environment* 9: 18–26.

Hsiang, S. M., K. C. Meng, and M. A. Cane. 2011. Civil conflicts are associated with the global climate. *Nature* 476: 438–441.˙

Izaurralde, R. C., A. M. Thomson, J. A. Morgan, P. A. Fay, H. W. Polley, and J. L. Hatfield. 2011. Climate impacts on agriculture: Implicatons for forage and rangeland production. *Agronomy Journal* 103: 371–381.

Janssen, R. H. H., M. B. J. Meinders, E. H. van Nes, and M. Scheffer. 2008. Microscale vegetation-soil feedback boosts hysteresis in a regional vegetation-climate system. *Global Change Biology* 14: 1104–1112.

Jørgensen, S. E. 2010. Ecosystem services, sustainability, and thermodynamic indicators. *Ecological Complexity* 7: 311–313.

Juang, J.-Y., G. G. Katul, A. Porporato, P. C. Stoy, M. S. Sequeira, M. Detto, H.-S. Kim, and R. Oren. 2007. Eco-hydrological controls on summertime convective rainfall triggers. *Global Change Biology* 13: 887–896.

Keeling, C. D., T. P. Whorf, M. Wahlen, and J. van der Pilcht. 1995. Interannual extremes in the rate of rise of atmospheric carbon dioxide since 1980. *Science* 375: 666–670.

Kizlinski, M. L., D. A. Orwig, R. C. Cobb, and D. R. Foster. 2002. Direct and indirect ecosystem consequences of an invasive pest on forests dominated by eastern hemlock. *Journal of Biogeography* 29: 1489–1503.

Klein, A.-M, B. E. Vaissière, J. H. Cane, I. Steffan-Dewenter, S. A. Cunningham, C. Kremen, and T. Tscharntke. 2007. Importance of pollinators in changing landscapes for world crops. *Proceedings of the Royal Society B* 274: 303–313.

Klock, G. O. and B. E. Wickman. 1978. Ecosystem effects. In *The Douglas-Fir Tussock Moth: A Synthesis*, M. H. Brookes, R. W. Stark, and R. W. Campbell, eds. 90–95. USDA Forest Service Technical Bulletin 1585. Washington, DC: USDA Forest Service.

Kogan, M. and P. Jepson. 2007. Ecology, sustainable development, and IPM: The human factor. In *Perspectives in Ecological Theory and Integrated Pest Management*, M. Kogan and P. Jepson, eds. 1–44. Cambridge, UK: Cambridge University Press.

Konishi, M. and Y. Itô. 1973. Early entomology in East Asia. In *History of Entomology*, R. F. Smith, T. E. Mittler, and C. N. Smith, eds. 1–20. Palo Alto, CA: Annual Reviews.

Krug, E. C. 2007. Coastal change and hypoxia in the northern Gulf of Mexico. Part 1. *Hydrology and Earth System Sciences* 11: 180–190.

Leuschner, W. A. 1980. Impacts of the southern pine beetle. In *The Southern Pine Beetle*, R. C. Thatcher, J. L. Searcy, J. E. Coster, and G. D. Hertel, eds. 137–151. USDA Forest Service Technical Bulletin 1631. Washington, DC: USDA Forest Service.

Lorenz, E. N. 1993. *The Essence of Chaos*. Seattle: University of Washington Press.

Losey, J. E. and M. Vaughn. 2006. The economic value of ecological services provided by insects. *BioScience* 56: 311–323.

MacEvilly, C. 2000. Bugs in the system. *Nutrition Bulletin* 25: 267–268.

Maser, C. 2009. *Earth in Our Care: Ecology, Economy, and Sustainability*. New Brunswick, NJ: Rutgers University Press.

Maser, C. 2010. *Social-Environmental Planning: The Design Interface Between Everyforest and Everycity*. Boca Raton, FL: CRC Press/Taylor & Francis.

Mattson, W. J. and N. D. Addy. 1975. Phytophagous insects as regulators of forest primary production. *Science* 190: 515–522.

Mehner T., J. Ihlau, H. Dörner, M. Hupfer, and F. Hölker. 2005. Can feeding of fish on terrestrial insects subsidize the nutrient pool of lakes? *Limnology and Oceanography* 50: 2022–2031.

Menninger, H. L., M. A. Palmer, L. S. Craig, and D. C. Richardson. 2008. Periodical cicada detritus impacts stream ecosystem metabolism. *Ecosystems* 11: 1306–1317.

Millenium Ecosystem Assessment. 2005. *Ecosystems and Human Well-Being: Biodiversity Synthesis*. Washington, DC: World Resources Institute.

Momose, K., T. Yumoto, T. Nagamitsu, M. Kato, H. Nagamasu, S. Sakai, R. D. Harrison, T. Itioka, A. A. Hamid, and T. Inoue. 1998. Pollination biology in a lowland dipterocarp forest in Sarawak, Malaysia. I. Characteristics of the plant-pollinator community in a lowland dipterocarp forest. *American Journal of Botany* 85: 1477–1501.

Novotny, V., Y. Basset, S. E. Miller, G. D. Weiblen, B. Bremer, L. Cizek, and P. Drozd. 2002. Low host specificity of herbivorous insects in a tropical forest. *Nature* 416: 841–844.

Nowlin, W. H., M. J. González, M. J. Vanni, M. H. H. Stevens, M. W. Fields, and J. J. Valenti. 2007. Allochthonous subsidy of periodical cicadas affects the dynamics and stability of pond communities. *Ecology* 88: 2174–2186.

Odum, H. T. 1957. Trophic structure and productivity of Silver Springs, Florida. *Ecological Monographs* 27: 55–112.

Odum, H. T. 1970. Summary: An emerging view of the ecological system at El Verde. In *A Tropical Rain Forest*, H. T. Odum and R. F. Pigeon, eds. I191–I289. Washington, DC: U.S. Atomic Energy Commission.

Oerke, E.-C. 2006. Centenary review: Crop losses to pests. *Journal of Agricultural Science* 144: 31–43.

Orwig, D. A 2002. Ecosystem to regional impacts of introduced pests and pathogens: Historical context, questions, and issues. *Journal of Biogeography* 29: 1471–1474.

Pedigo, L. P., S. H. Hutchins, and L. G. Higley. 1986. Economic injury levels in theory and practice. *Annual Review of Entomology* 31: 341–368.

Plapp, F. W. 1976. Biochemical genetics of insecticide resistance. *Annual Review of Entomology* 21: 179–197.

Pray, C. L, W. H. Nowlin, and M. J. Vanni. 2009. Deposition and decomposition of periodical cicadas (Homoptera: Cicadidae: Magicicada) in woodland aquatic ecosystems. *Journal of the North American Benthological Society* 28: 181–195.

Ricketts, T. H., J. Regetz, I. Steffen-Dewenter, S. A. Cunningham, C. Kremen, A. Bogdanski, B. Gemmill-Herren, S. S. Greenleaf, A. M. Klein, M. M. Mayfield, L. A. Morandin, A. Ochieng, and B. F. Viana. 2008. Landscape effects on crop pollinator services: Are there general patterns? *Ecology Letters* 11: 499–515.

Riley, C. V. 1878. *First Annual Report of the United States Entomological Commission for the Year 1877 Relating to the Rocky Mountain Locust and the Best Methods of Preventing Its Injuries and of Guarding Against Its Invasions, in Pursuance of an Appropriation Made by Congress for This Purpose.* Washington, DC: U.S. Department of Agriculture.

Roland, J. 1993. Large-scale forest fragmentation increases the duration of tent caterpillar outbreak. *Oecologia* 93: 25–30.

Roussel, J. S. and D. F. Clower 1957. Resistance to the chlorinated hydrocarbon insecticides in the boll weevil. *Journal of Economic Entomology* 50: 463–468.

Rural Sociological Society Task Force on Persistent Rural Poverty 1993. *Persistent Poverty in Rural America.* Boulder, CO: Westview Press.

Sanders, N. J., N. J. Gotelli, N. E. Heller, and D. M. Gordon. 2003. Community disassembly by an invasive species. *Proceedings of the National Academy of Sciences USA* 100: 2474–2477.

Schlesinger, W. H., J. F. Reynolds, G. L. Cunningham, L. F. Huenneke, W. M. Jarrell, R. A. Virginia, and W. G. Whitford. 1990. Biological feedbacks in global desertification. *Science* 247: 1043–1048.

Schowalter, T. D. 2011. *Insect Ecology: An Ecosystem Approach*, 3rd ed. San Diego, CA: Elsevier/Academic.

Schowalter, T. D. 2012. Insect responses to major landscape-level disturbance. *Annual Review of Entomology* 57: 1–20.

Smith, R. H. 2007. *History of the Boll Weevil in Alabama.* Alabama Agricultural Experiment Station Bulletin 670. Auburn, AL: Auburn University.

Soboll, A., M. Elbers, R. Barthel, J. Schmude, A. Ernst, and R. Ziller. 2011. Integrated regional modelling and scenario development to evaluate future water demand under global change conditions. *Mitigation and Adaptation Strategies for Global Change* 16: 477–498.

Soderlund, D. M. and J. R. Bloomquist. 1990. Molecular mechanisms of insecticide resistance. In *Pesticide Resistance in Arthropods*, R. T. Roush and B. E. Tabashnik, eds. 58–96. New York: Chapmann & Hall.

Stark, R. W. 1978. Introduction. In *The Douglas-Fir Tussock Moth: A Synthesis*, M. H. Brookes, R. W. Stark, and R.W. Campbell, eds.1–5. USDA Forest Service Technical Bulletin 1585. Washington, DC: USDA.

Steffen-Dewenter, I. and T. Tscharntke. 1999. Effects of habitat isolation on pollinator communities and seed set. *Oecologia* 121: 432–440.

Stige, L. C., K.-S. Chan, Z. Zhang, D. Frank, and N. C. Stenseth. 2007. Thousand-year-long Chinese time series reveals climatic forcing of decadal locust dynamics. *Proceedings of the National Academy of Sciences USA* 104: 16188–16193.

Thomas, J. A., M. G. Telfer, D. B. Roy, C. D. Preston, J. J. D. Greenwood, J. Asher, R. Fox, R. T. Clarke, and J. H. Lawton. 2004. Comparative losses of British butterflies, birds, and plants and the global extinction crisis. *Science* 303: 1879–1881.

Thompson, C. G. 1978. Nuclear polyhedrosis epizootiology. In *The Douglas-Fir Tussock Moth: A Synthesis*, M. H. Brookes, R. W. Stark, and R. W. Campbell, eds.136–140. USDA Forest Service Technical Bulletin 1585. Washington, DC: USDA.

Trumble, J. T., D. M. Kolodny-Hirsch, and I. P. Ting. 1993. Plant compensation for arthropod herbivory. *Annual Review of Entomology* 38: 93–119.

Turner, M. G. 1989. Landscape ecology: The effect of pattern on process. *Annual Review of Ecology and Systematics* 20: 171–197.

Tyndale-Biscoe, M. and W. G. Vogt. 1996. Population status of the bush fly, *Musca vetustissima* (Diptera: Muscidae), and native dung beetles (Coleoptera: Scarabaeinae) in south-eastern Australia in relation to establishment of exotic dung beetles. *Bulletin of Entomological Research* 86: 183–192.

Veblen, T. T., K. S. Hadley, E. M. Nel, T. Kitzberger, M. Reid, and R. Villalba. 1994. Disturbance regime and disturbance interactions in a Rocky Mountain subalpine forest. *Journal of Ecology* 82: 125–135.

Vitousek, P. M., H. A. Mooney, J. Lubchenco, and J. M. Melillo. 1997. Human domination of Earth's ecosystems. *Science* 277: 494–499.

Vittor, A. Y., R. H. Gilman, J. Tielsch, G. Glass, T. Shields, W. S. Lozano, V. Pinedo-Cancino, and J. A. Patz. 2006. The effect of deforestation on the human-biting rate of *Anopheles darlingi*, the primary vector of falciparum malaria in the Peruvian Amazon. *American Journal of Tropical Medicine and Hygiene* 74: 3–11.

Wall, D. H. 2007. Global change tipping points: Above- and below-ground biotic interactions in a low diversity ecosystem. *Philosophical Transactions of the Royal Society B* 362: 2291–2306.

Watts, J. G., E. W. Huddleston, and J. C. Owens. 1982. Rangeland entomology. *Annual Review of Entomology* 27: 283–311.

White, P. S. and S. T. A. Pickett. 1985. Natural disturbance and patch dynamics: An introduction. In *Ecology of Natural Disturbance and Patch Dynamics*, S. T. A. Pickett and P. S. White, eds. 3–13. New York: Academic Press.

White, W. B. and N. F. Schneeberger. 1981. Socioeconomic impacts. In *The Gypsy Moth: Research Toward Integrated Pest Management*, C. C. Doane and M. L. McManus, eds. 681–694. USDA Forest Service Technical Bulletin 1584. Washington, DC: USDA Forest Service.

Whitford, W. G. 1986. Decomposition and nutrient cycling in deserts. In *Pattern and Process in Desert Ecosystems*, W. G. Whitford, ed. 93–117. Albuquerque: University of New Mexico Press.

Willig, M. R. and L. R. Walker. 1999. Disturbance in terrestrial ecosystems: Salient themes, synthesis, and future directions. In *Ecosystems of the World: Ecosystems of Disturbed Ground*, L.R. Walker, ed. 747–767. Amsterdam, The Netherlands: Elsevier Science.

Wilson, E. O. 1987. The little things that run the world (the importance and conservation of invertebrates). *Conservation Biology* 1: 344–346.

Wilson, E. O. 1992. *The Diversity of Life*. Cambridge, MA: Harvard University Press.

Zenk, M. H. and M. Juenger. 2007. Evolution and current status of the phytochemistry of nitrogenous compounds. *Phytochemistry* 68: 2757–2772.

Zhang, D. D., P. Brecke, H. F. Lee, Y. Q. He, and J. Zhang. 2007. Global climate change, war, and population decline in recent human history. *Proceedings of the National Academy of Sciences USA* 104: 19214–19219.

2

Humans versus Insects: The Good, the Bad, and the Ugly

> The three great causes of famine in China are placed as flood, drought, and locusts.
>
> **Charles Valentine Riley (1878)**

Humans have had a long and complex history of interaction with insects, but insects have not always been viewed as "pests" requiring control. The influence of insects on people's lives is universal and pervasive. Even urbanites who have never seen a cow or chicken have dealt with cockroaches, termites, mosquitoes, or filth flies. Locust plagues are among the oldest recorded natural disasters, with records extending back 3,000 years (Riley 1878, 1883; Ma 1958; Konishi and Itô 1973). Widespread entomophobia (especially among Western societies) testifies to the impact of negative associations. However, many insect species have interacted positively with humans. Honey has been a valued food for millennia (Crane 1999), and insects themselves are eaten in many cultures, provide medically or industrially important products, or are important cultural symbols (Clausen 1954).

Most insect species have no more than incidental association with humans, and even those that transmit diseases have important ecological functions that contribute to ecosystem services. For example, of the more than 3,400 known species of mosquitoes worldwide (Foley et al. 2007; Rueda 2008), only 10% are medically important (Manguin and Boëte 2011), most are not even associated with mammals, and some are predators that prey on other mosquito species (Borer et al. 1989; Romoser and Stoffolano 1998). From a study along the Front Range in Colorado, Eisen et al. (2008) reported that only 12 of 27 mosquito species attracted to CO_2-baited traps were potential vectors of West Nile virus. Mosquito control campaigns do not kill only medically important species; other animals that feed on mosquitoes also are affected. Although it is tempting to wish for a world without mosquitoes, were we to be successful in eliminating mosquitoes, we would also eliminate the primary food supply for dragonflies, insectivorous songbirds, and major fisheries, among others.

Insects unquestionably transmit devastating diseases that cause millions of illnesses and deaths each year and are capable of causing billions of dollars in crop losses, resulting in famine and demographic upheaval (Riley 1878; Smith 1954; Pfadt and Hardy 1987; Acuña-Soto et al. 2002; Smith 2007;

Brouqui 2011). However, harsh control measures often are implemented in response to public demand, with little consideration of alternatives or the unintended consequences that may outweigh the benefits. Demand for pest-free produce and pest-free homes frequently drives decisions to apply pesticides at the same time that public concern about effects of pesticides generates controversy. For example, DDT is defended by many as a necessary tool to reduce malaria, but target mosquito species are resistant to DDT; DDT exposure itself causes serious human health problems; and alternative methods to reduce human exposure to mosquitoes are available but often disregarded (Goodman and Mills 1999; Penagos et al. 2004; Vittor et al. 2006).

Only recently have we started separating facts from myths surrounding many interactions between humans and insects (Clausen 1954). The myth of spontaneous generation, of maggots arising de novo in spoiling meat, for example, was dispelled by Francesco Redi's experiments in 1668 (Curtis 1968). Although records of disease epidemics date from at least 5,500 years ago, the identity of many diseases and discovery of their transmission by insects date only from the late 1800s (Manson 1879, 1898; Bray 1996; Chaves-Carballo 2005). Only in the last few decades have we begun to recognize that insect outbreaks in natural ecosystems may represent regulatory feedback that stabilizes ecosystem conditions, rather than a destructive disturbance (Mattson and Addy 1975; Schowalter 1981, 2011).

Humans have used insects as sources of food, as medical and industrial products, as cultural icons, and as indicators of environmental change since prehistoric times. Some insects have always been nuisances. Insect vectors of human and livestock diseases and locusts and other crop "pests" probably have plagued humans primarily since the advent of agriculture led to population concentration. More recently, insects have been employed as instruments of warfare. Various ways in which humans and insects have interacted are described as follows.

2.1 Honey Bees (the "Good")

Honey bees appear in the fossil record at least by the end of the Oligocene 25 million years ago (Crane 1999). A fossilized piece of honeycomb from Malaysia dates from the late Tertiary or early Quaternary period (Stauffer 1979). Early hominids certainly coexisted with honey-producing bees and shared a general primate fondness for honey (Crane 1999). An early cave painting at Altamira in Spain (from about 15,500 years ago) may represent honey hunting (Crane 1999). Mesolithic rock art from Spain, South Africa, and India more clearly records honey hunting, often depicting the use of ladders to climb trees or rock faces (Figure 2.1) (Clausen 1954; Dams 1978; Crane 1999). Unambiguous representation of honey bees and honey collection date from the Egyptian First Dynasty

FIGURE 2.1
Cave art depicting a prehistoric honey hunt. Hunters are shown climbing a ladder to a hive with flying bees at the top. (From Dams, L. R., *Bee World*, 59, 45–53, 1978. With permission.)

5,000 years ago (Crane 1999). Opportunistic honey hunting using similar techniques has continued to the present in many regions.

Beekeeping in trees has been practiced for several thousand years (Crane 1999). Beekeepers excavated artificial cavities, fitted a removable rectangular door with a smaller flight entrance for the bees, collected swarms, and protected hives. Climbing equipment or steps cut into trees facilitated access (Crane 1999). Construction of cavities in stone walls has been practiced in the Mediterranean and Central Asian regions since at least Roman times (Crane 1999). Mesolithic rock art appears to depict use of smoke to drive bees out of nests prior to honey collection (Crane 1999).

Egyptians of the First Dynasty developed artificial hives of clay at least 4,400 years ago (Clausen 1954; Crane 1999). The earliest artificial hives were cylinders made of sun-dried mud or pottery laid horizontally and stacked. Similar hives appeared somewhat later in Mesopotamia and China. Clay cylinders were used into the Roman period but woven, wicker, and wooden hives made from plant materials became common during Roman times. Rome also established laws concerning colony and swarm ownership and penalties for theft (Crane 1999). Upright log hives were introduced in northern Europe some 1,700 years ago and coiled- or woven-straw skeps in eastern Europe some 900 years ago. Protective clothing was developed about 600 years ago.

Coiled straw skeps and wooden-frame hives gradually replaced earlier versions in Europe. These light materials facilitated transport of hives among flower-producing crops to increase honey production. Subsequently, honey bees were transported to the New World with early colonists. A variety of novel designs were introduced during the last few hundred years. However, in 1851, Langstroth (1853) introduced the moveable frame hive that remains in widespread use today. This design permitted harvest of honey and beeswax from removable frames without destroying the colony, as was necessary with earlier designs. This improvement greatly facilitated maintenance of bee colonies and eliminated the need for sufficient growth of new colonies each year before honey production could occur.

Honey has remained among the most important sources of food and medical and industrial products (see sections 2.3 and 2.4) and a commercial trade product for millennia. Ownership of forests with bee trees often included rights to revenue from honey harvest (Crane 1999). Beeswax often has been more valuable than honey because of its pliability and flammability. Honey and beeswax have been preferred or acceptable payment for taxes, tolls, tithes, and trade goods.

More recently, pollination by honey bees and other insects has become more valuable than honey and other products. Pollination by insects has an estimated global value of US$120 billion dollars per year (Costanza et al. 1997). The first commercial introduction of beehives into apple orchards for the purpose of pollination occurred in 1926, increasing the apple harvest by 40% (Clausen 1954). Many crop producers now depend on rented bees to pollinate fruit and vegetable crops, and pollination service by mobile beekeepers, transporting their hives from south to north as the season progresses, has become a multibillion dollar industry in the United States (Figure 2.2). Decimation of native pollinators as a result of the intensified agricultural practices and insecticide use has increased demands on beekeepers to maintain healthy colonies for pollination services (Kremen et al. 2002, 2004; Klein et al. 2007). Current threats to hive survival by introduced Varroa mites, *Varroa destructor*, and the as yet unexplained colony collapse disorder (CCD) jeopardize the production of sufficient food resources globally (Genersch 2010).

2.2 Other Insects Used as Food or Cosmetics

Insects have represented valuable food resources in many parts of the world (Clausen 1954). Although the consumption of insects as food (entomophagy) is not practiced intentionally in Europe or North America (primarily due to

FIGURE 2.2 (SEE COLOR INSERT.)
Pollination services by honey bees and other insects are necessary for 35% of global crop production.

Biblical restrictions in Leviticus 11: 22-23, although grasshoppers and beetles are allowed), Pandora moth, *Coloradia pandora,* larvae and pupae were harvested as food by Native Americans in pine forests of Oregon and northern California (Clausen 1954; Furniss and Carolin 1977). Interestingly, the difficulty of separating insects and insect parts from harvested grain ensures that most people consume insects unintentionally.

Grasshoppers, crickets, cicadas, caterpillars, beetles, termites, and other insects have been valued as food in many cultures (Clausen 1954). Biblical records include several references to eating locusts. Insects continue to make up 5–10% of dietary protein in some cultures (Ramos-Elorduy 2009; Yen 2009). Edible caterpillars (primarily two saturniids, *Gynanisa maja* and *Gonimbrasia zambesina*) represent major food value to indigenous cultures in Zambia, and caterpillar harvest is ritually regulated (Mbata et al. 2002), demonstrating their importance. In a unique study of costs and benefits of insect consumption versus control, Cerritos and Cano-Santana (2008) calculated that the harvest of grasshoppers for sale during an outbreak in Mexico substantially reduced grasshopper damage and provided US$3,000 in revenue per family, compared with a cost of US$150 per family for insecticide treatment.

Insect products also are important food additives. Carmine dye from cochineal scale insects (see following section) is used to color a variety of foods and has become more popular as concerns grow over synthetic food additives. However, the U.S. Food and Drug Administration recently required that all foods or cosmetics containing either cochineal (extract from crushed scale insects) or carmine dye (a more purified dye made from cochineal) to declare this on their ingredient label (Food and Drug Administration 2009).

A number of insect products have been used in cosmetics. Beeswax is commonly used in lip balms, face masks, and hand and body moisturizers (Crane 1999). Carmine (cochineal) dye has been used historically in lipstick and other cosmetics (Vigueras G. and Portillo 2001).

2.3 Medical Uses

Many insects provide materials that have been used widely in medicine (Singh and Jayasomu 2002). Honey has been used in folk medicine to prevent or treat infection for at least 4,000 years (Clausen 1954; Crane 1999). More recently, honey has been demonstrated to have antimicrobial activity against human pathogens (Cooper and Molan 1999; Cooper et al. 1999). Honey was used as early as 4,300 years ago by ancient Akkadians and Egyptians to preserve corpses (Crane 1999).

A number of insects have pharmaceutical value in folk medicine (Clausen 1954; Namba et al. 1988; Pemberton 1999). Bee and wasp venom has been used since ancient times to treat arthritis and rheumatism (Crane 1999). Cantharidin, a defensive alkaloid produced by blister beetles, is used in modern medicine to remove warts (Epstein and Kligman 1948). Furthermore, insects can be used to identify plants with pharmaceutically active compounds (Helson et al. 2009).

Blow fly (*Lucilia* spp.) maggots have been used since antiquity as a wound treatment (Whitaker et al. 2007). Their medical value lies in their selective feeding on necrotic tissue, leaving clean tissue when they depart (Baer 1931; Sherman et al. 2007; Whitaker et al. 2007). Maggots were used for wound healing by Maya Indians, Australian Aborigines, and ancient Burmese over thousands of years (Greenberg 1973; Sherman and Pechter 1988).

More recently, soldiers with wounds colonized by maggots survived and suffered less tissue damage than did soldiers whose wounds were not colonized (Baer 1931; Sherman et al. 2000; Whitaker et al. 2007). During Napoleon's Egyptian campaign in Syria, 1798–1801, Baron Dominique Larrey, the army's general surgeon, reported that certain species of fly destroyed only dead tissue and had a positive effect on wound healing (Sherman et al. 2000). During the American Civil War, surgeons in both armies noted the beneficial effects of blow fly larvae in neglected wounds, which otherwise often became gangrenous (Whitaker et al. 2007). The first therapeutic use of maggots is credited to Dr. J. F. Zacharias, a Confederate medical officer who reported that, "Maggots...in a single day would clean a wound much better than any agents we had at our command...I am sure I saved many lives by their use." He recorded a high survival rate in patients he treated with maggots (Sherman et al. 2000).

During World War I, Dr. William Baer, an orthopedic surgeon, recognized the efficacy of maggot colonization for healing wounds of soldiers who had been left on the battlefield for several days before receiving medical treatment. Later, in 1929, Dr. Baer experimentally introduced maggots into 21 patients with untreatable chronic osteomyelitis and observed rapid debridement, reduction in the number of pathogenic organisms, reduced odor levels, alkalinization of wound beds, and ideal rates of healing (Baer 1931). All 21 patients' open lesions healed completely after two months of maggot therapy. Subsequently, maggot therapy became acceptable for wound treatment, particularly in the United States (McKeever 1933). More than 300 American hospitals employed maggot therapy through the 1940s (Whitaker et al. 2007). Maggot therapy declined after World War II, following the discovery of penicillin. However, maggot therapy is receiving renewed medical attention for wound debridement, as surgical treatments increasingly risk infection by antibiotic-resistant pathogens (Sherman et al. 2000, 2007; Kerridge et al. 2005).

Silk has been used to stitch wounds (Clausen 1954), and the silkworm played a role in discovery of the germ theory of disease. In 1865, Louis Pasteur was asked by the French government for help in solving the "silkworm disease" epidemic that threatened the French silk industry. Eventually, Pasteur discovered that the disease was caused by germs that were contagious but could be controlled, an important step in medical science (Clausen 1954).

Finally, ants have provided antibiotics (e.g., formalin) and surgical material (Clausen 1954). Use of ant mandibles to stitch wounds or surgical incisions was documented in India as early as 1000 B.C. and continued in some areas at least into the early 1900s (Gudger 1925). After live ants are induced to bite the pinched sides of a wound or incision, the ant bodies are removed, and the mandibles remain fastened until the wound heals (Gudger 1925).

2.4 Industrial Uses

Insects have been the source of a number of economically important industrial materials or ideas. Silkworms, *Bombyx mori*, have provided silk for garments and surgical stitching for millennia (Figure 2.3). They and a few

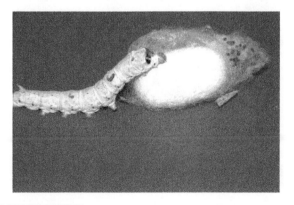

FIGURE 2.3 (SEE COLOR INSERT.)
Silkworm caterpillar and cocoon. Silkworms remain the primary source of commercial silk production.

other moth species remain the only source of commercial silk (Clausen 1954). The economic value of silk supported the historic Silk Road that connected Europe, the Middle East, and China for at least 500 years and was responsible for the introduction of the gypsy moth, *Lymantria dispar*, and other silk-producing species into North America; such efforts to establish silk production in the United States were unsuccessful (Andrews 1868; Forbush and Fernald 1896; Anelli and Prischman-Voldseth 2009).

Carmine (cochineal) dye is one of the oldest organic pigments. The dye is produced from the crushed bodies of scale insects, especially *Kermes vermilio* from the Near East and southern Europe and *Dactylopius coccus* from Mesoamerica (Figure 2.4). Carmine dye was used by ancient Egyptians, Greeks, and Persians. Pre-Columbian civilizations in the New World used the dye as early as 2,000 years ago (Greenfield 2005). The Aztecs raised cochineal scales and exacted them as tribute from their subjects (Clausen 1954). Cochineal was once the most valuable export from colonial Mexico, next to silver (Donkin 1977; Greenfield 2005; Anelli and Prischman-Voldseth 2009), resulting in transport of cactus and cochineal insects to other colonized territories (e.g., Australia) with disastrous results (Clausen 1954). Because this method of providing carmine dye is labor intensive, the scale insects were replaced in the twentieth century with synthetic dyes, but some cochineal production continues, primarily for use by indigenous people (Chávez-Moreno et al. 2009).

Scale insects, *Laccifer lacca*, also were the principle source of commercial lac products, including shoe polishes, electrical insulation, sealing waxes, glazes, phonographic records, and shellac varnishes for wood finishing (Clausen 1954). These insects remain a source of some commercial shellac.

Beeswax has been among the most valuable trade commodities in the past, often valued more highly than honey (Crane 1999). Beeswax was the

FIGURE 2.4 (SEE COLOR INSERT.)
Cochineal scale insects on a prickly pear cactus. The scales remain an important source of carmine dye used in cosmetics and food coloring.

basis for the lost-wax method of casting metals as early as 5,500 years ago (Crane 1999). A beeswax model was coated with clay to form a mold that was allowed to dry. The mold was then heated to melt the wax, which was drained out (lost) through one or more holes. Molten metal was poured into the mold through the hole(s) and allowed to solidify, after which the mold was removed. Beeswax was used to make the earliest candles, invented some 3,500 years ago (Crane 1999). Candles represented an improvement over the earlier oil lamps because the wick could be inserted into the wax candle and a pottery vessel was not required. Beeswax was used to make models of divinities, humans, and animals that were placed within wrappings of Egyptian mummies to accompany the deceased (Crane 1999). Later, the Romans used beeswax for seals on deeds and legal documents. Beeswax has been used as a surface finish or polish for stonework, pottery, wood, and leather for at least 5,500 years. Beeswax also was used as an ingredient in paints to provide luster.

More recently, insects have provided inspiration for technological advancements. During the 1930s, Joe Cox revolutionized the timber harvest industry in the Pacific Northwest with a new saw chain design. Cox's inspiration was the alternating cuts made by the curved mandibles of the cerambycid beetle, *Ergates spiculatus*; while one is cutting, the other acts as a depth gauge (Fore 1970). The "C" shaped cutters alternating sides along the length of the chain improved wood cutting efficiency and reduced maintenance requirements (Fore 1970).

Insects also have led to advances in walking or flying robots (Delcomyn 2004; Aktakka et al. 2011). Insects are uniquely capable of moving efficiently and stably through complex terrain (Ritzmann et al. 2004). Incorporating insect body flexion, leg articulation, and righting ability, as well as sensory feedback mechanisms to avoid collisions, has improved the use of robots in volcanic research and exploration of other planets (Frantsevich 2004; Ritzmann et al. 2004; Webb et al. 2004). Aktakka et al. (2011) demonstrated electrical generation by the wing motion of flying beetles as a model for developing flying robots that could explore hostile environments more quickly.

Gut microorganisms that facilitate the digestion of lignin and cellulose by wood-feeding insects are being explored for novel and inexpensive enzymes for wood degradation. Such biocatalysts may permit economically feasible biofuel production from wood (Cook and Doran-Peterson 2010).

2.5 Cultural Icons

Insects have been valued symbols in many cultures (Clausen 1954). They are depicted in art; used in dances, religious ceremonies and sporting events; or kept as pets.

Termites are used in a purification ceremony among headhunters of the Amazon River region (Clausen 1954). A widow seeking to remarry must be purified by sitting by a termite nest with her head and the nest covered by a tent-like cloth. The nest is set on fire and the widow must inhale the smoke until nearly suffocated in order to be purified and released from the marriage taboo associated with the death of her former husband. Dragonflies are symbols of victory in Japan and depicted during ceremonies to commemorate past successes (Clausen 1954). Some giant silk moth species spin large, tough cocoons that are used ceremonially by Native Americans in the western United States and Mexico and by Zulus in South Africa. The cocoons are split open, the pupae are removed (often eaten, see earlier), a few pebbles are placed in each cocoon, and the cocoons are stitched closed and tied singly or in groups to make rattles and musical instruments (Figure 2.5).

Scarab beetles were the ancient Egyptian symbol of eternal life and held sacred from prehistoric times, as demonstrated by preserved beetles and images carved in jade, malachite, and even emeralds (Clausen 1954). Their image symbolized the sun god, Khepera, creator and highest god to the Egyptians. Scarabs symbolized resurrection when used in burial ceremonies and were often inserted in place of the removed heart of the deceased (Clausen 1954). These Egyptian customs were later carried to Rome, and Roman soldiers customarily wore images of the sacred scarab as a talisman.

FIGURE 2.5 (SEE COLOR INSERT.)
Giant silk moth cocoon rattles worn by Zulu dancer in South Africa. The caterpillars provide valued food, and the pebble-filled cocoons are an important cultural instrument. (From Schowalter, T. D., *Insect Ecology: An Ecosystem Approach*, Elsevier, San Diego, 2011. With permission.)

Children in equatorial Africa tie a string to the leg of a goliath beetle, *Goliathus goliathus*, and use it as a toy, letting the beetle fly noisily (Clausen 1954). Currently, many of the metallic-green "jewel beetle" species of South America are highly prized by collectors worldwide (Cave 2001).

Cicadas symbolized resurrection to the Oraibi Indians of North America and the ancient Chinese, as a result of the adults' sudden appearance out of the ground. In China, a carved jade amulet in the image of a cicada was placed on the tongue of the deceased before burial—the cicada representing resurrection and jade symbolizing the triumph of good over evil (Clausen 1954). The Chinese continue to hold cicadas in high esteem for their songs and keep cicadas in special cages as pets.

The Chinese also value crickets for their singing and keep males in bamboo cages as pets. In addition, cricket fighting is popular in China, with large sums of money wagered on combat and high values for champions (Clausen 1954). Special, often highly ornate, cages are used to house these special insects (Figure 2.6).

FIGURE 2.6 (SEE COLOR INSERT.)
Chinese cricket box to hold fighting crickets. Note the ornate inlay, hidden sliding doors to each of three cricket compartments, and plug for holes used to insert rice grains as food into each compartment. The box is the size of a deck of playing cards for ease of transport in a shirt pocket.

During the early twentieth century, butterfly collecting became a popular social pastime, with amateur societies established to promote interest (Comstock 1911). Some tropical butterflies, including morphos (*Morpho* spp.) and birdwings (especially *Ornithoptera* spp.), have been collected so intensively that they are now endangered and their trade is prohibited by international law (Clausen 1954). Jewel beetles and dragonflies also represent colorful groups enjoyed and traded by avid amateur societies (Cave 2001). These insect orders are the focus of many popular field guides (Pyle 1981; White 1983; Ferro et al. 2010).

2.6 Indicators of Environmental Change

Insects have provided humans with signs of environmental change for millennia, although most of this evidence has been based on superstition (Clausen 1954). The Zuni Indians of the southwestern United States believe that the early appearance of butterflies is an indication of fair weather. White butterflies signify the onset of summer, and white butterflies flying from the southwest forecast rain. Folk beliefs in the eastern United States claim that "woolly bear" caterpillars (the common Arctiid species, *Isia Isabella*) predict local winter weather by the width of their reddish-brown band—the narrower the band, the colder and longer the winter. Similarly, unusually large stores of honey are believed by some to forecast a severe winter (Clausen 1954).

More recent improvement in understanding of factors that influence insect populations has led to identification of species that provide evidence of change in particular environmental conditions, although typically not of local weather forecasts. Because of their short life spans, rapid reproductive rates, and sensitivity to biochemical changes in their resources, insects can provide early warning of environmental changes that are not visible yet in the condition or abundance of larger, longer-lived plants or vertebrates, usually favored as bioindicators (Balanyá et al. 2006; Menéndez 2007).

Water yield and water quality are particularly important ecosystem services that are critical for support of urban and agricultural production (Bonada et al. 2006). Aquatic insects are sensitive to changes in water quality (especially concentrations of oxygen and pollutants), making them particularly useful indicators of water quality (Hawkins et al. 2000; Bonada et al. 2006). For example, replacement of chironomid species characterizing oligo-mesotrophic conditions by species characterizing eutrophic conditions provided early indication of pollution in Lake Balaton, Hungary (Dévai and Moldován 1983; Ponyi et al. 1983).

Ant associations are used as indicators of ecosystem integrity and the status of restoration efforts in Australia (Andersen and Majer 2004). Persistence of species known to be sensitive to adverse environmental conditions provides evidence of suitable conditions. Similarly, grasshoppers, dung beetles, ground beetles, and xylophagous (wood-feeding) beetles can be used to assess ecosystem integrity and recovery status (Klein 1989; Niemelä et al. 1992; Niemelä and Spence 1994; Fielding and Brusven 1995; Grove 2002; Maleque et al. 2009). The sensitivity of insect herbivores to changes in plant biochemistry can be used to identify plant stress before visible chlorosis or other symptoms appear.

The succession of insect species in decomposing carcasses has been applied by law enforcement agencies to determine time and circumstances of death (Smith 1986; Goff 2000; Byrd and Castner 2001; Watson and Carlton 2003). For example, fly colonization rate in a corpse differs between exposed or protected locations. Research on the sequence and timing of colonization by various insect species on corpses under different environmental conditions provides critical evidence in criminal cases.

Insects also provide useful indications of global climate change. Responses to increased global temperatures include measurable shifts in species' geographic ranges toward higher elevations and latitudes, earlier activity in spring as degree-day requirements are met earlier, and measurable change in frequencies of genes conferring temperature tolerance. Menéndez (2007) analyzed distribution data for 1,700 species of plants, insects, and vertebrates and found a significant range shift averaging 6.1 km per decade toward the poles (or 6.1 m per decade upward in elevation) and significant advancement of spring events by two to three days per decade. Balanyá et al. (2006) compared genetic change and climate change for 26 populations of *Drosophila subobscura*, a cosmopolitan species, for which genetic composition has been known for, on average, 24 years. Over this period, 22 of these populations experienced measurable warming, and 21 showed a shift toward the lower latitude (warm adapted) genotype. Temperature and genetic shifts were equivalent to a one-degree shift in latitude toward the equator.

2.7 Nuisances (the "Bad")

A variety of insects are attracted to humans, livestock, or their abodes. While seeking moisture from sweat, gnats and other small flies may annoy people or livestock. One of the Biblical "plagues of Egypt" was of flies.

A number of commensal or parasitic species are closely associated with humans, their clothing, or their structures. Human abodes offer stable temperature and moisture conditions and an abundance of unprotected food, human and animal waste, and other detrital resources that attract and sustain many insect species that would not survive unprotected outside of human habitations. Urban temperatures typically are 10°C higher than the surrounding landscape (Arnfield 2003) and provide large concentrations of resources for a variety of insects.

Bedbugs probably spread with humans moving out of central Asia. These insects generally are considered nuisances, rather than medical problems, because they do not (yet) transmit any human diseases. However, serious infestation of bedbugs causes cimicosis, an allergic condition characterized by an intensely itchy cutaneous response to the bites that can include blisters and papules that fuse into large, scabby lesions. Fleas originated at least as early as the mid-Jurassic and the rounded form of fossils from this period indicate early feeding on reptiles with a later switch to mammals and birds (Huang et al. 2012). The laterally flatted bodies of modern fleas are adaptations for feeding between hair or feathers. These insects are not well-adapted to feeding on nearly hairless humans, but they feed on humans who live in close association with rat, dog, or cat hosts and often transmit diseases from those hosts (see section 2.9). Lice are adapted to the haired parts of humans and can be a serious problem, causing intense itching as well as transmitting serious diseases (see section 2.9).

Human habitations also support various species of cockroaches, silverfish, moths, beetles, flies, ants, and other insects, as well as mites, that feed on clothing, wood, or stored products, or on human debris. Cockroaches, filth flies, and ants spread human disease organisms while crawling from unsanitary refuges to unsealed food or cooking utensils, requiring proper sanitation to avoid illness, but are not specific vectors of disease (Clausen 1954). These insects also can trigger allergies and asthma among humans in close association (Curtis and Davies 2001). A large proportion of urban asthma sufferers are sensitive to cockroach allergens (Santos et al. 1999).

Many people suffer from unexplained itching ascribed to insects or mites although no cause can be identified. This dermatological condition, termed delusory parasitosis, often persists after repeated pesticide applications. This condition may be caused by allergies to various materials, including dust mite or insect allergens (see earlier) or household products—such as soap, laundry detergent, and commonly used organophosphate and pyrethroid insecticides (Victor et al. 2010). In fact, an estimated 16% of the general population may be allergic to pesticides (Penagos et al. 2004).

Termites, carpenter ants, and wood-feeding beetles are serious problems in wooden structures (Figure 2.7). Damage by these insects requires constant repair amounting to billions of dollars per year (Guillot et al. 2010). Public demand for protection from these insects, as well as nuisance pests (see earlier), has fueled an urban pest management industry worth billions of dollars. Use of insecticides in urban environments, especially improper use and disposal by homeowners, may cause as many or more unintended environmental problems than does the more regulated use in agricultural systems.

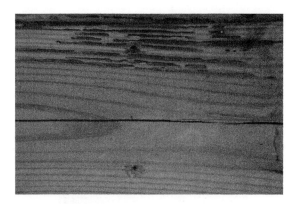

FIGURE 2.7 (SEE COLOR INSERT.)
Termite damage to wood beams.

2.8 Venomous Insects and Other Arthropods

Venomous insects represent serious health threats to humans. A number of insects and other arthropods produce venoms, primarily peptides, including phospholipases, histamines, proteases, and esterases, but also formic and acetic acids (Habermann 1972; Schmidt 1982; Meinwald and Eisner 1995). These chemicals are used for predation and defense but can cause severe reactions or death in humans and other vertebrates.

Both neurotoxic and hemolytic venoms are represented among insects. Phospholipases are particularly well-known because of their high toxicity and their strong antigen activity capable of inducing life-threatening allergy. Venoms are most common among bees, wasps, and ants and consist of a variety of enzymes, biogenic amines (such as histamine and dopamine), epinephrine, norepinephrine, and acetylcholine. This combination produces severe pain and affects cardiovascular, central nervous, and endocrine systems in vertebrates (Schmidt 1982). Melittin, found in honey bee venom, disrupts erythrocyte membranes (Habermann 1972). The toxicity of these compounds rivals that of more-feared snakes (Figure 2.8).

Some venoms include nonpeptide components. For example, the venom of red imported fire ants, *Solenopsis invicta*, contains piperidine alkaloids with hemolytic, insecticidal, and antibiotic effects (Lai et al. 2008).

Although stinging Hymenoptera are the most familiar venomous insects, a variety of other insects also can cause painful stings. Many caterpillars produce proteolytic and histaminergic venoms that can cause symptoms ranging from urticarial dermatitis to life-threatening renal failure or intracerebral hemorrhaging (Diaz 2005; Battisti et al. 2011). Unlike Hymenoptera, which have a stinger attached to a venom gland, envenomation by caterpillars

FIGURE 2.8 (SEE COLOR INSERT.)
Trail sign in Fushan Botanical Garden, Taiwan, warning hikers to be alert for venomous giant centipedes, hornets, and snakes.

occurs primarily through contact with urticating setae ("hairs") or spines (Figure 2.9) (Battisti et al. 2011). However, defensive posturing or flailing by the caterpillar can increase the degree of contact and injection of its venom. A number of Hemiptera, Diptera, Neuroptera, and Coleoptera produce orally derived venoms that facilitate prey capture, as well as defense, but also cause extremely painful bites to humans or other vertebrates (Schmidt 1982).

Spiders, scorpions, and centipedes are important predators that feed on insects, mites, or other arthropods. All species are venomous and bites may

FIGURE 2.9 (SEE COLOR INSERT.)
The saddleback caterpillar, *Sibene stimulea*, is perhaps the most potent stinging caterpillar in North America.

be painful, but most species are too small to pierce human skin and their venom too weak to threaten humans. However, a few species—such as black and brown widows, *Latrodectus* spp., and brown recluse, *Loxosceles reclusa*, in the United States; the Brazilian wandering spider, *Phoneurtria fera*, in southern Brazil; funnel-web spiders, *Atrax* spp., in Australia; the Durango scorpion, *Centruroides suffuses*, in Mexico; and larger species of tropical and subtropical *Scolopendra* centipedes can cause particularly painful bites, tissue necrosis, and death in susceptible patients (Harwood and James 1979).

2.9 Vectors of Human and Livestock Diseases

Insect parasites and vectors of disease organisms certainly have plagued humans and their livestock from earliest times, although the role of insects in disease epidemiology was not understood until relatively recently. Insect vectors and agents of major vertebrate diseases have been documented from as early as the Middle Jurassic (Boucot and Poinar 2010; Huang et al. 2012). Poinar and Telford (2005) detected malarial parasites in a ceratopogonid biting midge in Burmese amber (Cretaceous). Poinar and Poinar (2004a, b) found *Paleoleishmania proterus*, the causal agent of leishmaniasis, in sand flies (Psychodidae) in Burmese amber and Dominican amber (Miocene). Vector relationships often are highly species specific, requiring particular adaptations by the pathogen to permit survival in the vector and transmission into the host. For example, yellow fever is spread primarily by one mosquito species, *Aedes aegypti*, that, unfortunately, has been introduced around the globe via human transportation. However, some pathogens, such as the West Nile virus, are more opportunistic and can be transmitted by a variety of mosquito species, a factor that explains its rapid spread across the United States from its point of introduction in New York City in 1999 (Allan et al. 2009; Schowalter 2011).

Although few insect parasites or human diseases leave traces on bones (Tanke and Rothschild 1997), Egyptian, Chinese, and Peruvian mummies provide evidence of early association with insect parasites and vectors of diseases. Head lice are present on Peruvian Inca mummies (Ewing 1926), and human fleas and body lice were present on humans and sheep in Viking-age Greenland, 1,000 years ago (Sadler 1990). The Old Testament of the Bible provides the earliest record of human wound infection by fly larvae (Zumpt 1965).

The advent of animal husbandry may have facilitated host switching from domesticated animals to humans by some pathogens. For example, smallpox likely originated from close association between humans and cattle infected with cowpox (Diamond 1999). Some diseases still are transmitted between humans and animal reservoirs, for example, bubonic plague and West Nile virus (Peterson 1995; Bray 1996; Allan et al. 2009). However, epidemics probably did not occur until humans began living in dense communities following the advent of agriculture about 6,000 years ago (Diamond 1999).

Insects vector some of the most devastating epidemic diseases, including malaria, yellow fever, bubonic plague, and typhus, all of which have documented effects on recent human history, including the success of military campaigns (Peterson 1995; Diamond 1999). Unfortunately, early descriptions of epidemics often do not permit identification of the specific disease (Bray 1996). Nevertheless, we can assume that many of the devastating epidemics that occurred in ancient times, especially during warfare, were caused by the same insect-transmitted pathogens that have caused more recent epidemics under crowded conditions in military or refugee camps. Of the major human diseases (malaria, bubonic plague, yellow fever, typhus, smallpox, cholera, and influenza), four are vectored by insects. Currently, mosquito-vectored diseases kill more than one million people per year and sicken more than 500 million, mostly in poor, undeveloped countries (Snow et al. 2005). More than two billion people may be exposed to malaria each year (Snow et al. 2005).

A eukaryotic parasite, *Plasmodium falciparum*, transmitted by *Anopheles gambiae* and other mosquitoes, is the agent of human malaria. Evidence of malaria has been found in Egyptian mummies from 5,200 years ago (Miller et al. 1994). Angel (1966) found symptoms of malaria in skeletal remains of early farmers in marshy areas, but not in dry areas or in remains of the latest Paleolithic hunters, in the eastern Mediterranean region as early as 8,500 years ago. The evolution of the sickle cell trait, which confers resistance to malaria in tropical Africa (Friedman 1978), is evidence of long association between human populations and this disease. Dr. Ronald Ross, a surgeon-major in the British Army, was the first to identify mosquitoes as the vector of malarial parasites (Manson 1898). Mosquito control subsequently became the focus of efforts to reduce the incidence of malaria. Though malaria cases have been greatly reduced in Europe and the New World—initially through application of DDT (see next section)—it continues to kill an estimated one million infants and an additional one million others each year, primarily in Africa and Asia (Bray 1996). Deforestation, encroachment of humans into previously unpopulated areas, and altered distribution of aquatic habitats have resulted in a resurgence of malaria in some areas (Vittor et al. 2006).

Bubonic plague, caused by a bacterium (*Yersinia pestis*) transmitted by rat fleas (*Xenopsylla cheopsis*) was endemic to the Mediterranean and eastern Asia regions and already had determined the course of battles and the fall of civilizations before the great epidemics of the Middle Ages (1368–1720). However, in 1342, events triggered an epidemic that killed more than 25% of Europe's population before it ended in 1450 (Bray 1996; Lockwood 2008). In that year, a Mongol army under Jani Beg Khan besieged Kaffa, a strategic Genoese seaport on the Crimean Peninsula. Although the Mongol army was relentless, the city walls were impregnable, and the Genoese dominance of sea power ensured adequate supply to the city. After three years of siege, the

Mongol army was devastated by an outbreak of bubonic plague that forced the withdrawal of their forces. However, before leaving, the Khan ordered that human cadavers be catapulted into the city. The Khan clearly meant to spread the disease to his enemies, but as the disease spread in Kaffa, the Genoese evacuated to their ships and fled to sea, stopping at various ports along the Mediterranean. At each port, infected humans, rats, and fleas disembarked, spreading the disease along the coast of southern Europe. The disease spread inland, and by 1350 the pandemic covered all of Europe. In addition to its devastating effects on European populations, the disease had farther-reaching consequences. The mortality and displacement of serfs caused wages to increase for farm labor and generated labor-saving inventions that ended feudalism and initiated the Renaissance (Bray 1996). Bubonic plague also is credited with halting Napoleon's campaign against the Turks (Lockwood 2008).

Yellow fever, caused by a ribonucleic acid (RNA) virus transmitted primarily by *Aedes aegypti*, originated in Africa and was introduced to the Americas in the 1600s by African slaves (Barrett and Higgs 2007). Periodic epidemics caused major mortality and disruption in the New World from the time of colonization until its elimination in 1905 (Barrett and Higgs 2007). General George Washington was forced to flee Philadelphia during the epidemic of 1793, which killed thousands (Crosby 2006). Nearly 8,000 residents of New Orleans died during an epidemic in 1853, and 20,000 died during an epidemic covering the lower Mississippi Valley in 1878 (Barrett and Higgs 2007). The French effort to build the Panama Canal during 1882–1889 was ended by casualties (about one-third of the workers) resulting from yellow fever (Crosby 2006). In 1881, Dr. Carlos Finley, a Cuban doctor and scientist, proposed that yellow fever is transmitted by mosquitoes rather than direct human contact (Chaves-Carballo 2005). Unfortunately, his work was ignored for 20 years before Dr. Walter Reed, a U.S. Army surgeon, and his team confirmed Dr. Finley's work in 1903 during U.S. construction of the Panama Canal (Crosby 2006). Subsequently, mosquito control was implemented as a major tactic during an outbreak in New Orleans in 1905. Mortality during this epidemic was much lower than during previous epidemics, and no subsequent epidemics have occurred in the United States. Nevertheless, yellow fever continues to enter the United States via travelers to South America, and future outbreaks could occur if an *Aedes aegypti* mosquito fed on an infected person before being diagnosed (Crosby 2006).

Typhus, caused by a bacterium (*Rickettsia prowazekii*) transmitted by human lice (*Pediculus humanus*) is commonly associated with war, famine, and overcrowding (Bray 1996; Brouqui 2011). Typhus unquestionably has killed more soldiers than has combat (Diamond 1999). The origins of this disease are not clear. No convincing description of disease that can unambiguously be identified as typhus is available until about 1,000 years ago, with the first recorded epidemic beginning in Spain in 1489. This epidemic spread to Italy and caused the decimation of French troops under Francis I and the cessation of their siege of Naples (Bray 1996). Typhus subsequently was responsible for determining the outcome of every war in Europe, especially the outcome of the Thirty Years' War (1618–1648), during which Germany lost half of its population, and Napoleon's retreat from Moscow (1812), during which his army was reduced from 655,000 to 93,000 and made vulnerable to eventual defeat (Bray 1996; Lockwood 2008). After the Austrian attack on Serbia in 1914, typhus forced an Austrian retreat, but 60,000 typhus-infected Austrian prisoners spread the disease to Serbian troops. Typhus killed an estimated 25% of the Serbian army and eliminated Serbia as a warring nation (Bray 1996; Lockwood 2008). During World War II, the successful control of lice on troops using DDT prevented typhus from becoming a major factor (Lockwood 2008). However, lice and typhus have continued to be major medical challenges in refugee camps.

In addition to specific insect vectors of diseases, a number of human diseases are transmitted mechanically by insects associated with human or animal wastes. For example, filth flies that breed in dung or spoiling detritus can contaminate human food or cooking utensils as they travel between food and breeding resources (Eldridge and Edman 2003; Lockwood 2008; Mullen and Durden 2009).

2.10 Crop Pests

Insect associations with plants predate human agriculture, but herbivorous insect populations typically are regulated at small sizes by low availability and antiherbivore defenses of host plants (Schowalter 2011). When humans began planting crop species at higher densities than occurred naturally, one major factor regulating insect herbivore populations was relaxed, allowing host-adapted insects to increase in abundance on crops. Breeding efforts to improve crop flavor and production reduced the concentration of distasteful plant defenses (Michaud and Grant 2009) and increased nutritional quality for insects, relaxing a second major factor regulating insect herbivore populations. Records of crop-destroying locust outbreaks go back at least 3,000 years, according to Biblical and ancient Chinese records (Riley 1883; Ma 1958; Konishi and Itô 1973; Stige et al. 2007). Currently, a vast majority of humans rely on agricultural production for their food supply, but insects destroy about 18% of global crop production and represent a major impediment to producing sufficient food for a growing human population (Oerke 2006).

Insect outbreaks are capable of causing devastating losses of food or other resources and have led to substantial expenditures for pest control. For example, locust outbreaks have destroyed entire crop and rangeland production over large areas and caused massive human migration from devastated areas or have required extraordinary economic aid to sustain farmers or ranchers on their land (Riley 1878; Smith 1954; Pfadt and Hardy 1987). Even during average years, 21–23% loss of available range vegetation due to grasshopper feeding represents an economic loss of about US$393 million (Pfadt and Hardy 1987). Locust laws represent the oldest known pest management practices. During the Song Dynasty, in 1075, Emperor Shen-Tzoon ordered district governors and their assistants to be present at locations where locusts were expected to occur and to invite their people to gather locusts in exchange for established bounties of grain or equivalent money (Riley 1883). Gathered locusts were to be burned. Locusts in the open were to be driven by villagers into ditches where they could be burned. Failure to properly implement these practices was punishable by whipping (Riley 1883). A number of western states in the United States enacted locust laws that provided for payment of bounties for grasshoppers collected during the devastating plague of Rocky Mountain grasshoppers, *Melanoplus spretus*, in the 1870s (Riley 1878). On the other hand, the locusts themselves could often be consumed in place of the lost crops (Cerritos and Cano-Santana 2008), as described earlier.

During the late 1800s and early 1900s, the spread of boll weevils (*Anthonomus grandis*) from sparsely distributed native cotton (*Gossypium* spp.) throughout the U.S. Cotton Belt devastated the southern economy, ended the reign of cotton as the dominant crop in the South, and led to a massive demographic shift away from bankrupt farms and communities (Smith 2007). Tree mortality caused by gypsy moth defoliation resulted in an economic loss of more than US$104 million over a three-year period in Pennsylvania (Ticehurst and Finley 1988); more than US$194 million was spent on monitoring and control of gypsy moths in the United States during 1985–2004 (Johnson et al. 2006).

Crop losses to insects have generated focused efforts to control insect populations, including coordinated efforts to collect and destroy target insects, invention of various tools to improve control (such as hopper dozers and various chemical application devices), and application of various chemicals to kill targeted insects (Riley 1883, 1885; Jones 1917). The earliest control methods were primarily physical removal of insects from crops (Riley 1878; 1883; Vincent et al. 2003). Collected insects could be destroyed or used commercially (Riley 1878, 1883; Cerritos and Cano-Santana 2008; Ramos-Elorduy 2009). Crop breeding for insect resistance and crop rotation were recognized methods for reducing losses to insects by the late 1700s (Painter 1951; Smith 2005). Pyrethrum, prepared from the powdered flower heads of *Pyrethrum roseum*, was used at least as early as 1800 to control insects south of the Caucasus Mountains (Riley 1885). However, pyrethrum was effective only in direct contact with insects and became ineffective within an hour of application. More rapid and effective measures were demanded. Consequently, arsenical compounds became widely used in the late 1800s and early 1900s, followed by DDT and other chlorinated hydrocarbons during and after World War II.

Insect adaptation to selection by insecticides was recognized as early as the 1920s (Melander 1923) and evidence of resistance to DDT by the mid 1950s (Clausen 1954; Roussel and Clower 1957; see Chapter 6). Accumulating evidence of nontarget environmental effects and target pest resistance to chemical controls has led to increasing efforts to manage insect populations with more target-specific methods (Schowalter 2011). However, any insecticide applied consistently results in resistance among target insects within a decade. The Integrated Pest Management (IPM) concept that developed during the 1950s–1970s emphasizes adherence to ecological principles and a combination of multiple preventative (routinely used) and remedial (used as necessary) tactics to minimize or delay resistance and minimize nontarget effects (see Chapter 9).

2.11 Instruments of Warfare (the "Ugly")

A less laudatory association between humans and insects involves the use of insects in warfare. The Mayans developed bee grenades and booby traps as early as 2,600 years ago (Lockwood 2008). Around 2,300 years ago, the Greeks recorded the release of bees and wasps into tunnels being excavated by enemy forces to undermine a fortified position (Crane 1999). Nests of stinging bees, wasps, or ants could be enclosed in clay pots, sealed with grass plugs, and thrown into enemy forces or over fortifications (Lockwood 2008). During the Middle Ages, defenders of attacked fortresses would hurl beehives down on invading soldiers; in some cases the attackers were so badly stung that they abandoned their attack (Crane 1999; Lockwood 2008). Beehives also were launched into besieged fortresses by catapults as early as the Roman period (Lockwood 2008). Bees were used as weapons of defense or offense into the eighteenth century (Lockwood 2008). Insect toxins have been used for the preparation of poisoned arrows (Lockwood 2008).

During World War II, Japanese scientists of Unit 731 developed methods for mass rearing fleas infected with particularly virulent forms of bubonic plague, using human guinea pigs from conquered Manchuria and POW camps (Lockwood 2008). Infected fleas were packed in porcelain Uji bombs that would explode 100–200 m above the ground, raining fleas over a large area and initiating disease outbreaks (Lockwood 2008). An attack on Quzhou in 1940 initiated an outbreak that continued for six years, even after the end of the war, eventually killing 50,000 people. Over the next several years, the Japanese launched more than a dozen such attacks, causing more than 100,000 casualties as well as general panic (Lockwood 2008).

Also during World War II, German, French, and American scientists worked on weaponization of the Colorado potato beetle to destroy enemy crops and cause famine (Lockwood 2008). Others pursued research on livestock pests and transmission of weaponized diseases by native filth flies from aerially dropped, contaminated dung (Lockwood 2008).

The possibility that the West Nile virus outbreak in the United States, starting in 1999, was a bioterrorist attack by Iraq was considered by bioterrorism experts. If this had represented an attack, the United States responded slowly, and only quickly launched mosquito control prevented a worse outcome (Lockwood 2008). Bioterrorism has been discounted, but the origin of this mosquito-borne virus has not been explained (perhaps an infected bird or mosquito introduced into the United States); several countries had explored this disease earlier as an instrument of war (Lockwood 2008). West Nile disease is relatively mild (<10% death rate), but its rate of spread across the United States (via a combination of bird movement and a variety of mosquito vectors) and the public reaction are reminders that a disease need not be deadly to disrupt daily life (Lockwood 2008; Schowalter 2011). The prospect of future attacks by weaponized insects should not be discounted, but the real danger may be panic spread by the mere threat of such attacks (Lockwood 2008).

2.12 Summary

Insects and humans have interacted since prehistoric times. Insects are a far older group, and parasitic species certainly had adapted to human hosts as they had to their predecessors. Insects also have served as vectors of human and livestock diseases from earliest times, but the role of insects in transmission of disease has been known only since the late 1800s. Many species are simply nuisances. Humans undoubtedly used insects as food from prehistoric times although fossil evidence of this would be difficult to find.

Over time, a variety of other associations has developed. Insects have provided medical and industrial products and served as important cultural icons and as indicators of environmental changes. Silk and some cosmetic dyes and wood finishing materials are derived solely or primarily from insects. Insect diversity and morphology have inspired art and new technologies. With the advent of agriculture and the corresponding concentration of humans and crops into sedentary societies, crop pests and epidemic, insect-vectored, and other diseases began to plague humans. More recently, humans have used insects as instruments of warfare. Although entomophobia is widespread among Western societies, and some insects continue to cause serious crop loss and human disease, most insects pose no threat to humans; many provide useful products or contribute to other ecosystem services. Factors that affect their abundance and effects on ecosystem services are the subjects of the following chapters.

References

Acuña-Soto, R., D. W. Stahle, M. K. Cleaveland, and M. D. Therrell. 2002. Megadrought and megadeath in 16th century Mexico. *Emerging Infectious Diseases* 8: 360–362.

Aktakka, E. E., H. Kim, and K. Najafi. 2011. Energy scavenging from insect flight. *Journal of Micromechanics and Microengineering* 21: 095016.

Allan, B. F., R. B. Langerhans, W. A. Ryberg, W. J. Landesman, N. W. Griffin, R. S. Katz, B. J. Oberle, M. R. Schutzenhofer, K. N. Smyth, A. de St. Maurice, et al. 2009. Ecological correlates of risk and incidence of West Nile virus in the United States. *Oecologia* 158: 699–708.

Andersen, A. N. and J. D. Majer. 2004. Ants show the way Down Under: Invertebrates as bioindicators in land management. *Frontiers in Ecology and the Environment* 2: 291–298.

Andrews, W. V. 1868. The Cynthia silk-worm. *American Naturalist* 2: 311–320.

Anelli, C. M. and D. A. Prischmann-Voldseth. 2009. Silk batik using beeswax and cochineal dye: An interdisciplinary approach to teaching entomology. *American Entomologist* 55: 95–105.

Angel, J. L. 1966. Porotic hyperostosis, anemias, malaria, and marshes in the prehistoric eastern Mediterranean. *Science* 153: 760–763.

Arnfield, A. J. 2003. Two decades of urban climate research: A review of turbulence, exchanges of energy and water, and the urban heat island. *International Journal of Climatology* 23: 1–26.

Baer, W. S. 1931. The treatment of chronic osteomyelitis with the maggot (larvae of the blowfly). *Journal of Bone and Joint Surgery* 13: 438–475.

Balanyá, J., J. M. Oller, R. B. Huey, G. W. Gilchrist, and L. Serra. 2006. Global genetic change tracks global climate warming in *Drosophila subobscura*. *Science* 313: 1773–1775.

Barrett, A. D. T. and S. Higgs. 2007. Yellow fever: A disease that has yet to be conquered. *Annual Review of Entomology* 52: 209–229.

Battisti, A., G. Holm, B. Fagrell, and S. Larsson. 2011. Urticating hairs in arthropods: Their nature and medical significance. *Annual Review of Entomology* 56: 203–220.

Bonada, N., N. Prat, V. H. Resh, and B. Statzner. 2006. Developments in aquatic insect biomonitoring: A comparative analysis of recent approaches. *Annual Review of Entomology* 51: 495–523.

Borer, D. J., C. A. Triplehorn, and N. F. Johnson. 1989. *An Introduction to the Study of Insects*, 6th ed. Orlando, FL: Harcourt Brace.

Boucot, A. J. and G. O. Poinar, Jr. 2010. *Fossil Behavior Compendium*. Boca Raton, FL: CRC Press/Taylor & Francis.

Bray, R. S. 1996. *Armies of Pestilence: The Impact of Disease on History*. New York: Barnes and Noble.

Brouqui, P. 2011. Arthropod-borne diseases associated with political and social disorder. *Annual Review of Entomology* 56: 357–374.

Byrd, J. H. and J. L. Castner, eds. 2001. *The Utility of Arthropods in Legal Investigations*. Boca Raton, FL: CRC Press/Taylor & Francis.

Cave, R. D. 2001. Jewel scarabs. *National Geographic* 199: 52–61.

Cerritos, R. and Z. Cano-Santana. 2008. Harvesting grasshoppers *Sphenarium purpurascens* in Mexico for human consumption: A comparison with insecticidal control for managing pest outbreaks. *Crop Protection* 27: 473–480.

Chaves-Carballo., E. 2005. Carlos Finley and yellow fever: triumph over adversity. *Military Medicine* 170: 881–885 .

Chávez-Moreno, C. K., A. Tecante, and A. Casas. 2009. The *Opuntia* (Cactaceae) and *Dactylopius* (Hemiptera: Dactylopiidae) in Mexico: A historical perspective of use, interaction, and distribution. *Biodiversity Conservation* 18: 3337–3355.

Clausen, L. W. 1954. *Insect Fact and Folklore*. New York: MacMillan.

Comstock, A. B. 1911. *The Handbook of Nature Study*. Ithaca, NY: Comstock.

Cook, D. M. and J. Doran-Peterson. 2010. Mining diversity of the natural biorefinery housed within *Tipula abdominalis* larvae for use in an industrial biorefinery for production of lignocellulosic ethanol. *Insect Science* 13: 303–312.

Cooper, R. A. and P. C. Molan. 1999. The use of honey as an antiseptic in managing *Pseudomonas* infection. *Journal of Wound Care* 8: 161–164.

Cooper, R. A., P. C. Molan, and K. G. Harding. 1999. Antibacterial activity of honey against strains of *Staphylococcus aureus* from infected wounds. *Journal of the Royal Society of Medicine* 92: 283–285.

Costanza, R., R. d'Arge, R. de Groot, S. Farger, M. Grasso, B. Hannon, K. Limburg, S. Naeem, R. V. O'Neill, et al. 1997. The value of the world's ecosystem services and natural capital. *Nature* 387: 253–260.

Crane, E. 1999. *The World History of Beekeeping and Honey Hunting*. New York: Routledge.

Crosby, M. C. 2006. *The American Plague: The Untold Story of Yellow Fever, the Epidemic that Shaped our History*. New York: Berkeley Books.

Curtis, C. F. and C. R. Davies. 2001. Present use of pesticides for vector and allergen control and future requirements. *Medical and Veterinary Entomology* 15: 231–235.

Curtis, H. 1968. *Biology*. New York: Worth Publishers.

Dams, L. R. 1978. Bees and honey-hunting scenes in the Mesolithic rock art of eastern Spain. *Bee World* 59: 45–53.

Delcomyn, F. 2004. Insect walking and robotics. *Annual Review of Entomology* 49: 51–70.

Dévai, G. and J. Moldován. 1983. An attempt to trace eutrophication in a shallow lake (Balaton, Hungary) using chironomids. *Hydrobiologia* 103: 169–175.

Diamond, J. 1999. *Guns, Germs, and Steel: The Fates of Human Societies*. New York: W.W. Norton.

Diaz, J. H. 2005. The evolving global epidemiology, syndromic classification, management, and prevention of caterpillar envenoming. *American Journal of Tropical Medicine and Hygiene* 72: 347–357.

Donkin, R. A. 1977. An ethnogeographical study of cochineal and the opuntia cactus. *Transactions of the American Philosophical Society* 67: 1–84.

Eisen, L., B. G. Bolling, C. D. Blair, B. J. Beaty, and C. G. Moore. 2008. Mosquito species richness, composition, and abundance along habitat-climate-elevation gradients in the northern Colorado Front Range. *Journal of Medical Entomology* 45: 800–811.

Eldridge, B. F. and J. D. Edman. 2003. *Medical Entomology: A Textbook on Public Health and Veterinary Problems Caused by Arthropods*, 2nd ed. Dordrecht, The Netherlands: Springer.

Epstein W. L. and A. M. Kligman. 1958. Treatment of warts with cantharidin. *American Medical Association Archives of Dermatology* 77: 508–511.

Ewing, H. E. 1926. A revision of the American lice of the genus *Pediculus*, together with a consideration of the significance of their geographical and host distribution. *Proceedings of the United States National Museum* 68: 1–30.

Ferro, M. L., K. A. Parys, and M. L. Gimmel. 2010. *Dragonflies and Damselflies of Louisiana*. Baton Rouge: Louisiana State Arthropod Museum.

Fielding, D. J. and M. A. Brusven. 1995. Ecological correlates between rangeland grasshopper (Orthoptera: Acrididae) and plant communities of southern Idaho. *Environmental Entomology* 24: 1432–1441.

Foley, D. H., L. M. Rueda, and R. C. Wilkerson. 2007. Insight into global mosquito biogeography from country species records. *Journal of Medical Entomology* 44: 554–567.

Food and Drug Administration. 2009. Listing of color additives exempt from certification; food, drug, and cosmetic labeling: Cochineal extract and carmine declaration; confirmation of effective date. *Federal Register* 74: 10483.

Forbush, E. H. and C. H. Fernald. 1896. *The Gypsy Moth*. Boston: Massachusetts Board of Agriculture.

Fore, T. 1970. Joe Cox and the beetle bug. *Oregonizer* 93: 1–3.

Frantsevich, L. 2004. Righting kinematics in beetles (Insecta: Coleoptera). *Arthropod Structure and Development* 33: 221–235.

Friedman, M. J. 1978. Erythrocytic mechanism of sickle cell resistance to malaria. *Proceedings of the National Academy of Sciences USA* 75: 1994–1997.

Furniss, R. L. and V. M. Carolin. 1977. *Western Forest Insects*. USDA Forest Service Misc. Publ. 1339. Washington, DC: USDA Forest Service.

Gensersch, E. 2010. Honey bee pathology: Current threats to honey bees and beekeeping. *Applied Microbiology and Biotechnology* 87: 87–97.

Goff, M. L. 2000. *A Fly for the Prosecution: How Insect Evidence Helps Solve Crimes*. Harvard University Press, Cambridge, MA.

Goodman, C. A. and A. J. Mills. 1999. The evidence base on the cost-effectiveness of malaria control measures in Africa. *Health Policy and Planning* 14: 301–312.

Greenberg, B. 1973. *Flies and Disease*. Princeton, NJ: Princeton University Press.

Greenfield, A. B. 2005. *A Perfect Red: Empire, Espionage, and the Quest for the Color of Desire*. New York: Harper Collins.

Grove, S. J. 2002. Saproxylic insect ecology and the sustainable management of forests. *Annual Review of Ecology and Systematics* 33: 1–23.

Gudger, E. W. 1925. Stitching wounds with the mandibles of ants and beetles. *Journal of the American Medical Association* 84: 1861–1864.

Guillot, F. S., D. R. Ring, A. R. Lax, A. Morgan, K. Brown, C. Riegel, and D. Boykin. 2010. Area-wide management of the Formosan subterranean termite, *Coptotermes formosanus* Shiraki (Isoptera: Rhinotermitidae), in the New Orleans French Quarter. *Sociobiology* 55: 311–338.

Habermann, E. 1972. Bee and wasp venoms. *Science* 177: 314–322.

Harwood, R. F. and M. T. James. 1979. *Entomology in Human and Animal Health*, 7th ed. New York: Macmillan.

Hawkins, C. P., R. H. Norris, J. N. Hogue, and J. W. Feminella. 2000. Development and evaluation of predictive models for measuring the biological integrity of streams. *Ecological Applications* 10: 1456–1477.

Helson, J. E., T. L. Capson, T. Johns, A. Aiello, and D. M. Windsor. 2009. Ecological and evolutionary bioprospecting: Using aposematic insects as guides to rainforest plants active against disease. *Frontiers in Ecology and the Environment* 7: 130–134.

Huang, D., M. S. Engel, C. Cai, H. Wu, and A. Nel. 2012. Diverse transitional giant fleas from the Mesozoic era of China. *Nature* 483: 201–204.

Johnson, D. M., A. M. Liebhold, P. C. Tobin, and O. N. Bjørnstad. 2006. Allee effects and pulsed invasion by the gypsy moth. *Nature* 444: 361–363.

Jones, C. R. 1917. Grasshopper control. *Colorado Agricultural Experiment Station Bulletin* 233, Ft. Collins, CO: Colorado State University.

Kerridge, A., H. Lappin-Scott, and J. R. Stevens. 2005. Antibacterial properties of larval secretions of the blowfly, *Lucilia sericata*. *Medical and Veterinary Entomology* 19: 333–337.

Klein, A.-M, B. E. Vaissière, J. H. Cane, I. Steffen-Dewenter, S. A. Cunningham, C. Kremen, and T. Tscharntke. 2007. Importance of pollinators in changing landscapes for world crops. *Proceedings of the Royal Society B* 274: 303–313.

Klein, B. C. 1989. Effects of forest fragmentation on dung and carrion beetle communities in central Amazonia. *Ecology* 70: 1715–1725.

Konishi, M. and Y. Itô. 1973. Early entomology in East Asia. In *History of Entomology*, R. F. Smith, T. E. Mittler, and C. N. Smith, eds. 1–20. Palo Alto, CA: Annual Reviews.

Kremen, C., N. M. Williams, R. L. Bugg, J. P. Fay, and R. W. Thorp. 2004. The area requirements of an ecosystem service: Crop pollination by native bee communities in California. *Ecology Letters* 7: 1109–1119.

Kremen, C., N. M. Williams, and R. W. Thorp. 2002. Crop pollination from native bees as risk from agricultural intensification. *Proceedings of the National Academy of Sciences USA* 99: 16812–16816.

Lai, L.-C., R.-N. Huang, and W.-J. Wu. 2008. Venom alkaloids of monogyne and polygyne forms of the red imported fire ant, *Solenopsis invicta*, in Taiwan. *Insectes Sociaux* 55: 443–449.

Langstroth, L. L. 1853. *The Hive and the Honey-Bee: A Bee Keeper's Manual*. Northampton, MA: Hopkins, Bridgman.

Lockwood, J. A. 2008. *Six-Legged Soldiers: Using Insects as Weapons of War*. New York: Oxford University Press.

Ma, S.-C. 1958. The population dynamics of the oriental migratory locust (*Locusta migratoria manilensis* Mayen) in China. *Acta Entomologica Sinica* 8: 1–40.

Maleque, M. A., K. Maeto, and H. T. Ishii. 2009. Arthropods as bioindicators of sustainable forest management, with a focus on plantation forests. *Applied Entomology and Zoology* 44: 1–11.

Manguin, S. and C. Boëte. 2011. Global impact of mosquito biodiversity, human vector-borne diseases, and environmental change. In *The Importance of Biological Interactions in the Study of Biodiversity*, J. L. Pujol, ed. 27–50. Croatia: InTech, Rikela. http://www.intechopen.com/articles/show/title/global-impact-of-mosquito-biodiversity-human-vector-borne-diseases-and-environmental-change#reference. Boëte.

Manson, P. 1879. On the development of *Filaria sanguinis hominis*, and on the mosquito considered as a nurse. *Zoological Journal of the Linnaean Society* 14: 304–311.

Manson, P. 1898. Surgeon-Major Ronald Ross's recent investigations on the mosquito-malaria theory. *British Medical Journal* June 18: 1575–1577.

Mattson, W. J. and N. D. Addy. 1975. Phytophagous insects as regulators of forest primary production. *Science* 190: 515–522.

Mbata, K. J., E. N. Chidumayo, and C. M. Lwatula. 2002. Traditional regulation of edible caterpillar exploitation in the Kopa area of Mpika district in northern Zambia. *Journal of Insect Conservation* 6: 115–130.

McKeever, D. C. 1933. Maggots in treatment of osteomyelitis: A simple inexpensive method. *Journal of Bone and Joint Surgery* 15: 85–93.

Meinwald, J. and T. Eisner. 1995. The chemistry of phyletic dominance. *Proceedings of the National Academy of Sciences USA* 92: 14–18.

Melander, A.L. 1923. *Tolerance of San Jose Scale to Sprays*. Agricultural Experiment Station Bulletin 174. Pullman, WA: State College of Washington.

Menéndez, R. 2007. How are insects responding to global warming? *Tijdschrift voor Entomologie* 150: 355–365.

Michaud, J. P. and A. K. Grant. 2009. The nature of resistance to *Dectes texanus* (Col., Cerambycidae) in wild sunflower, *Helianthus annuus*. *Journal of Applied Entomology* 133: 518–523.

Miller, R. L., S. Ikraum, G. J. Armelagos, R. Walker, W. B. Harer, C. J. Shiff, D. Baggett, M. Carrigan, and S. M. Marel. 1994. Diagnosis of *Plasmodium falciparum* in mummies using the rapid manual ParaSightTM-F test. *Transaction of the Royal Society of Tropical Medicine and Hygiene* 88: 31–32.

Mullen, G. and L. Durden. 2009. *Medical and Veterinary Entomology*, 2nd ed. San Diego: Elsevier/Academic.

Namba, T., Y. H. Ma, and K. Inagaki. 1988. Insect-derived crude drugs in the Chinese Song Dynasty. *Journal of Ethnopharmacology* 24: 247–285.

Niemelä, J., D. Langor, and J. R. Spence. 1992. Effects of clear-cut harvesting on boreal ground beetle assemblages in western Canada. *Conservation Biology* 7: 551–561.

Niemelä, J. and J. R. Spence. 1994. Distribution of forest dwelling carabids: Spatial scale and concept of communities. *Ecography* 17: 166–175.

Oerke, E.-C. 2006. Centenary review: Crop losses to pests. *Journal of Agricultural Science* 144: 31–43.

Painter, R. H. 1951. *Insect Resistance in Crop Plants*. New York: Macmillan.

Pemberton, R. W. 1999. Insects and other arthropods used as drugs by Korean traditional medicine. *Journal of Ethnopharmacology* 65: 207–216.

Penagos, H., C. Ruepert, T. Partanen, and C. Wesseling. 2004. Pesticide patch test series for the assessment of allergic contact dermatitis among banana plantation workers in Panama. *Dermatitis* 15: 137–145.

Peterson, R. K. D. 1995. Insects, disease, and military history. *American Entomologist* 41: 147–160.

Pfadt, R. E. and D. M. Hardy. 1987. A historical look at rangeland grasshoppers and the value of grasshopper control programs. In *Integrated Pest Management on Rangeland*, J. L. Capinera, ed. 183–195, Boulder, CO: Westview Press.

Poinar, G., Jr., and R. Poinar. 2004a. *Palaeoleishmania proterus* n. gen., n. sp., (Trypanosomatidae: Kinetoplastida) from Cretaceous Burmese amber. *Protist* 155: 305–310.

Poinar, G., Jr., and R. Poinar. 2004b. Evidence of vector-borne disease of early Cretaceous reptiles. *Vector-Borne and Zoonotic Diseases* 4: 281–284.

Poinar, G., Jr., and S. R. Telford, Jr. 2005. *Paleohaemoproteus burmacis* gen. n., sp. n. (Haemospororida: Plasmodiidae) from an Early Cretaceous biting midge (Diptera: Ceratopogonidae). *Parasitology* 131: 79–84.

Ponyi, J. E., I. Tátrai, and A. Frankó. 1983. Quantitative studies on Chironomidae and Oligochaeta in the benthos of Lake Balaton. *Archiv für Hydrobiologie* 97: 196–207.

Pyle, R. M. 1981. *The Audubon Society Field Guide to North American Butterflies*. New York: Alfred A. Knopf.

Ramos-Elorduy, J. 2009. Anthro-entomophagy: Cultures, evolution, and sustainability. *Entomological Research* 39: 271–288.

Riley, C. V. 1878. *First Annual Report of the United States Entomological Commission for the Year 1877 Relating to the Rocky Mountain Locust and the Best Methods of Preventing Its Injuries and of Guarding Against Its Invasions, in Pursuance of an Appropriation Made by Congress for This Purpose.* Washington, DC: U.S. Department of Agriculture.

Riley, C. V. 1883. *Third Report of the United States Entomological Commission, Relating to the Rocky Mountain Locust, the Western Cricket, the Army-Worm, Canker Worms, and the Hessian Fly, Together with Descriptions of Larvae of Injurious Forest Insects, Studies on the Embryological Development of the Locust and of Other Insects, and on the Systematic Position of the Orthoptera in Relation to Other Orders of Insects.* Washington, DC: U.S. Department of Agriculture.

Riley, C. V. 1885. *Fourth Report of the United States Entomological Commission, Being a Revised Edition of Bulletin No. 3, and the Final Report on the Cotton Worm, Together with a Chapter on the Boll Worm.* Washington, DC: U.S. Department of Agriculture.

Ritzmann, R. E., R. D. Quinn, and M. S. Fischer. 2004. Convergent evolution and locomotion through complex terrain by insects, vertebrates, and robots. *Arthropod Structure and Development* 33: 361–379.

Romoser, W. S. and J. G. Stoffolano, Jr. 1998. *The Science of Entomology*, 4th ed. Boston: McGraw-Hill.

Roussel, J. S. and D. F. Clower 1957. Resistance to the chlorinated hydrocarbon insecticides in the boll weevil. *Journal of Economic Entomology* 50: 463–468.

Rueda, L. M. 2008. Global diversity of mosquitoes (Insecta: Diptera: Culicidae) in freshwater. *Hydrobiologia* 595: 477–487.

Sadler, J. P. 1990. Record of ectoparasites on humans and sheep from Viking Age deposits in the former Western Settlement on Greenland. *Journal of Medical Entomology* 27: 628–631.

Santos, A. B. R., M. D. Chapman, R. C. Aalberse, L. D.Vailes, V. P. L. Ferriani, C. Oliver, M. C. Rizzo, C. K. Naspitz, and L. K. Arruda. 1999. Cockroach allergens and asthma in Brazil: Identification of tropomyosin as a major allergen with potential crossreactivity with mite and shrimp allergens. *Journal of Allergy and Clinical Immunology* 104: 329–337.

Schmidt, J. O. 1982. Biochemistry of insect venoms. *Annual Review of Entomology* 27: 339–368.

Schowalter, T. D. 1981. Insect herbivore relationship to the state of the host plant: Biotic regulation of ecosystem nutrient cycling through ecological succession. *Oikos* 37: 126–130.

Schowalter, T. D. 2011. *Insect Ecology: An Ecosystem Approach.* San Diego: Elsevier/Academic.

Sherman, R. A., M. J. R. Hall, and S. Thomas. 2000. Medical maggots: An ancient remedy for some contemporary afflictions. *Annual Review of Entomology* 45: 55–81.

Sherman, R. A. and E. A. Pechter. 1988. Maggot therapy: A review of the therapeutic applications of fly larvae in human medicine, especially for treating osteomyelitis. *Medical and Veterinary Entomology* 2: 225–230.

Sherman, R. A., H. Stevens, D. Ng, and E. Iversen. 2007. Treating wounds in small animals with maggot debridement therapy: A survey of practitioners. *Veterinary Journal* 173: 138–143.

Singh, K. P. and R. S. Jayasomu. 2002. Bombyx mori—A review of its potential as a medicinal insect. *Pharmaceutical Biology* 40: 28–32.

Smith, C. M. 2005. *Plant Resistance to Arthropods: Molecular and Conventional Approaches.* Dordrecht, The Netherlands: Springer.

Smith, K. G. V. 1986. *A Manual of Forensic Entomology.* Ithaca, NY: Cornell University Press.

Smith, R. C. 1954. An analysis of 100 years of grasshopper populations in Kansas (1854 to 1954). *Transactions of the Kansas Academy of Science* 57: 397–433.

Smith, R. H. 2007. *History of the Boll Weevil in Alabama.* Alabama Agricultural Experiment Station Bulletin 670, Auburn, AL: Auburn University.

Snow, R. W., C. A. Guerra, A. M. Noor, H. Y. Myint, and S. I. Hay. 2005. The global distribution of clinical episodes of *Plasmodium falciparum* malaria. *Nature* 434: 214–217.

Stauffer, P. H. 1979. A fossilized honey bee comb from late Cenozoic cave deposits at Batu Caves, Malay Peninsula. *Journal of Paleontology* 53: 1416–1421.

Stige, L. C., K.-S. Chan, Z. Zhang, D. Frank, and N. C. Stenseth. 2007. Thousand-year-long Chinese time series reveals climatic forcing of decadal locust dynamics. *Proceedings of the National Academy of Sciences USA* 104: 16188–16193.

Tanke, D. H. and B. M. Rothschild. 1997. Paleopathology. In *Encyclopedia of Dinosaurs*, P. J. Currie and K. Padian, eds. 525–530. San Diego: Academic Press.

Ticehurst, M. and S. Finley. 1988. An urban forest integrated pest management program for gypsy moth: An example. *Journal of Arboriculture* 14: 172–175.

Victor, F. C., D. E. Cohen, and N. A. Soter. 2010. A 20-year analysis of previous and emerging allergens that elicit photoallergic contact dermatitis. *Journal of the American Academy of Dermatology* 62: 605–610.

Vigueras G., A.L. and L. Portillo. 2001. Uses of *Opuntia* species and the potential impact of *Cactoblastis cactorum* (Lepidoptera: Pyralidae) in Mexico. *Florida Entomologist* 84: 493–498.

Vincent, C., G. Hallman, B. Panneton, and F. Fleurat-Lessard. 2003. Management of agricultural insects with physical control methods. *Annual Review of Entomology* 48: 261–281.

Vittor, A. Y., R. H. Gilman, J. Tielsch, G. Glass, T. Shields, W. S. Lozano, V. Pinedo-Cancino, and J. A. Patz. 2006. The effect of deforestation on the human-biting rate of *Anopheles darlingi*, the primary vector of falciparum malaria in the Peruvian Amazon. *American Journal of Tropical Medicine and Hygiene* 74: 3–11.

Watson, E. J. and C. E. Carlton. 2003. Spring succession of necrophilous insects on wildlife carcasses in Louisiana. *Journal of Medical Entomology* 40: 338–347.

Webb, B., R. R. Harrison, and M. A. Willis. 2004. Sensorimotor control of navigation in arthropod and artificial systems. *Arthropod Structure and Development* 33: 301–329.

Whitaker, I. S., C. Twine, M. J. Whitaker, M. Welck, C. S. Brown, and A. Shandall. 2007. Larval therapy from antiquity to the present day: Mechanisms of action, clinical applications, and future potential. *Postgrad Medical Journal* 83: 409–413.

White, R. E. 1983. *Peterson Field Guide to the Beetles of North America.* Boston: Houghton Mifflin.

Yen, A. L. 2009. Entomophagy and insect conservation: Some thoughts for digestion. *Journal of Insect Conservation* 13: 667–670.

Zumpt, F. 1965. *Myiasis in Man and Animals in the Old World.* London: Butterworths.

3

Insect Responses to Environmental Changes

In 22 of 26 populations [of *Drosophila subobscura* on three continents over an average 24-yr period], climates warmed over the intervals, and genotypes characteristic of low latitudes (warm climates) increased in frequency in 21 of those 22 populations. Thus, genetic change in this fly is tracking climate warming and is doing so globally.

Balanyá et al. (2006)

As noted in Chapter 1 (see Figure 1.8), insects are the most successful group of organisms on the planet in terms of diversity and ecological function. Their success reflects adaptations to major changes in temperature, atmospheric chemistry, and geographic distribution of resources and habitats over 400 million years. They have survived extinction events that have eliminated other major groups. Their adaptive ability makes them extremely hard to control.

Several adaptations in particular have contributed to their evolutionary and ecological success in a constantly changing environment and affect their responses to anthropogenic changes and management efforts (see Chapter 6). Clearly, we need to understand how insects respond to environmental changes in order to anticipate their responses to future natural and anthropogenic changes and to develop ecosystem management approaches that mitigate these responses.

3.1 Adaptive Attributes

Small size, exoskeleton, metamorphosis, and flight represent a combination of characteristics that have ensured the survival of insects over millions of years of environmental changes. The contribution of each of these attributes is described here.

Small size (an attribute insects share with other invertebrates and microorganisms) has permitted exploitation of habitat and food resources at a microscopic scale. Insects find protection from adverse conditions in microsites too small for larger organisms (e.g., within individual leaves). Large numbers of insects can exploit the resources represented by a single leaf, by partitioning leaf resources, with some species feeding on cell contents, others on sap in leaf veins, some on top of the leaf, others on the underside, some internally. At the same time, small size makes insects particularly sensitive to changes in temperature, moisture, air or water chemistry, and other factors.

The exoskeleton (shared with other arthropods) provides protection against predation and desiccation or waterlogging (necessary for small organisms) and innumerable points of muscle attachment (for flexibility). However, the exoskeleton also limits the size attainable by arthropods. The increased weight of exoskeleton required to support larger body size would limit mobility. Early arthropods reached large sizes before the appearance of faster, more flexible vertebrate predators. Larger arthropods also occur in aquatic environments where water helps support their weight.

Metamorphosis is necessary for exoskeleton-limited growth but permits partitioning of habitats and resources among life stages. Immature and adult insects can differ dramatically in form and function and thereby live in different habitats and feed on different resources, reducing intraspecific competition. For example, dragonflies and mayflies live in aquatic ecosystems as immatures but in terrestrial ecosystems as adults. Many butterflies and beetles feed on foliage or wood resources as immatures and on nectar as adults. Among insects with complete metamorphosis (holometaboly), the quiescent, pupal stage facilitates survival during unfavorable environmental conditions (Figure 3.1). However, insects, as well as other arthropods, are particularly vulnerable to desiccation and predation during molting, when the old exoskeleton is shed and the new exoskeleton has not yet hardened.

Finally, insects were the first animals to fly and thereby gained a distinct advantage over other organisms. Flight permits rapid long-distance movement that facilitates discovery of new resources, as well as escape from predators or unfavorable conditions. The aerial acrobatics of many insects has

FIGURE 3.1 (SEE COLOR INSERT.)
Complete metamorphosis is exemplified by the monarch butterfly, *Danaus plexippus*, with nearly completed development of adult features visible through the pupal exoskeleton (chrysalis).

inspired awe and admiration. Flight remains a dominant feature of insect ecology.

3.2 Dispersal

Dispersal is the movement of individuals among habitat patches. This process minimizes the risk that an entire population will be destroyed by disturbance or resource depletion and maximizes the probability that some individuals will colonize and exploit new resources (Wellington 1980; Johnson 2004; Schowalter 2011). However, although many insects demonstrate capacity to disperse over large portions of the globe (e.g., locusts), dispersal also involves considerable risk to individuals and requires considerable energy expenditure (Rankin and Burchsted 1992). Torres (1988) documented cases of African insects being blown across the Atlantic Ocean by hurricane winds, to be deposited in Puerto Rico, including a swarm of desert locusts, *Schistocerca gregaria*. Most of these insects fail to find or reach suitable habitats and die without reproducing. A number of factors affect the probability of successful dispersal, that is arrival at suitable habitats, including life history strategy, crowding, nutritional status, habitat and resource conditions, and the mechanism of dispersal.

3.2.1 Life History Strategy

The degree of species adaptation to disturbance (see section 3.4) affects the predisposition of individuals to disperse. Species characterizing relatively stable, infrequently disturbed, habitats tend to disperse slowly (i.e., produce few offspring and move short distances). Infrequent disturbance and consistent resource availability provides little or no selection for greater dispersal ability. Many forest species (especially Lepidoptera and Coleoptera) are flightless, or at least poor fliers. By contrast, species (such as aphids) that characterize temporary, frequently disturbed, habitats produce large numbers of individuals with a high proportion of dispersers. Such traits are important adaptations for "weedy" species exploiting temporary, unstable, conditions (Janzen 1977).

3.2.2 Crowding

Crowding affects insects' tendency to disperse, and in some cases may stimulate morphological or physiological transformations that facilitate dispersal (Anstey et al. 2009). Survival and fecundity typically depend on population density (i.e., survival and fecundity decline as density increases). Therefore, dispersing individuals may achieve higher fitnesses than do nondispersing individuals at high population densities (Price 1997). For example, some bark beetle species oviposit their full complement of eggs in one tree under low density conditions, but only a portion of their eggs in one tree under high density conditions, leaving that tree and depositing remaining eggs in other trees (Wagner et al. 1981). If all eggs were laid in the first tree under crowded conditions, the large number of offspring could deplete resources before completing development.

Under crowded conditions, some insects spend more time eating and less time resting in order to maximize personal share of resources (Chapman 1998). Crowding increases the incidence of cannibalism in many species (Fox 1975a, b), encouraging dispersal. In addition, crowding can induce morphological changes that promote dispersal. Uncrowded desert locusts tend to repel one another and feed individually on clumps of vegetation. Under crowded conditions, locusts undergo a phase shift from smaller, solitary greenish nymphs that develop into greenish adults with short wings and long legs to larger, gregarious black nymphs that initiate group marching behavior and develop into long-winged, short-legged, black adults that migrate as immense swarms (Maeno and Tanaka 2008; Anstey et al. 2009; Matthews and Matthews 2010; Maeno et al. 2011). Tanaka (2012) demonstrated that this phase shift is induced by the visual stimulus of five to ten nearby insects.

3.2.3 Nutritional Status

Nutritional status affects the physiological condition, reproductive capacity, and endurance of dispersing insects. Most resources available to insects do not provide appropriate concentrations or balances of essential nutrients (Sterner and Elser 2002; Raubenheimer and Simpson 2003; Behmer 2009; Zehnder and Hunter 2009). Insects require a number of essential amino acids and vitamins in their food resources (Fraenkel and Blewett 1946; Rodriguez 1972; Mattson 1980; Chapman 1998; Klowden 2008) as well as sufficient carbohydrates to fuel metabolism for various activities, including dispersal. The balances of most of these essential nutrients are not optimal in most plant parts because of differences between plant needs and insect needs. Consequently, insects must select foods to provide improved balance, that is, to gain enough of the limited nutrient(s) without ingesting toxic amounts of others (Sterner and Elser 2002; Raubenheimer and Simpson 2003; Behmer 2009).

Plant defensive compounds reduce the net nutritional value of food for herbivorous insects and require specialized feeding behavior or production of expensive detoxification enzymes by species adapted to feed on particular plants. Populations of many insects show considerable variation in fat storage and vigor as a result of variation in food quality and the quantity and maternal partitioning of nutrient resources to progeny (Wellington 1980; Wagner et al. 1981). Many species exhibit obligatory flight distances that are determined by the amount of energy and nutrient reserves: Dispersing individuals respond to external stimuli only after depleting these reserves to a threshold level. Hence, less vigorous individuals tend to colonize neighboring habitats, whereas more vigorous individuals fly greater distances and colonize more distant habitats. Because crowding and nutritional status have opposite effects on dispersal, the per capita accumulation of adequate energy reserves and the number of dispersing individuals should peak at intermediate population densities at which crowding encourages dispersal and resource quality and quantity are still sufficient to fuel dispersal.

3.2.4 Mechanism of Dispersal

The mechanism of dispersal strongly affects the probability that suitable resources can be found and colonized. Three general mechanisms can be identified: random, phoretic, and directed (Matthews and Matthews 2010).

Random dispersal direction and path are typical of most small insects with little capacity to detect or orient toward environmental cues. Such insects are at the mercy of physical barriers or wind or water currents, and their direction and path of movement are determined by obstacles and patterns of air or water movement. For example, first instar nymphs of a *Pemphigus* aphid that lives on the roots of sea asters growing in salt marshes climb the sea asters and are set adrift on the rising tide. Sea breezes enhance movement, and successful nymphs are deposited at low tide on new mud banks where they must seek hosts (Kennedy 1975). Aquatic insects often are carried downstream during floods. Hatching gypsy moth, *Lymantria dispar*, and other tussock moth larvae (Lymantriidae), scale insect crawlers, and spiders disperse by launching themselves into the airstream (McClure 1990; Matthews and Matthews 2010). Lymantriid females and scale insect adults have poor (if any) flight capacity. Wind-aided dispersal by larval Lepidoptera and spiders is facilitated by extrusion of silk strands, a practice known as "ballooning." Western spruce budworm, *Choristoneura occidentalis*, adults aggregate in mating swarms above the forest canopy and are carried by wind currents to new areas (Wellington 1980).

The distance traveled by wind- or water-dispersed insects depends on several factors, including flow rate and insect size or mass. Even small insects and mites have some control over buoyancy and landing (Jung and Croft 2001).

The probability that at least some insects will arrive at suitable resources depends on the number of dispersing insects and the predictability of wind or water movement in the direction of new resources. Most individuals fail to colonize suitable sites, and many become part of the aerial or aquatic plankton that eventually "falls out" and becomes deposited in remote, unsuitable, locations. For example, Edwards and Sugg (1990) documented fallout deposition of many aerially dispersed insect species on montane glaciers in western Washington, where they become resources for a community of detritivorous and predaceous arthropods.

Phoretic dispersal is a special case in which a flightless insect or other arthropod hitches a ride on another animal (Figure 3.2). Phoresy is particularly common among wingless Hymenoptera and mites. For example, scelionid wasps ride on the backs of female grasshoppers, benefitting from transport and the eventual opportunity to oviposit on the grasshoppers' eggs. Wingless Mallophaga (lice) attach themselves to hippoboscid flies that parasitize the same bird hosts. Many species of mites attach themselves to dispersing adult insects that feed on the same dung or wood resources (Krantz and Mellott 1972; Stephen et al. 1993). Birds and mammals provide long-distance transport for hemlock woolly adelgids, *Adelges tsugae* (McClure 1990). The success of phoresy (as with wind- or water-aided dispersal) depends on the predictability of host dispersal. However, in the case of phoresy, success is enhanced by the association of both the hitchhiker and its mobile (and perhaps cue-directed) host with the same resource.

Directed dispersal provides the highest probability of successful colonization and is observed in larger, stronger fliers capable of orienting and directing movement toward suitable resources. Many wood-boring insects, such as wood wasps (Siricidae) and beetles (especially Buprestidae), are attracted to sources of smoke, infrared radiation, or volatile tree chemicals emitted from burned or injured trees over distances of up to 50 km (Wickman 1964; Evans 1966; Gara et al. 1984; Raffa et al. 1993; Schütz et al. 1999; Matthews and Matthews 2010). Attraction to suitable hosts often is significantly enhanced by mixing of host odors with pheromones emitted by early colonists. Visual, acoustic, or magnetic cues also aid orientation (Matthews and Matthews 2010). For example, masking the silhouette of tree boles with white

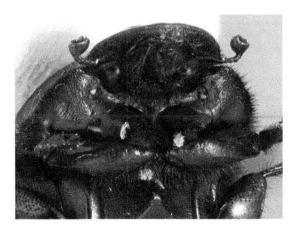

FIGURE 3.2 (SEE COLOR INSERT.)
Phoretic mites attached to legs of a scarab beetle. The flightless mites disperse to new dung pads by hitching a ride on dung beetles capable of dispersing to new dung pads over long distances. (Photo courtesy of A. Tishechkin. From Schowalter, T. D., *Insect Ecology: An Ecosystem Approach*, Elsevier, San Diego, 2011. With permission.)

paint substantially reduced numbers of attracted southern pine beetles, *Dendroctonus frontalis* (Strom et al. 1999), *Ips* engraver beetles, and some bark beetle predators (Goyer et al. 2004).

Migration is an active mass movement of individuals that functions to displace entire populations. Migration always involves females but not always males. Examples of migratory behavior in insects include locusts, monarch butterflies, and ladybird beetles. Desert locust and migratory locust, *Locusta migratoria*, migration depends, at least in part, on wind patterns. Locust swarms remain compact, not because of directed flight, but because randomly oriented locusts reaching the swarm edge reorient toward the body of the swarm. Swarms are displaced downwind into equatorial areas where converging air masses rise, leading to precipitation and vegetation growth favorable to the locusts (Matthews and Matthews 2010). In this way, migration displaces the swarm from an area of crowding and insufficient food to an area with more abundant food resources. Monarch butterfly and ladybird beetle migration occurs seasonally and displaces large numbers to and from overwintering sites—Mexico for North American monarch butterflies and sheltered sites for ladybird beetles. Merlin et al. (2009) reported that a gene-controlled circadian clock in monarch antennae provides the internal timing mechanism for a time-compensated sun compass that allows the insects to correct their flight direction relative to the position of the sun as it moves across the sky during the day.

3.2.5 Habitat and Resource Conditions

The likelihood that an insect will find a suitable landscape patch depends strongly on patch or resource size, ease of discovery, and proximity to insect population sources. Larger or more conspicuous habitats or resources are more likely to be perceived by dispersing insects or to be intercepted by a given direction of flight. Larger habitat patches intersect a longer arc centered on a given starting point. Insects dispersing in any direction have a higher probability of contacting larger patches than they do smaller patches. Courtney (1985, 1986) reported that the pierid butterfly, *Anthocharis cardamines*, preferentially oviposited on the most conspicuous (in terms of flower size) host species, which were less suitable for larval development than were less conspicuous hosts. This behavior by the adults represented a trade-off between the prohibitive search time required to find the most suitable hosts and the reduced larval survival on the most conspicuous hosts.

The probability of survival declines with distance as a result of depletion of metabolic resources and protracted exposure to various mortality factors (Pope et al. 1980). Hence, more insects are successful over shorter distances and reach closer resources or sites (Figure 3.3). Sartwell and Stevens (1975) and Schowalter et al. (1981a) reported that, under nonoutbreak conditions, probability of bark beetle, *Dendroctonus* spp., colonization of living pine trees declined with distance from currently attacked trees. Trees more than 6 m

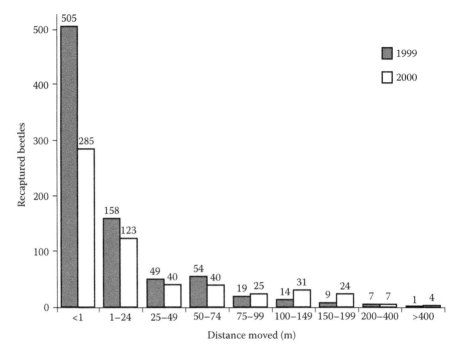

FIGURE 3.3

Range of dispersal distances from a population source for the weevil, *Rhyssomatus lineati-collis*, in the United States. (From St. Pierre, M. J. and S. D. Hendrix, 2003, *Ecological Entomology*, 28, 579–586, 2003. With permission.)

from currently colonized trees had negligible probability of colonization by sufficient numbers to successfully kill the tree. Under outbreak conditions, the effect of distance disappeared because sufficient numbers of beetles were likely to land on any tree within a larger radius of source trees (Schowalter et al. 1981a). Similarly, He and Alfaro (1997) reported that, under nonoutbreak conditions, colonization of white spruce by the white pine weevil, *Pissodes strobi*, depended on host condition and distance from trees colonized the previous year, but during outbreaks most trees were sufficiently near occupied trees to be colonized.

3.3 Environmental Variation

Abiotic conditions vary seasonally in most biomes and alter habitat and resource conditions to which insects, as well as other organisms, respond. Temperate ecosystems are characterized by obvious seasonality in temperature, with cooler winters and warmer summers, and also may show distinct seasonality in precipitation patterns, resulting from seasonal changes in the orientation of earth's axis relative to the sun. Although tropical ecosystems experience relatively consistent temperatures, precipitation often shows pronounced seasonal variation. Aquatic habitats show seasonal variation in water level and circulation patterns related to seasonal patterns of precipitation and evaporation. Intermittent streams and ponds may disappear during dry periods or when evapotranspiration exceeds precipitation.

Physical conditions vary over longer periods as a result of climate cycles resulting from changes in global circulation patterns of air or water. For example, the east-west gradient in surface water temperature in the southern Pacific diminishes in some years, altering oceanic and atmospheric currents globally; this is the El Niño/southern oscillation (ENSO) phenomenon (Rasmussen and Wallace 1983; Windsor 1990). Particularly strong El Niño years (e.g., 1982–1983 and 1997–1998) are characterized by extreme drought conditions in some tropical ecosystems and severe storms and wetter conditions in some higher latitude ecosystems. Seasonal patterns of precipitation can be reversed (i.e., drier wet season and wetter dry season). The year following an El Niño year may show a rebound, an opposite but less intense effect (La Niña). Windsor (1990) found a strong positive correlation between El Niño index and precipitation during the preceding year in Panama. Precipitation in Panama typically is lower than normal during El Niño years, in contrast to the greater precipitation accompanying El Niño in Peru and Ecuador (Windsor 1990; Zhou et al. 2002).

Insects are sensitive to the changes in temperature and moisture that accompany such events. Zhou et al. (2002) reported that extremely high populations of sand flies, *Lutzomyia verrucarum*, were associated with El Niño conditions in Peru, resulting in near doubling of human cases of bartonellosis, an emerging, vector-borne, highly fatal infectious disease in the region (Figure 3.4). Regional drying triggers outbreaks of many herbivorous species (Mattson and Haack 1987; Schowalter et al. 1999; Van Bael et al. 2004).

Terrestrial and aquatic biomes differ in the type and extent of variation in physical conditions. Terrestrial habitats are exposed to wide variation in air temperature, wind speed, relative humidity, and other atmospheric conditions. Aquatic habitats are relatively buffered from sudden changes in air temperature but are subject to changes in flow rate, depth, and chemistry, especially changes in pH and concentrations of dissolved gases, nutrients, and pollutants. Vegetation cover insulates the soil surface and reduces albedo, thereby reducing diurnal and seasonal variation in soil and near-surface temperatures (Foley et al. 2003). Hence, desert biomes with sparse vegetation cover typically show the widest diurnal and seasonal variation in physical conditions. Areas with high proportions of impervious surfaces (such as roads, roofs, parking lots) greatly alter conditions of terrestrial and aquatic systems by increasing albedo and precipitation runoff (Elvidge et al. 2004).

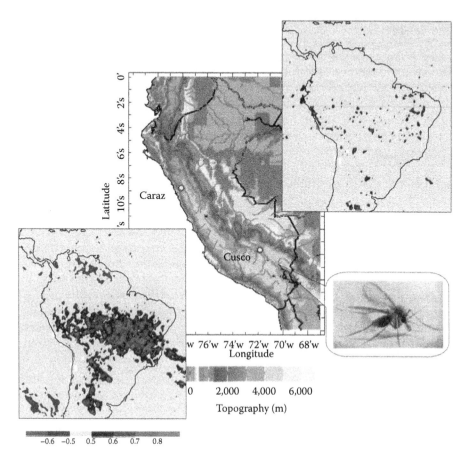

FIGURE 3.4 (SEE COLOR INSERT.)
Topography of Peru (center), comparison of TRMM TMI rainfall at Cuzco (lower left) and Caraz (upper right) relative to their surroundings and the sand fly, *Lutzomyia verrucarum*, vector of bartonellosis that shows increased spread associated with higher rainfall during El Niño events. (From Zhou, J., W. K.-M. Lau, P. M. Masuoka, R. G. Andre, J. Chamberlin, P. Lawyer, and L. W. Laughlin, *Eos, Transactions, American Geophysical Union*, 83, 157, 160–161, 2002. With permission.)

Physiological tolerances of organisms, including insects, generally reflect the normal range of physical conditions of the biomes in which they occur. Insects associated with the tundra biome tolerate a lower range of temperatures than do insects associated with tropical biomes. The upper threshold temperature for survival of a tundra species might be the lower threshold temperature for survival of a tropical species. Similarly, insects characterizing mesic or aquatic biomes generally have less tolerance for desiccation than do insects characterizing xeric biomes. However, species characterizing temporary streams or ponds may have adapted mechanisms for withstanding desiccation during dry periods (Batzer and Wissinger 1996). Some species show greater capacity than do others to adapt genetically to changing environmental conditions, especially rapid changes resulting from anthropogenic activity. Species adapted to frequent disturbance may be predisposed to adapt to rapid changes resulting from human activity.

Bradshaw and Holzapfel (2001) and Mathias et al. (2007) found that the genetically controlled photoperiodic response of the pitcher-plant mosquito, *Wyeomyia smithii*, has shifted toward shorter, more southern daylengths as growing seasons have become longer. This shift was detectable over time intervals as short as five years. Faster evolutionary response has occurred in northern populations where selection for response to seasonal change in photoperiod is stronger and genetic variation was greater than in southern populations. Therefore, *W. smithii* represents an example of genetic differentiation of a seasonality trait that is consistent with an adaptive evolutionary response to recent global warming.

3.4 Disturbance

Disturbances are relatively abrupt events in time and space that substantially alter ecosystem conditions and resource distribution across landscapes (White 1969; Wilson 1986; Waring and Price 1990; Willig and Walker 1999). The extreme change in conditions created by disturbances is among the most significant factors affecting insects, as well as other organisms, and induces dramatic responses by species and communities (Schowalter 2012). Each disturbance event is characterized by a unique combination of type, magnitude, frequency, and extent that determines its effect on various organisms (White 1969; Waring and Price 1990). Superimposing a sequence of events on the landscape creates a mosaic of patches that differ in their disturbance histories (see Figure 1.6). For example, over a 20-year period a rainforest in Puerto Rico experienced two major hurricanes (Hugo in 1989 and Georges in 1998) that broke or toppled trees on the windward sides of slopes over large areas and caused numerous landslides, several moderate hurricanes (Luis and Marilyn in 1995, Bertha and Hortense in 1996, Erika in 1997, Jose in 1999, and Jeanne in 2004) that caused substantial defoliation and flooding, a number of minor hurricanes and tropical storms, a major drought (1994–1995, during which precipitation was only 41% of annual mean) and several minor droughts (1991, 1996, 2001, and 2003), as well as an overall drying trend of 2 mm yr^{-1} since 1988 (Heartsill-Scalley et al. 2007). The particular sequence of disturbances affects insect species responses (Schowalter et al. 2003; Ballinger et al. 2007; Chase 2007). A species adapted to post-storm conditions, but eliminated by a previous fire, would not be represented in the post-storm community. Responses of litter arthropod and canopy Lepidoptera to canopy-opening disturbance in a conifer forest can reflect filtering of species composition by previous harvests as much as 60 years earlier (Schowalter et al. 2003; Taylor and MacLean 2009).

3.4.1 Type

Disturbance types vary in the conditions they impose on organisms, including insects. Severe storms, especially tropical cyclones, dislodge or injure insects and initiate landslides and flooding that redistribute sediments and bury organisms. Fires and volcanic eruptions impose extreme high temperatures, whereas ice storms impose extreme low temperatures. Fires and volcanic eruptions also fill the air with ash and caustic gases, and lava or ash deposition burns and/or buries organisms. The eruption of Mt. St. Helens in 1980, for example, deposited up to 30 kg ash m^{-2} at depths of 1–8 cm over an area of 54,000 km^2 (Cook et al. 1981).

Insect outbreaks often are perceived as disturbances because they seem to occur suddenly and kill plants and alter distribution of biomass over large areas (MacLean 2004; Breshears et al. 2005; Kurz et al. 2008; Brown et al. 2010). Insect outbreaks often predispose ecosystems to subsequent fire or storm disturbances (Tisdale and Wagner 1990; McCullough et al. 1998; Parker et al. 2006; Jenkins et al. 2008; Taylor and McLean 2009). However, unlike abiotic disturbances, insect outbreaks represent population responses to relatively predictable changes in environmental conditions, especially to high density and/or stress of host plants as a result of fire, drought, flooding, storms, or anthropogenic management (Schowalter 2011; see Chapter 4).

3.4.2 Magnitude

Disturbances vary in intensity and severity. Intensity often may be difficult to measure (e.g., actual temperature during a fire or velocity of wind gusts during a storm). Magnitude is measured more often by the severity of effects on species and ecosystems. A low intensity ground fire affects primarily surface-dwelling organisms, many of which may be adapted to this level of disturbance, whereas the intense heat of a fire storm (created by convection during a catastrophic wildfire) penetrates more deeply into soil and wood and kills a larger proportion of the community (Wikars and Schimmel 2001; Van Bael et al. 2004). A minor flood slowly filling a floodplain for a few days affects fewer insects than does a major flood that scours and inundates the landscape for weeks (White and Pickett 1985). Dead vegetation deprives many insects of food resources and alters microclimatic conditions that affect habitat quality (Willig and Walker 1999; Koptur et al. 2002). However, some species thrive under the altered conditions (Van Straalen and Verhoef 1997; Schowalter et al. 1999).

3.4.3 Frequency

Disturbances vary in their return interval. Fire intense enough to kill most vegetation occurred, on average, every 200 years since 1633 in a montane forest landscape in western North America (Veblen et al. 1994), whereas hurricanes of this magnitude recur on average every 60 years at sites in the Caribbean region (Heartsill-Scalley et al. 2007). Frequency, with respect to generation times, of a particular disturbance type affects the rate of selection for adaptive traits that confer tolerance (resistance) to disturbance. Infrequently disturbed ecosystems, such as tropical forests, may be affected most by disturbances because dominant species in these ecosystems experience too few disturbances to drive selection for tolerance. Long-lived organisms, especially perennial plants, are more likely to experience disturbances consistently among generations, leading to stronger selection for adaptive traits.

Insects, with short life spans relative to disturbance return intervals, experience less selective pressure for adaptation to disturbance than do longer-lived species. Nevertheless, some insect species show apparent adaptations to disturbance. Some wood-boring insects, with life spans of two to ten years, are attracted to sources of heat or smoke, indicative of dead wood resources in fire-killed trees (Evans 1966; Schütz et al. 1999). Avoidance of dense overstory and accumulated litter by ground-pupating larvae of the pandora moth, *Coloradia pandora*, may reflect adaptation to minimize mortality during fire, which occurs every five years (on average) compared with a two-year life cycle for this moth (Miller and Wagner 1984). Fire ants, *Solenopsis invicta*, and other ants characteristic of floodplain habitats respond to flooding by forming floating mats of ants and larvae that disperse downstream (Anderson et al. 2002; Mlot et al. 2011). Some floodplain insects are capable of tolerating extended periods of inundation (Hoback and Stanley 2001; Zerm and Addis 2003; see section 3.5.1.2). Dispersal ability is an important adaptation for species exploiting temporary, unstable, conditions (Janzen 1977). Many species (especially Lepidoptera and Coleoptera) that characterize infrequently disturbed forests are flightless, or at least weak dispersers, whereas species that characterize temporary, frequently disturbed, habitats (such as many aphids) produce large numbers of strong dispersers (see earlier).

Disturbances of greater severity typically occur at lower frequency. Increasing disturbance magnitude or frequency generally reduces species diversity (Ballinger et al. 2007; Haddad et al. 2008; Mertl et al. 2009) because fewer species are able to tolerate more extreme changes in habitat or resource conditions. Hanula and Wade (2003) reported that most forest floor species decreased in abundance, but some species increased, with increasing fire frequency. De Mazancourt et al. (2008) suggested that high biodiversity increases the likelihood that some species have genotypes that are adapted to altered conditions. Arrival of adapted colonists of various species from other areas augments community recovery (see section 3.5.2).

Timing of disturbances, relative to insect developmental stage, also affects insect responses. Winged adults may be able to escape as conditions become intolerable, whereas exposed pupae would be most vulnerable to disturbance. On the other hand, Martin-R. et al. (1999) reported that experimental fires set during different developmental stages of spittlebug, *Aeneolamia albofas-ciata*, in buffelgrass, *Cenchrus ciliaris*, grassland in Sonora, Mexico, eliminated spittlebugs for at least four years after burning, regardless of developmental stage at the time of burning.

3.4.4 Extent

Disturbances range in extent from a few hectares to continental, but the magnitude of particular disturbances varies across landscapes due to variation in topography, substrate condition or vegetation (e.g., intervening hills, bedrock outcrops, bodies of water, or patches of vegetation that are resistant to particular disturbances). Veblen et al. (1994) reported that fire has affected 59% and snow avalanches 9% of a montane forest landscape in western North America since 1633, compared with 39% by outbreaks of spruce beetle, *Dendroctonus rufipennis*. The 1998 flood in Bangladesh inundated nearly 70% (about 100,000 km^2) of the country (Kundzewicz et al. 1998), leaving few refuges for insects intolerant of immersion. Heavy rains producing such floods typically saturate soils over a much larger area. Lava and ash from volcanic eruptions can cover thousands of square kilometers (Cook et al. 1981), but volcanic gases can affect atmospheric chemistry and climate globally (McCormick et al. 1995). El Niño events in the southern Pacific Ocean also affect climate and insects globally (Zhou et al. 2002; Stapp et al. 2004; Van Bael et al. 2004). Insect outbreaks themselves are capable of consuming or killing most host plants over thousands of square kilometers (Pfadt and Hardy 1987; Schowalter 2011).

The extent of disturbed area affects insect responses. Populations restricted to areas smaller than the disturbed area are likely to disappear. This threat creates a challenge for conservation biologists who must work to conserve large enough areas, or sufficient distribution of increasingly isolated refuges, to maintain target species vulnerable to large-scale disturbance (Hanski and Simberloff 1997). Colonization typically progresses inward from the edges of disturbed areas, as dispersing individuals from population sources find such "sinks," so smaller areas can be colonized more quickly than larger areas (Shure and Phillips 1991; Antunes et al. 2009). Insects with limited mobility may require considerable time to colonize large areas (Knight and Holt 2005). More extensive disturbances or disturbances that result in greater contrast between disturbed and undisturbed patches (e.g., Figure 1.6d) create steeper gradients in post-disturbance temperature and relative humidity between disturbed and undisturbed patches, with sharper boundaries relative to insect tolerance ranges (Shure and Phillips 1991). However, edges of disturbed areas may provide unique resources for insects, for example, species that exploit forest plants that are stressed by exposure or are favored by higher light available at the edge of a disturbed area (Roland and Kaupp 1995; Knight and Holt 2005).

3.5 Insect Responses

Individual insects have specific tolerance ranges to abiotic conditions that dictate their ability to survive exposure to extreme temperatures, water availability, chemical concentrations, or other factors during and after a disturbance. Variable ecosystem conditions typically select for wider tolerance ranges than do more stable conditions. Although changes in abiotic conditions during disturbances can affect insects directly (e.g., burning, drowning, particle blocking of spiracles), disturbances also affect insects indirectly through changes in resource quality and availability and in exposure to predation or parasitism (Alstad et al. 1982; Miller and Wagner 1984; Mattson and Haack 1987; Shure and Wilson 1993; Roland and Kaupp 1995; Van Straalen and Verhoef 1997; Hunter and Forkner 1999). Population size and degree of genetic heterogeneity affects survival during disturbance (Ballinger et al. 2007). As habitat conditions change, intolerant individuals or species disappear, but tolerant species may be favored by reduced predation or improved resource conditions following disturbance (Roland and Kaupp 1995; Schowalter 1995, 2012; Schowalter et al. 1999). Because survival and reproduction of individual insects determine population size, distribution, and subsequent effects on recovery of ecosystem processes and services, this section focuses on factors that affect insect responses directly and indirectly.

3.5.1 Direct Effects of Abiotic Changes

Disturbances alter abiotic conditions to varying degrees depending on disturbance type and severity. Insects are particularly vulnerable to changes in temperature, water availability, and air or water chemistry because of their relatively large surface area to volume ratio and limited homeostatic ability. Although some habitats may protect insects from disturbances, and some insects may be able to escape as conditions approach tolerance limits (Wilson 1986; Johnson 2004), survival of many depends on physiological tolerance ranges relative to the extreme environmental conditions experienced during or after disturbances.

3.5.1.1 Temperature Extremes

Fires and volcanic eruptions, in particular, create lethal temperatures for insects unable to escape. Disturbances that reduce vegetation cover subsequently expose surviving or colonizing insects to elevated surface temperatures. Ulyshen et al. (2010) found that fire reduced abundances, but not species representation, of wood-boring beetles in coarse woody debris. Beetles were relatively protected in larger diameter logs. Bark beetles in subcortical tissues, however, are vulnerable to heat mortality (Wikars and Schimmel 2001). Small, flightless litter species only need to move a few millimeters vertically within the soil profile to avoid lethal temperatures and desiccation during fire or canopy-opening disturbances (Seastedt and Crossley 1981). Nevertheless, Hanula and Wade (2003) found that abundances of most forest floor species (especially predators) were reduced by prescribed burning, and reduced more by annual than by biennial or quadrennial burning, but a few species (especially detritivores) increased in abundance with more frequent burning.

Survival at high temperatures requires high body water content or access to water because desiccation at low relative humidity causes death (Hadley 1994). Disturbances that reduce riparian canopy cover significantly increase water temperature, and reduce oxygen levels, of aquatic patches especially in the summer (Kiffney et al. 2003; Rykken et al. 2007). A distinct riparian fauna may be vulnerable to canopy opening disturbance within 30 m of streams (Rykken et al. 2007). However, stream grazers may respond positively to increased primary production resulting from higher light level when riparian canopies are opened (Kiffney et al. 2003).

Conversely, unseasonable ice storms or extreme cold periods also kill exposed insects. Ability to survive depends on prior preconditioning to sublethal temperatures (Kim and Kim 1997) or physiological mechanisms, such as production of cryoprotectants (antifreeze) to prevent intracellular ice formation and voiding the gut to prevent food particles from serving as nuclei for ice formation (Hadley 1994).

3.5.1.2 Precipitation Extremes

Water availability becomes particularly limited during drought or excessive during floods, but high temperatures during fire or volcanic eruption may reduce relative humidity severely. Maintenance of water balance becomes a challenge for small organisms such as insects, but some insects are capable of minimizing water loss or tolerating dehydration (Gibbs and Matzkin 2001). The exoskeleton is an important mechanism for control of water exchange. Larger, more heavily sclerotized arthropods are less susceptible to desiccation or waterlogging than are smaller, more delicate species (Alstad et al. 1982; Kharboutli and Mack 1993).

Extreme dehydration may trigger the onset of anhydrobiosis, a physiological state characterized by an absence of free water and of measurable metabolism (Whitford 1992; Hadley 1994). Survival during anhydrobiosis requires stabilization of membranes and enzymes by compounds other than water, for example, glycerol and trehalose, whose synthesis is stimulated by dehydration (Hadley 1994). Among insects, only some larval Diptera and adult Collembola have been shown to undergo anhydrobiosis (Hadley 1994). Hinton (1960a, b) reported that a chironomid fly, *Polypedilum vanderplancki*, found in temporary pools in central Africa, withstands repeated dehydration to 8% of body water content. At 3% body water content, this midge is capable of surviving temperatures from –270°C to 100°C.

On the other hand, insects subjected to flooding must contend with excess water. Subterranean termites can survive short periods of inundation by entering a quiescent state; relative abilities of species to withstand periods of flooding correspond to their utilization of aboveground or belowground wood resources (Forschler and Henderson 1995). Litter-dwelling ants are vulnerable to seasonal flooding in Amazonian forests (Mertl et al. 2009). Specialist predators were virtually eliminated by flooding; one *Hypoponera* species was adapted to a high degree of flooding, increasing in abundance with the frequency and duration of flooding. Webb and Pullin (1998) found that pupae of a wetland butterfly, *Lycaena dispar batavus*, could tolerate 28 days of submergence, but survival was negatively correlated with duration of submergence between 28 and 84 days. However, inundation affects oxygen availability (see section 3.5.1.4), as well as water balance.

3.5.1.3 Wind Speed and Water Flow

High wind speeds and water flow during storms dislodge and displace exposed insects, as well as sediment and organic debris, crushing or injuring many individuals. Some may survive and be able to move around and colonize new habitats to which they are relocated, but immature and sedentary insects most likely perish. Torres (1992) documented a large number of insect species, including a swarm of desert locusts that were blown across the Atlantic Ocean from North Africa to Puerto Rico by hurricane winds.

3.5.1.4 Air and Water Quality

Some disturbances alter atmospheric and water quality. Fire and volcanic eruptions, in particular, release toxic abiotic and biotic gases and particulate materials, but high winds during storms and high river levels during flooding also increase the amount of dissolved and suspended materials that may affect exposed insects.

Oxygen supply is especially critical to survival but may become limiting during or after some disturbances, such as soil saturation during flooding or burial by sedimentation or ash fall. Many insects can tolerate short periods of anoxia, but prolonged periods result in reduced survival and developmental abnormalities (Hoback and Stanley 2001). Adult alder leaf beetles, *Agelastica alni*, that overwinter in frequently waterlogged or flooded riparian soil, showed a reduction in metabolic activity after three days to 2% of normal metabolic activity (Kölsch et al. 2002). Larval tiger beetles, *Phaeoxantha klugiis*, found in central Amazonian floodplains are able to tolerate anoxic conditions in flooded soils for up to 3.5 months at 29°C (Zerm and Adis 2003). This exceptional degree of anoxia tolerance appeared to require several days of induction as water levels rose, suggesting vulnerability to more rapid inundation. Brust and Hoback (2009) found that tolerance to hypoxia among several tiger beetles, *Cicindela* spp., was not related to likelihood of immersion.

Increased concentrations of atmospheric CO_2 that result from fire, volcanic eruption, or other causes appear to have little direct effect on insects or other arthropods. Fluorides, sulfur compounds, nitrogen oxides, ozone, and other toxic fumes affect many insect species directly, although the physiological mechanisms of toxicity are not well-known (Alstad et al. 1982; Heliövaara and Väisänen 1986, 1993). Disruption of epicuticular or spiracular tissues by these reactive chemicals may be involved.

Soil and water pH affects a variety of chemical reactions, including enzymatic activity. Changes in pH resulting from deposition of ash and release of caustic gases from fire or volcanic eruptions affect osmotic exchange, gill or spiracular surfaces, and digestive processes. Changes in pH often are correlated with other chemical changes, such as increased concentrations of nitrogen or sulfur compounds, and effects of pH change may be difficult to separate from other factors. Van Straalen and Verhoef (1997) found that several species of soil collembolans and oribatid mites varied in their responses to acidic or alkaline soil conditions.

Dust and ash from volcanic eruptions or fires kill many insects, apparently because they absorb and abrade the thin epicuticular wax-lipid film that is the principal barrier to water loss, causing death by desiccation (Cook et al. 1981; Alstad et al. 1982; Marske et al. 2007). Insects exposed to volcanic debris also can suffer gut epithelial stress from accumulation of heavy metals (Rodrigues et al. 2008). Ash accumulation and retention by aquatic insects following eruption of Mount St. Helens were affected by exoskeletal sculpturing, armature, and pubescence (Gersich and Brusven 1982). Substantial accumulation was noted on respiratory structures, potentially interfering with respiration. Ash-covered insects showed increased activity and orientation upstream, which successfully washed ash off within 24 hours. However, ash coating over cobbles, pebbles, and sand significantly inhibited colonization of these substrates (Brusven and Hornig 1984).

3.5.2 Indirect Effects of Post-Disturbance Changes

Insects that survive the direct effects of disturbance must contend with altered habitat conditions and resource availability. Disturbances destroy some habitats and resources and alter distribution or quality of others for a period, during which ecosystem conditions recover to a semblance of pre-disturbance conditions (ecological succession). Some insect species respond positively, others negatively, to these changes in community and ecosystem conditions, based on adaptive characteristics and trophic interactions (Evans 1988; Willig and Camilo 1991; Paquin and Coderre 1997; Schowalter et al. 1999; Hanula and Wade 2003; Cleary and Grill 2004; Mertl et al. 2004; Buddle et al. 2006; Gandhi et al. 2007; Hirao et al. 2008; Moretti and Legg 2009). Species that increase in abundance following disturbance typically are favored by exposed conditions, stressed or rapidly growing plants, or detrital resources. In addition, some may be promoted by decoupling of predator-prey relationships in patches of enemy-free space.

3.5.2.1 Exposure

Disturbances that remove vegetation cover and/or litter expose insects to a wider range of ambient temperature and relative humidity (Chen et al. 1995; Rykken et al. 2007) and predators (Björklund et al. 2003). Arboreal beetle responses to cyclone disturbance in tropical rainforest in Australia reflected species' adaptations to moisture, with more xerophilic species increasing in abundance and mesophilic species decreasing in abundance following canopy opening and general drying of the forest (Grimbacher and Stork 2009). Furthermore, reproduction, especially egg hatch, may be reduced at high temperature and low relative humidity (Tisdale and Wagner 1990).

Some ant species (e.g., *S. invicta* and *Atta laevigatta*) preferentially colonize bare soil habitats over soil covered by vegetation or litter (Stiles and Jones 1998; Vasconcelos et al. 2006). However, other species may be unable to survive the extreme temperatures of exposed sites. Meisel (2006) reported that the army ant, *Eciton burchellii*, is restricted to forest fragments in Costa Rica because workers survived less than three minutes at 51°C (the midday temperature of surrounding pastures) and only 18 minutes at 43°C.

3.5.2.2 Resource Abundance and Quality

BOX 3.1   PLANT DEFENSIVE CHEMISTRY

Plants produce a remarkable variety of compounds that constitute effective defenses against herbivores. Plant chemical defenses can be classified as nonnitrogenous, nitrogenous, and elemental to indicate their ecological costs to the plant, in terms of limited nitrogen, or as toxins, repellents, feeding deterrents, insect hormone analogues, or predator attractants to indicate their cost to the insect. Many of these compounds are broadly effective against invertebrate and vertebrate (including human) herbivores and are the source of important pharmaceutical compounds.

Phenolics, also known as tannins or flavenoids, are distributed widely among terrestrial plants and are likely among the oldest plant secondary (i.e., nonmetabolic) compounds (Figure B3.1). Although phenolics are perhaps best known as defenses against herbivores and plant pathogens, they also protect plants from damage by ultraviolet radiation, provide support for vascular plants (lignins), compose pigments that determine flower color for angiosperms, and play a role in plant nutrient acquisition by affecting soil chemistry. These compounds are distasteful, typically bitter and astringent, and act as feeding deterrents for many herbivores. When ingested, they bind N-bearing molecules to form indigestible complexes (Feeny 1969) and oxidize in the herbivore gut to form semiquinones and other highly oxidative compounds that damage cell membranes and DNA (Barbehenn et al. 2008). Insects that cannot catabolize these compounds or prevent binding of proteins suffer gut damage and are unable to assimilate nitrogen from their food. Herbivores with high gut pH can prevent these effects (Feeny 1969; Fox and Macauley 1977).

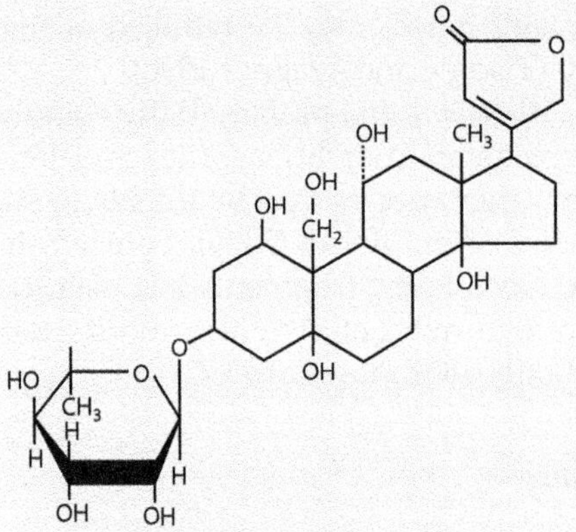

Terpenoid cardiac glycoside, ouabain, from *Acokanthera ouabaio*

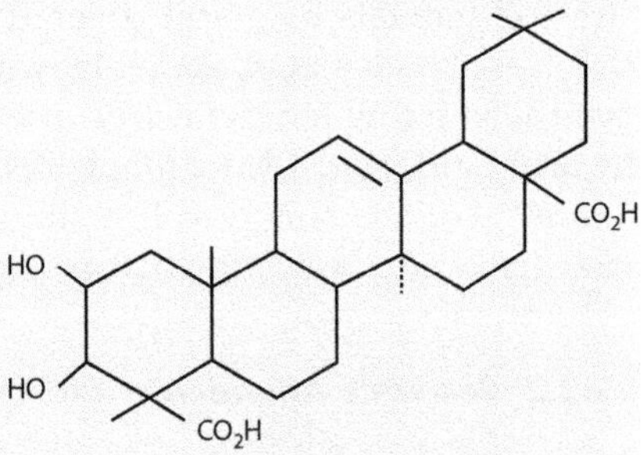

Terpenoid saponin, medicagenic acid, from *Medicago sativa*

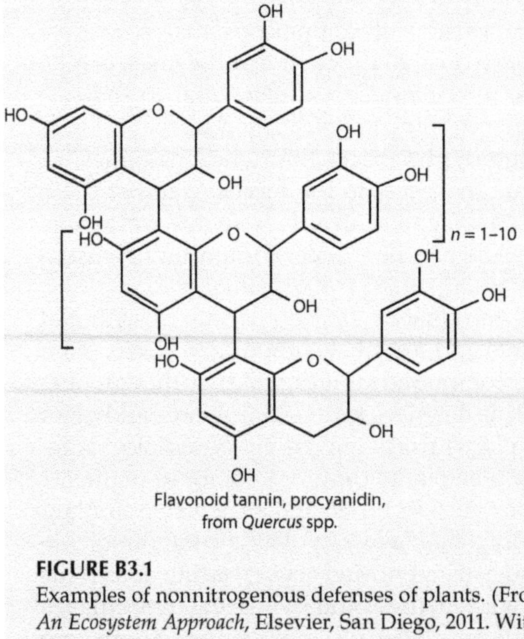

Flavonoid tannin, procyanidin, from *Quercus* spp.

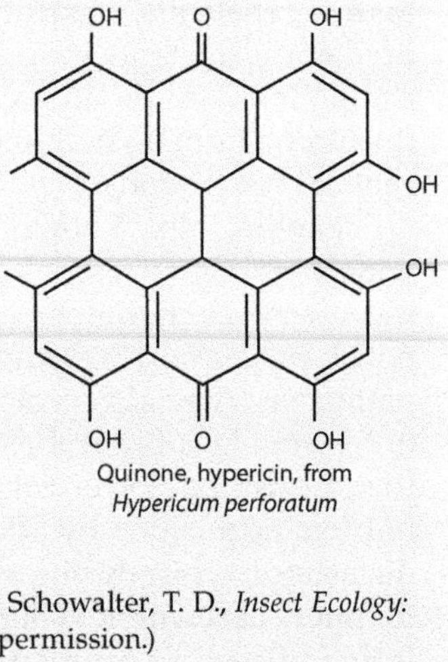

Quinone, hypericin, from *Hypericum perforatum*

FIGURE B3.1
Examples of nonnitrogenous defenses of plants. (From Schowalter, T. D., *Insect Ecology: An Ecosystem Approach*, Elsevier, San Diego, 2011. With permission.)

Terpenoids also are widely represented among plant groups. Monoterpenes and sesquiterpenes with low molecular weights are highly volatile compounds that are important floral and foliar scents that attract pollinators and herbivores, and often associated predators and parasites, to their hosts. Terpenoids with higher molecular weights include plant resins, cardiac glycosides, saponins, and latex (Figure B3.1). Terpenoids typically are distasteful or toxic to herbivores. Cardiac glycosides (cardenolides) are the terpenoids best known as the milkweed compounds sequestered by monarch butterflies, *Danaus plexippus*. Ingestion of these compounds by vertebrates either induces vomiting or results in cardiac arrest. The butterflies thereby gain protection against predation by birds (Brower et al. 1968). However, monarch caterpillars suffer high mortality on milkweed species (e.g., *Asclepias humistrata*) that have high concentrations of cardiac glycosides (Zalucki et al. 2001).

Photooxidants, such as quinones (Figure B3.1) and furanocoumarins, increase epidermal sensitivity to solar radiation. Assimilation of these compounds results in severe sunburn, necrosis of the skin, and other epidermal damage upon exposure to sunlight. Feeding on furanocoumarin-producing plants in daylight can cause 100% mortality to insects, whereas feeding in the dark causes only 60% mortality (Harborne 1994). Adapted insects circumvent this defense by becoming leaf rollers or nocturnal feeders (Harborne 1994) or by sequestering antioxidants (Blum 1992).

Plants produce a variety of insect hormone analogues that disrupt insect development, typically preventing maturation or producing imperfect and sterile adults (Harborne 1994). Insect development is governed primarily by molting hormone (ecdysone) and juvenile hormone (Figure B3.2). The relative concentrations of these two hormones dictate the timing of molting and the subsequent stage of development. Some phytoecdysones are as much as 20 times more active than are ecdysones produced by insects and resist inactivation by insect enzymes (Harborne 1994). Plants also produce juvenile hormone analogues (primarily juvabione) and compounds that interfere with juvenile hormone activity (primarily precocene, Figure B3.2). The antijuvenile hormones typically cause precocious development.

Some plants produce insect alarm pheromones that induce rapid departure of colonizing insects. For example, wild potato, *Solanum berthaultii*, produces (E)-β-farnesene, the major component of alarm pheromones for many aphid species. This compound is released from glandular hairs on the foliage at sufficient quantities to induce avoidance by host-seeking aphids and departure of settled colonies (Gibson and Pickett 1983).

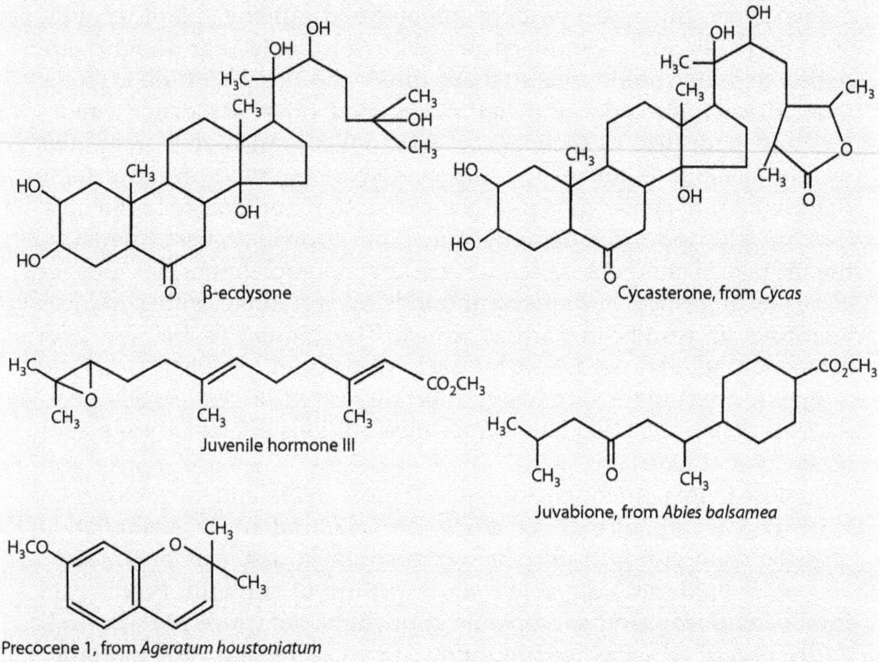

FIGURE B3.2
Insect developmental hormones and examples of their analogues in plants. (From Schowalter, T. D., *Insect Ecology: An Ecosystem Approach*, Elsevier, San Diego, 2011. With permission.)

Pyrethroids (Figure B3.3) are an important group of plant toxins and among the earliest recognized plant defenses (Riley 1885). Many synthetic pyrethroids are widely used as contact insecticides (i.e., absorbed through the exoskeleton) because of their rapid effect on insect pests.

Aflatoxins (Figure B3.3) are highly toxic fungal compounds. Aflatoxins produced by mutualistic endophytic or mycorrhizal fungi augment defense by host plants (Clay et al. 1985, 1993; Carroll 1988; Clay 1990; Harborne 1994; van Bael et al. 2009). Endophytic and mycorrhizal fungi also may induce host plants to increase production of defensive compounds when injured by herbivores (Hartley and Gange 2009).

Nonprotein amino acids are analogues of essential amino acids (Figure B3.4). Their substitution for essential amino acids in proteins results in improper configuration, loss of enzyme function, and inability to maintain physiological processes critical to survival. Others, such as 3,4-dihyrophenylalanine (L-DOPA), interfere with tyrosinase (an enzyme critical to hardening of the insect cuticle). Over 300 nonprotein

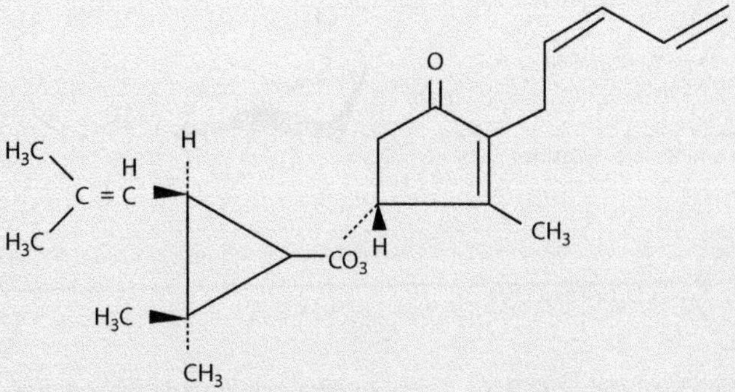

Pyrethrin I, from *Chrysanthemum cinearifolium*

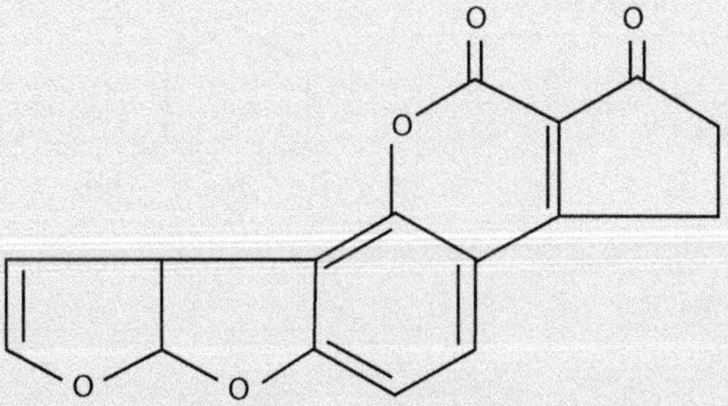

Aflatoxin B, from *Aspergillus flavus*

FIGURE B3.3
Examples of pyrethroid and aflatoxin defenses. (From Schowalter, T. D., *Insect Ecology: An Ecosystem Approach*, Elsevier, San Diego, 2011. With permission.)

amino acids are known, primarily from seeds of legumes (Harborne 1994). Proteinase inhibitors, produced by a variety of plants, interfere with insect digestive enzymes (Thaler et al. 2001; Kessler and Baldwin 2002).

Cyanogenic glycosides are distributed widely among plant families (Figure B3.4). These compounds are inert in plant cells, but when crushed plant cells enter the herbivore gut, the glycoside is hydrolyzed by glucosidases into glucose and hydrogen cyanide (Zenk and Juenger 2007). Hydrogen cyanide is toxic to most organisms because of its inhibition of cytochromes in the electron transport system (Harborne 1994).

Glucosinolates, such as glucobrassicin from cabbage and related plants (Figure B3.4), have been shown to deter feeding and reduce growth in a variety of herbivores (Renwick 2002; Strauss et al. 2004).

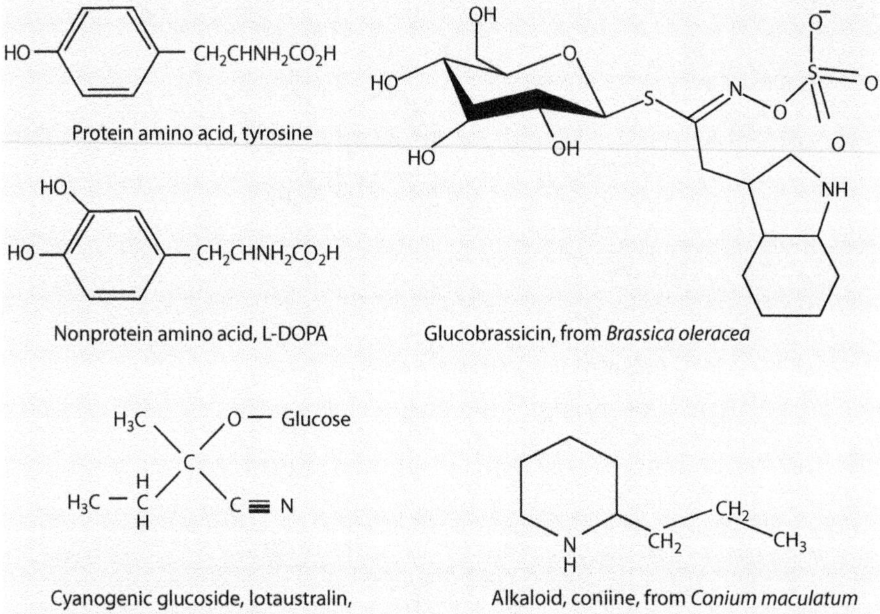

Protein amino acid, tyrosine

Nonprotein amino acid, L-DOPA

Glucobrassicin, from *Brassica oleracea*

Cyanogenic glucoside, lotaustralin, from *Lotus corniculatus*

Alkaloid, coniine, from *Conium maculatum*

FIGURE B3.4
Examples of nitrogenous defenses of plants. (From Schowalter, T. D., *Insect Ecology: An Ecosystem Approach*, Elsevier, San Diego, 2011. With permission.)

Intact glucosinolates confer some resistance to herbivores, but damaged plant cells release the enzyme myrosinase that converts glucosinolates to toxic isothiocyanates, nitriles, and oxazolidinethiones that are more toxic (Hopkins et al. 2009).

Alkaloids include more than 5,000 known structures from about 20% of higher plant families (Harborne 1994), ranging in molecular size and complexity from the relatively simple coniine (Figure B3.4) to multicyclic compounds such as solanine. These compounds are highly toxic and teratogenic, even at relatively low concentrations, because they interfere with major physiological processes, especially cardiovascular and nervous system functions, but include important pharmaceuticals, such as epinephrine and caffeine.

Some plants accumulate and tolerate high concentrations of toxic elements, including Se, Mn, Cu, Ni, Zn, Cd, Cr, Pb, Co, Al, and As (Boyd 2004, 2007, 2009; Trumble and Sorensen 2008). In some cases, foliage concentrations of these metals can exceed 2% (Jhee et al. 1999). Although the function of such hyperaccumulation remains unclear, high concentrations in some plants confer protection against herbivores (Pollard and Baker 1997; Boyd and Moar 1999; Vickerman et al. 2002; Boyd 2004, 2007; Galeas et al. 2008).

In addition to compounds that are normally present in plant tissues (constitutive defenses), plants are able to induce production of new defenses in response to particular injury (Ralph et al. 2006). Induced responses may be new compounds, elevated concentrations of compounds such as those described earlier, or proteinase inhibitors (Schmelz et al. 2006, 2007; Little et al. 2007). Plant ability to produce these defenses depends on availability of water and nutrients, especially nitrogen in the case of compounds including this element, and may be impeded under adverse conditions that limit resource availability (Schowalter 2011).

Insects that feed on particular plant species have evolved genetic mechanisms to avoid or detoxify plant defenses (Becerra 1997). Many compounds have similar modes of action and differ only in the structure and composition of attached radicals, making related compounds vulnerable to desensitization or detoxification by the same insect enzymes. Because many insecticides share chemical structure with plant defensive compounds, they are similarly vulnerable to insect desensitization or detoxification mechanisms (see Chapter 6). However, insect adaptations to plant defenses also involve fitness costs (Voelckel et al. 2001) that favor plants when other factors impose stronger selection pressure on insect herbivores.

Insects that depend on lost resources may disappear, but some insects flourish on surviving hosts that are stressed and less capable of defense or on new hosts that exploit reduced competition or predation. Sap-sucking hemipterans are favored by rapid growth of early successional plants (Schowalter et al. 1981b; Schowalter 1995; Schowalter and Ganio 2003). Other species also respond to rapid growth of early successional plants (Bishop 2002; Ulyshen et al. 2010). Carabid beetle abundance and species richness increased in riparian forests subject to periodic flooding, compared with nonflooded sites, indicating that flooding contributed to habitat suitability for these beetles (Cartron et al. 2003; Lambeets et al. 2008).

Disturbances that create large amounts of coarse woody debris (fires and storms) and/or stressed trees (droughts and storms) are typical triggers for bark beetle and wood borer outbreaks (Mattson and Haack 1987; Breshears et al. 2005; Gandhi et al. 2007; Raffa et al. 2008). Acoustic cues from cavitating cell walls may attract bark beetles to water-stressed trees (Mattson and Haack 1987). Many wood-boring insects, such as wood wasps (Siricidae) and beetles (especially Buprestidae) are attracted to sources of smoke, infrared radiation, or volatile tree chemicals emitted from burned or injured trees over distances of up to 50 km (Evans 1966; Gara et al. 1984; Raffa et al. 1993;

Schmitz 1997; Schütz et al. 1999; Wikars 2002). These cues signal the avail-ability of dead trees, typically rare in undisturbed forests, that are suitable for reproduction. Interestingly, locust outbreaks appear to be triggered by either drought or flooding disturbances. A 1,000-year record of locust out-breaks in China indicated that outbreaks typically originated in floodplain refuges, which are characterized by adequate vegetation and suitable ovipo-sition sites, during drought years and years after flooding (Stige et al. 2007). Droughts increase the availability of suitable oviposition sites—as well as stressed vegetation—as water recedes, whereas similar conditions occur in formerly flooded areas in the year after flooding.

On the other hand, many herbivorous insects become less abundant on stressed host plants (Waring and Price 1990; Price 1991; Schowalter et al. 1999). A major drought in Pacific Northwestern North America virtually eliminated the dominant folivore, a budmoth *Zeiraphera hesperiana*, and favored its replacement by western spruce budworm, *C. occidentalis*, and bal-sam fir sawfly, *Neodiprion abietis*; following the drought, *Z. hesperiana* recov-ered its dominance, and *C. occidentalis* and *N. abietis* disappeared (Figure 3.5) (Schowalter 1995, 2011). Schowalter et al. (1999) reported that some herbivo-rous species increased, whereas others decreased, in abundance on creosote bushes, *Larrea tridentata*, subjected to an experimental moisture gradient. A review by Koricheva et al. (1998) also revealed that response to plant water status varies widely among herbivorous insects.

Variation in response to plant stress is often, but not always, associated with changes in plant defensive chemistry (see Box 3.1). Hale et al. (2005) reported that drought-stressed poplars increased production of phenolic glycoside concentrations, with differing effects on gypsy moths, *L. dis-par*, and white marked tussock moths, *Orgyia leucostigma*. Forest canopy-opening disturbances often result in increased production of phenolics by early successional plants growing under conditions of higher light avail-ability (Shure and Wilson 1993; Hunter and Forkner 1999). However, despite higher foliar phenolic concentrations, many herbivores increased in abun-dance following Hurricane Opal in southeastern North America (Hunter and Forkner 1999). Nutrient limitation, especially of nitrogen, may limit plant production of nitrogenous defensive compounds, such as those repre-sented in Figure B3.4.

Even within families and genera, individual species respond quite differ-ently to disturbances. Among Hemiptera, some scale insect species increased in abundance, and others decreased, during forest canopy recovery from hurricanes in Puerto Rico (Schowalter and Ganio 2003). Root bark beetles (e.g., *Hylastes nigrinus*) are attracted to chemicals emanating from exposed stump surfaces that advertise suitable conditions for brood development and become more abundant following forest thinning (Witcosky et al. 1986), whereas stem-feeding bark beetles (e.g., *Dendroctonus* spp.) are sensitive to tree spacing and become less abundant in thinned forests (Sartwell and Stevens 1975; Amman et al. 1988; Schowalter and Turchin 1993).

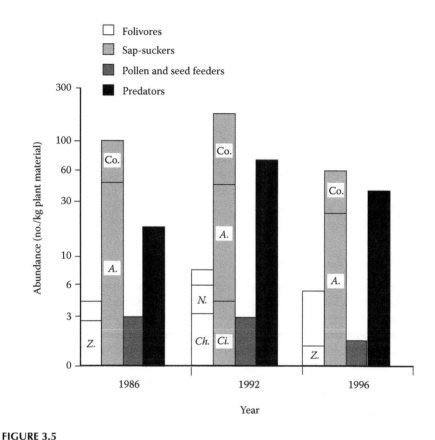

FIGURE 3.5

Temporal change in arthropod abundances in old-growth Douglas-fir canopies at the H. J. Andrews Experimental Forest in western Oregon; 1986 and 1996 were relatively wet years, 1992 was in the middle of an extended drought period (1987–1993). Z. = *Zeiraphera hesperiana* (bud moth); Ch. = *Choristoneura occidentalis* (western spruce budworm), N. = *Neodiprion abietis* (fir sawfly); Ci. = *Cinara* spp. (giant conifer aphids); A. = *Adelges cooleyi* (Cooley spruce gall adelgid); Co. = Coccoidea (scale insects, 4 spp.). Note the log scale of abundance. Data from Schowalter (1989, 1995, and unpublished data). (From Schowalter, T. D., *Insect Ecology: An Ecosystem Approach*, Elsevier, San Diego, 2011. With permission.)

Responses also vary among disturbance types. Paquin and Coderre (1997) compared forest floor arthropod responses to forest clearing versus fire. Decomposers were less abundant, whereas predators were more abundant in cleared plots, relative to undisturbed plots. Arthropod abundance overall was reduced 96% following experimental fire, but some organisms survived due to occurrence in deeper soil levels or to the patchy effect of fire. Abundances of some species differed between cleared and burned plots.

Following disturbance, populations and ecosystems may recover to their predisturbance condition at rates that reflect the extent of change and the size of the disturbed area. Recovery can be as quick as a few months for rapidly reproducing species or assemblages, such as many insects, or for patches in landscape matrices that facilitate dispersal (Haynes and Cronin 2003; Gandhi et al. 2007) or many years-to-centuries for long-lived, slowly reproducing species or assemblages. Factors that delay recovery of plant communities also influence recovery of habitat conditions and rates of insect recovery. Insects often influence rates of community recovery. Herbivorous species can accelerate replacement of earlier successional host plant species by later successional nonhost plant species (Torres 1992; Davidson 1993). If the disturbance-free interval is shorter than the time needed for recovery, then earlier-successional communities may persist. Clearly, anthropogenic activities such as crop planting and harvesting, land conversion, or elimination of species that are instrumental in community development will retard or prevent succession.

3.5.2.3 Predator/Parasite Abundance and Foraging Activity

Disturbances affect abundances and foraging activity of predators and their prey differently, creating areas of concentrated predation or of enemy-free space (Lancaster 1996; Thies et al. 2003; Hance et al. 2007). Insects at higher trophic levels appear to be particularly susceptible to disturbances (Paquin and Coderre 1997; Schowalter and Ganio 2003; Thies et al. 2003; Hance et al. 2007; Gandhi et al. 2008), although some carabid beetle species increase in abundance in burned sites (Gandhi et al. 2008). Parasitoids commonly have lower temperature tolerance than their hosts, and different thermal tolerances affect the temporal synchronization of parasitoid and host during the season (Hance et al. 2007). Beuzelin et al. (2009) reported a threefold reduction in red imported fire ants in areas inundated by storm surge during Hurricane Rita. Entomopathogen abundance may be reduced in disturbed areas by exposure to UV radiation (Roland and Kaupp 1995). Reduced predation or parasitism in disturbed areas may permit prey species to increase in abundance.

The variety of insect defensive strategies is evidence of the importance of predation in directing adaptation (Schowalter 2011). Many insects are cryptically colored to blend with their food plants (Figure 3.6a and b). Others produce venoms or sequester defensive toxins from their host plants to deter predators (Figures 2.9 and 3.6c). Many toxic or venomous insects advertise their threat to would-be predators, whereas other nondangerous insects similarly warn predators away by mimicking dangerous species (Figure 3.6d).

3.6 Rate of Adaptation by Insects

As a result of their small size, insects are sensitive to environmental changes, and as a result of their high reproductive capacity, they can respond quickly

FIGURE 3.6 (SEE COLOR INSERT.)
Representative insect defenses against predators. (a) Cryptic coloration (camouflage) by a grasshopper, *Syrbula admirabilis*, from western North America. (b) Deception by a Taiwanese moth, *Phalera angustipennis*, that resembles a piece of broken wood, and (c) combination of camouflage, warning stripes, and venomous spines in the Io moth caterpillar, *Automeris io*, from eastern North America. (d) Brightly colored insects, such as the red and black cicada, *Huechys sanguinea*, from Southeast Asia attract our attention and appreciation, but such coloration either advertises toxic or distasteful properties or mimics a distasteful model to would-be predators. [Photo (d) from Schowalter, T. D., *Insect Ecology: An Ecosystem Approach*, Elsevier, San Diego, 2011. With permission.]

to changes in environmental factors that occur consistently among generations. Insects have survived the major extinction events of the past that led to extinction of other major groups. They have adapted to a broad spectrum of plant defenses that include physical barriers, feeding deterrents, toxic elements and compounds, digestion inhibitors, and growth hormone analogues

FIGURE 3.6 (CONTINUED)
Representative insect defenses against predators. (a) Cryptic coloration (camouflage) by a grasshopper, *Syrbula admirabilis*, from western North America. (b) Deception by a Taiwanese moth, *Phalera angustipennis*, that resembles a piece of broken wood, and (c) combination of camouflage, warning stripes, and venomous spines in the Io moth caterpillar, *Automeris io*, from eastern North America. (d) Brightly colored insects, such as the red and black cicada, *Huechys sanguinea*, from Southeast Asia attract our attention and appreciation, but such coloration either advertises toxic or distasteful properties or mimics a distasteful model to would-be predators. [Photo (d) from Schowalter, T. D., *Insect Ecology: An Ecosystem Approach*, Elsevier, San Diego, 2011. With permission.]

that interfere with normal development (see Box 3.1). In some cases, patterns of speciation within herbivorous insect genera directly mirror speciation within their host plant genera (Becerra 1997), supporting the concept of a "coevolutionary arms race" (Whittaker and Feeny 1971).

Introduction of novel insecticides and other toxins into the environment has demonstrated how quickly adaptation can occur in insects. DDT was introduced into agricultural practice following World War II. By the mid-1950s (less than a decade later), many target insect species were sufficiently resistant to the effects of DDT that they could no longer be controlled

(Clausen 1954; Roussel and Clower 1957; Soderlund and Bloomquist 1990; Feyereisen 1997). Felland et al. (1990) reported that the soybean looper, *Chrysodeixis includens*, became resistant to permethrin (mortality declining from 99% to 60%) within five years of first application in soybeans in Mississippi. Currently, more than 500 insect species are sufficiently resistant to insecticides that they can no longer be controlled effectively (Soderlund and Bloomquist 1990; Hsu et al. 2004).

This brings into question the reliance on insecticides for long-term control of insects given the often serious undesirable consequences of insecticides for nontarget species and processes that affect the sustainability of ecosystem services (see Chapter 7). Even transgenic crops that incorporate genes for toxin production from *Bacillus thuringiensis* (Bt), introduced in the 1990s to minimize many of the undesirable consequences of insecticides, can select for resistant insects (Tabashnik et al. 1996, 1997). Although producers were required to maintain refuges of nontransgenic cultivars in order to delay increase in resistance within populations, several insect species have shown the ability to adapt to transgenic cultivars with single Bt toxins, perhaps aided by pollen contamination, adventitious volunteer Bt plants within refuges, or contamination of nontransgenic seed lots with transgenic seed (Tabashnik et al. 1996, 1997; Ottea and Leonard 2006; Huang et al. 2007; Heuberger et al. 2008 a, b). Consequently, transgenic crops with multiple genes for different Bt toxins have been introduced to delay resistance development. The effectiveness of this strategy depends on all toxins being produced at levels capable of killing all target insects throughout the growing season (Heuberger et al. 2011). In fact, this does not occur (Kranthi et al. 2005; Mahon and Olsen 2009; Showalter et al. 2009). Toxin levels may decline as a result of plant stress, and not all plants produce all toxins, as a result of independent assortment of unlinked genes among seeds. Heuberger et al. (2011) modeled the time to resistance for scenarios in which all toxin genes are linked on the same chromosomes or unlinked. Linked genes (on the same chromosome) would result in all seeds being capable of producing all toxins. Simulated time to insecticide resistance could be delayed by decades if genes for all toxins were linked.

3.7 Summary

Insects possess a number of attributes that make them remarkably adaptive to environmental changes. Small size permits use of abundant microscopic resources and protection in habitats too small for penetration by many chemicals or larger predators. The exoskeleton limits size and speed but permits flexibility and protection from water loss or penetration by many chemicals. Metamorphosis is necessary for growth but also permits partitioning of habitats and resources among life stages for many insects, improving life history efficiency and reducing intraspecific competition. Flight is arguably the most important attribute. Insects were the first organisms to fly and have achieved remarkable aerial speed and agility. This attribute permits rapid movement during foraging and escape from predators but also permits escape from adverse conditions or rapid colonization of new habitats created by disturbances.

Disturbances are pervasive among ecosystems, differing only in type, magnitude, frequency of occurrence, and severity in terms of effects on habitat and resource conditions. Some disturbances occur with sufficient frequency in some ecosystems to select for specific insect adaptations. For example, many floodplain insects can withstand hypoxia for long periods during inundation, and many ants can form living rafts that exploit floods for dispersal to new habitats. Insects characterizing frequently disturbed ecosystems generally have wider tolerances to variation in temperature, moisture, or chemical conditions. However, the direct effects of disturbances can eliminate most insect populations locally. Many insect adaptations primarily permit survival in the postdisturbance environment. In particular, many herbivorous insects exploit plants that either grow rapidly in the absence of competition following disturbance or are stressed by disturbance and become less capable of producing chemical defenses. Under such conditions, insect populations can grow exponentially, reaching outbreak levels in short periods of time, as described in the next chapter.

Introduction of novel insecticides and other toxins in the environment in recent decades has permitted measurement of the rate at which insects can adapt to new selective factors. Most target species have shown sufficient adaptation of resistance within a decade, and often within five years, to make new chemicals ineffective for control. This not only brings into question the wisdom of many control efforts but also makes insects prime examples of evolution.

References

Alstad, D. N., G. F. Edmunds, Jr., and L. H. Weinstein. 1982. Effects of air pollutants on insect populations. *Annual Review of Entomology* 27: 369–384.

Amman, G. D., M. D. McGregor, R. F. Schmitz, and R. D. Oakes. 1988. Susceptibility of lodgepole pine to infestation by mountain pine beetles following partial cutting of stands. *Canadian Journal of Forest Research* 18: 688–695.

Anderson, C., G. Theraulaz, and J.-L. Deneubourg. 2002. Self-assemblage in insect societies. *Insectes Sociaux* 49: 99–110.

Anstey, M. L., S. M. Rogers, S. R. Ott, M. Burrows, and S. J. Simpson. 2009. Serotonin mediates behavioral gregarization underlying swarm formation in desert locusts. *Science* 323: 627–630.

Antunes, S. C., N. Curado, B. B. Castro, and F. Gonçalves. 2009. Short-term recovery of soil functional parameters and edaphic macro-arthropod community after a forest fire. *Journal of Soils and Sediments* 9: 267–278.

Balanyá, J., J. M. Oller, R. B. Huey, G. W. Gilchrist, and L. Serra. 2006. Global genetic change tracks global climate warming in *Drosophila subobscura*. *Science* 313: 1773–1775.

Ballinger, A., P. S. Lake, and R. M. MacNally. 2007. Do terrestrial invertebrates experience floodplains as landscape mosaics? Immediate and longer-term effects of flooding on ant assemblages in a floodplain forest. *Oecologia* 152: 227–238.

Barbehenn, R., Q. Weir, and J.-P. Salminen. 2008. Oxidation of ingested phenolics in the tree-feeding caterpillar *Orgyia leucostigma* depends on foliar chemical composition. *Journal of Chemical Ecology* 34: 748–756.

Batzer, D. P. and S. A. Wissinger. 1996. Ecology of insect communities in nontidal wetlands. *Annual Review of Entomology* 41: 75–100.

Becerra, J. X. 1997. Insects on plants: Macroevolutionary chemical trends in host use. *Science* 276: 253–256.

Behmer, S. T. 2009. Insect herbivore nutrient regulation. *Annual Review of Entomology* 54: 165–187.

Beuzelin, J. M., T. E. Reagan, W. Akbar, H. J. Cormier, J. W. Flanagan, and D. C. Blouin. 2009. Impact of Hurricane Rita storm surge on sugarcane borer (Lepidoptera: Crambidae) management in Louisiana. *Journal of Economic Entomology* 102: 1054–1061.

Bishop, J. G. 2002. Early primary succession on Mount St. Helens: Impact of insect herbivores on colonizing lupines. *Ecology* 83: 191–202.

Björklund, N., G. Nordlander, and H. Bylund. 2003. Host-plant acceptance on mineral soil and humus by the pine weevil, *Hylobius abietis* (L.). *Agricultural and Forest Entomology* 5: 61–65.

Blum, M. S. 1992. Ingested allelochemicals in insect wonderland: a menu of remarkable functions. *American Entomologist* 38: 222–234.

Boyd, R. S. 2004. Ecology of metal hyperaccumulation. *New Phytologist* 162: 563–567.

Boyd, R. S. 2007. The defense hypothesis of elemental hyperaccumulation: Status, challenges, and new directions. *Plant and Soil* 293: 153–176.

Boyd, R. S. 2009. High-nickel insects and nickel hyperaccumulator plants: A review. *Insect Science* 16: 19–31.

Boyd, R. S. and W. J. Moar. 1999. The defensive function of Ni in plants: Response of the polyphagous herbivore *Spodoptera exigua* (Lepidoptera: Noctuidae) to hyperaccumulator and accumulator species of *Streptanthus* (Brassicaceae). *Oecologia* 118: 218–224.

Bradshaw, W. E. and C. M. Holzapfel. 2001. Genetic shift in photoperiodic response correlated with global warming. *Proceedings of the National Academy of Sciences USA* 98: 14509–14511.

Breshears, D. D., N. S. Cobb, P. M. Rich, K. P. Price, C. D. Allen, R. G. Balice, W. H. Romme, J. H. Kastens, M. L. Floyd, J. Belnap, J. J. Anderson, O. B. Meyers, and C. W. Meyer. 2005. Regional vegetation die-off in response to global-change-type drought. *Proceedings of the National Academy of Sciences USA* 102: 15144–15148.

Brower, L. P., W. N. Ryerson, L. L. Coppinger, and S. C. Glazier. 1968. Ecological chemistry and the palatability spectrum. *Science* 161: 1349–1351.

Brown, M., T. A. Black, Z. Nesic, V. N. Foord, D. L. Spittlehouse, A. L. Fredeen, N. J. Grant, P. J. Burton, and J. A. Trofymow. 2010. Impact of mountain pine beetle on the net ecosystem production of lodgepole pine stands in British Columbia. *Agricultural and Forest Meteorology* 150: 254–264.

Brust, M. L. and W. W. Hoback. 2009. Hypoxia tolerance in adult and larval *Cicindel* tiger beetles varies by life history but not habitat association. *Annals of the Entomological Society of America* 102: 462–466.

Brusven, M. A. and C. E. Hornig. 1984. Effects of suspended and deposited volcanic ash on survival and behavior of stream insects. *Journal of the Kansas Entomological Society* 57: 55–62.

Buddle, C. M., D. W. Langor, G. R. Pohl, and J. R. Spence. 2006. Arthropod responses to harvesting and wildfire: Implications for emulation of natural disturbance in forest management. *Biological Conservation* 128: 346–357.

Carroll, G. 1988. Fungal endophytes in stems and leaves: From latent pathogen to mutualistic symbiont. *Ecology* 69: 2–9.

Cartron, J.-L. E., M. C. Molles, Jr., J. F. Schuetz, C. S. Crawford, and C. N. Dahm. 2003. Ground arthropods as potential indicators of flooding regime in the riparian forest of the middle Rio Grande, New Mexico. *Environmental Entomology* 32: 1075–1084.

Chapman, R. F. 1998. *The Insects: Structure and Function*, 4th ed. Cambridge, U.K.: Cambridge University Press.

Chase, J. M. 2007. Drought mediates the importance of stochiastic community assembly. *Proceedings of the National Academy of Sciences USA* 104: 17430–17434.

Chen, J., J. F. Franklin, and T. A. Spies. 1995. Growing-season microclimatic gradients from clearcut edges into old-growth Douglas-fir forests. *Ecological Applications* 5: 74–86.

Clausen, L.W. 1954. *Insect Fact and Folklore*. New York: MacMillan.

Clay, K. 1990. Fungal endophytes of grasses. *Annual Review of Ecology and Systematics* 21: 275–297.

Clay, K., T. N. Hardy, and A. M. Hammond. 1985. Fungal endophytes of grasses and their effects on an insect herbivore. *Oecologia* 66: 1–5.

Clay, K., S. Marks, and G. P. Cheplick. 1993. Effects of insect herbivory and fungal endophyte infection on competitive interactions among grasses. *Ecology* 74: 1767–1777.

Cleary, D. F. R. and A. Grill. 2004. Butterfly response to severe ENSO-induced forest fires in Borneo. *Ecological Entomology* 29: 666–676.

Cook, R.J., J. C. Barron, R. I. Papendick, and G. J. Williams III. 1981. Impact on agriculture of the Mount St. Helens eruptions. *Science* 211: 16–22.

Courtney, S. P. 1985. Apparency in coevolving relationships. *Oikos* 44: 91–98.

Courtney, S. P. 1986. The ecology of pierid butterflies: Dynamics and interactions. *Advances in Ecological Research* 15: 51–131.

Davidson, D. W. 1993. The effects of herbivory and granivory on terrestrial plant succession. *Oikos* 68: 23–35.

de Mazancourt, C., E. Johnson, and T. G. Barradough. 2008. Biodiversity inhibits species' evolutionary responses to changing environments. *Ecology Letters* 11: 380–388.

Edwards, J. S. and P. Sugg. 1990. Arthropod fallout as a resource in the recolonization of Mt. St. Helens. *Ecology* 74: 954–958.

Elvidge, C. D., C. Milesi, J. B. Dietz, B. T. Tuttle, P. C. Sutton, R. Nemani, and J. E. Vogelmann. 2004. U.S. constructed area approaches the size of Ohio. *EOS, Transactions, American Geophysical Union* 85: 233.

Evans, E. W. 1988. Community dynamics of prairie grasshoppers subjected to periodic fire: Predictable trajectories or random walks in time? *Oikos* 52: 283–292.

Evans, W. G. 1966. Perception of infrared radiation from forest fires by *Melanophila acuminata* de Geer (Buprestidae, Coleoptera). *Ecology* 47: 1061–1065.

Feeny, P. P. 1969. Inhibitory effect of oak leaf tannins on the hydrolysis of proteins by trypsin. *Phytochemistry* 8: 2119–2126.

Felland, C. M., H. N. Pitre, R. G. Luttrell, and J. L. Hamer. 1990. Resistance to pyrethroid insecticides in soybean looper (Lepidoptera: Noctuidae) in Mississippi. *Journal of Economic Entomology* 83: 35–40.

Feyereisen, R. 1999. Insect P450 enzymes. *Annual Review of Entomology* 44: 507–533.

Foley, J. A., M. H. Costa, C. Delire, N. Ramankutty, and P. Snyder. 2003. Green surprise? How terrestrial ecosystems could affect earth's climate. *Frontiers in Ecology and the Environment* 1: 38–44.

Forschler, B. T. and G. Henderson. 1995. Subterranenan termite behavioral reaction to water and survival of inundation: Implications for field populations. *Environmental Entomology* 24: 1592–1597.

Fox, L. R. 1975a. Cannibalism in natural populations. *Annual Review of Ecology and Systematics* 6: 87–106.

Fox, L. R. 1975b. Some demographic consequences of food shortage for the predator, *Notonecta hoffmanni. Ecology* 56: 868–880.

Fox, L. R. and B. J. Macauley. 1977. Insect grazing on eucalyptus in response to variation in leaf tannins and nitrogen. *Oecologia* 29: 145–162.

Fraenkel, G. and M. Blewett. 1946. Linoleic acid, vitamin E, and other fat-soluble substances in the diet of certain insects, *Ephestria kuehniella, E. elutella, E. cautella,* and *Plodia interpunctella* (Lepidoptera). *Journal of Experimental Biology* 22: 172–190.

Galeas, M. L., E. M. Klamper, L. E. Bennett, J. L. Freeman, B. C. Kondratieff, C. F. Quinn, and E. A. H. Pilon-Smits. 2008. Selenium hyperaccumulation reduces plant arthropod loads in the field. *New Phytologist* 177: 715–724.

Gandhi, K. J. K., D. W. Gilmore, S. A. Katovich, W. J. Mattson, J. R. Spence, and S. J. Seybold. 2007. Physical effects of weather events on the abundance and diversity of insects in North American forests. *Environmental Review* 15: 113–152.

Gandhi, K. J. K., D. W. Gilmore, S. A. Katovich, W. J. Mattson, J. C. Zasada, and S. J. Seybold. 2008. Catastrophic windstorm and fuel-reduction treatments alter ground beetle (Coleoptera: Carabidae) assemblages in a North American subboreal forest. *Forest Ecology and Management* 256: 1104–1123.

Gara, R. I., D. R. Geiszler, and W. R. Littke. 1984. Primary attraction of the mountain pine beetle to lodgepole pine in Oregon. *Annals of the Entomological Society of America* 77: 333–334.

Gersich, F. M. and M. A. Brusven. 1982. Volcanic ash accumulation and ash-voiding mechanisms of aquatic insects. *Journal of the Kansas Entomological Society* 55: 290–296.

Gibbs, A. G. and L. M. Matzkin. 2001. Evolution of water balance in the genus *Drosphila. Journal of Experimental Biology* 204: 2331–2338.

Gibson, R. W. and J. A. Pickett. 1983. Wild potato repels aphids by release of aphid alarm pheromone. *Nature* 302: 608–609.

Goyer, R. A., G. J. Lenhard, and B. L. Strom. 2004. The influence of silhouette color and orientation on arrival and emergence of *Ips* pine engravers and their predators in loblolly pine. *Forest Ecology and Management* 191: 147–155.

Grimbacher, P. S. and N. E. Stork. 2009. How do beetle assemblages respond to cyclonic disturbance of a fragmented tropical rainforest landscape? *Oecologia* 161: 591–599.

Haddad, N. M., M. Holyoak, T. M. Mata, K. F. Davies, B. A. Melbourne, and K. Preston. 2008. Species' traits predict the effects of disturbance and productivity on diversity. *Ecology Letters* 11: 348–356.

Hadley, N. F. 1994. *Water Relations of Terrestrial Arthropods*. San Diego: Academic.

Hale, B. K., D. A. Herms, R. C. Hansen, T. P. Clausen, and D. Arnold. 2005. Effects of drought stress and nutrient availability on dry matter allocation, phenolic glycosides, and rapid induced resistance of poplar to two lymantriid defoliators. *Journal of Chemical Ecology* 31: 2601–2620.

Hance, T., J. van Baaren, P. Vernon, and G. Boivin. 2007. Impact of extreme temperatures on parasitoids in a climate change perspective. *Annual Review of Entomology* 52: 107–126.

Hanski, I. and D. Simberloff. 1997. The metapopulation approach, its history, conceptual domain, and application to conservation. In *Metapopulation Biology: Ecology, Genetics and Evolution*, I. A. Hanski and M. E. Gilpin, eds. 5–26. San Diego: Academic.

Hanula, J. L. and D. D. Wade. 2003. Influence of long-term dormant-season burning and fire exclusion on ground-dwelling arthropod populations in longleaf pine flatwoods ecosystems. *Forest Ecology and Management* 175: 163–184.

Harborne, J. B. 1994. *Introduction to Ecological Biochemistry*, 4th ed. London: Academic.

Hartley, S. E. and A. C. Gange. 2009. Impacts of plant symbiotic fungi on insect herbivores: Mutualism in a multitrophic context. *Annual Review of Entomology* 54: 323–342.

Haynes, K. J. and J. T. Cronin. 2003. Matrix composition affects the spatial ecology of a prairie planthopper. *Ecology* 84: 2856–2866.

He, F. and R. I. Alfaro. 1997. White pine weevil (Coleoptera: Curculionidae) attack on white spruce: Spatial and temporal patterns. *Environmental Entomology* 26: 888–895.

Heartsill-Scalley, T., F. N. Scatena, C. Estrada, W. H. McDowell, and A. E. Lugo. 2007. Disturbance and long-term patterns of rainfall and throughfall nutrient fluxes in a subtropical wet forest in Puerto Rico. *Journal of Hydrology* 333: 472–485.

Heliövaara, K. and R. Väisänen. 1986. Industrial air pollution and the pine bark bug, *Aradus cinnamomeus* Panz. (Het., Aradidae). *Zeitschrift für angewandte Entomologie* 101: 469–478.

Heliövaara, K. and R. Väisänen. 1993. *Insects and Pollution*. Boca Raton, FL: CRC Press.

Heuberger, S., C. Ellers-Kirk, C. Yafuso, A. J. Gassmann, B. E. Tabashnik, T. J. Dennehy, and Y. Carriére. 2008a. Effects of refuge contamination by transgenes on Bt resistance in pink bollworm (Lepidoptera: Gelichiidae). *Journal of Economic Entomology* 101: 504–514.

Heuberger, S., C. Yafuso, G. DeGrandi-Hoffman, B. E. Tabashnik, Y. Carriére, and T. J. Dennehy. 2008b. Outcrossed cottonseed and adventitious Bt plants in Arizona refuges. *Environmental Biosafety Research* 7: 87–96.

Heuberger, S. M., D. W. Crowder, T. Brevault, B. E. Tabashnik, and Y. Carriére. 2011. Modeling the effects of plant-to-plant gene flow, larval behavior, and refuge size on pest resistance to Bt cotton. *Environmental Entomology* 40: 484–495.

Hinton, H. E. 1960a. A fly larva that tolerates dehydration and temperatures of −270° to +102° C. *Nature* 188: 336–337.

Hinton, H. E. 1960b. Cryptobiosis in the larva of *Polypedilum vanderplanki* Hint. (Chironomidae). *Journal of Insect Physiology* 5: 286–315.

Hirao, T., M. Murakami, J. Iwamoto, H. Takafumi, and H. Oguma. 2008. Scale-dependent effects of windthrow disturbance on forest arthropod communities. *Ecological Research* 23: 189–196.

Hoback, W. W. and D. W. Stanley. 2001. Insects in hypoxia. *Journal of Insect Physiology* 47: 533–542.

Hopkins, R. J., N. M. van Dam, and J. J. A. van Loon. 2009. Role of glucosinolates in insect-plant relationships and multitrophic interactions. *Annual Review of Entomology* 54: 57–83.

Hsu, J.-C., H.-T. Feng, and W.-J. Wu. 2004. Resistance and synergistic effects of insecticides in *Bactrocera dorsalis* (Diptera: Tephritidae) in Taiwan. *Journal of Economic Entomology* 97: 1682–1688.

Huang, F., B. R. Leonard, and D. A. Andow. 2007. Sugarcane borer (Lepidoptera: Crambidae) resistance to transgenic *Bacillus thuringiensis* maize. *Journal of Economic Entomology* 100: 164–171.

Hunter, M. D. and R. E. Forkner. 1999. Hurricane damage influences foliar polyphenolics and subsequent herbivory on surviving trees. *Ecology* 80: 2676–2682.

Janzen, D. H. 1977. What are dandelions and aphids? *American Naturalist* 111: 586–589.

Jenkins, M. J., E. Herbertson, W. Page, and C. A. Jorgensen. 2008. Bark beetles, fuels, fires, and implications for forest management in the intermountain west. *Forest Ecology and Management* 254: 16–34.

Jhee, E. M., K. L. Dandridge, A. M. Christy, Jr., and A. J. Pollard. 1999. Selective herbivory on low-zinc phenotypes of the hyperaccumulator *Thlaspi caerulescens* (Brassicaceae). *Chemoecology* 9: 93–95.

Johnson, D. M. 2004. Life history and demography of *Cephaloleia fenestrate* (Hispinae: Chrysomelidae: Coleoptera). *Biotropica* 36: 352–361.

Jung, C. and B. A. Croft. 2001. Aerial dispersal of phytoseiid mites (Acari: Phytoseiidae): Estimating falling speed and dispersal distance of adult females. *Oikos* 94: 182–190.

Kennedy, J. S. 1975. Insect dispersal. In *Insects, Science, and Society*, D. Pimentel, ed. 103–119. New York: Academic.

Kessler, A. and I. T. Baldwin. 2002. Plant responses to herbivory: The emerging molecular analysis. *Annual Review of Plant Biology* 53: 299–328.

Kharboutli, M. S. and T. P. Mack. 1993. Tolerance of the striped earwig (Dermaptera: Labiduridae) to hot and dry conditions. *Environmental Entomology* 22: 663–668.

Kiffney, P. M., J. S. Richardson, and J. P. Bull. 2003. Responses of periphyton and insects to experimental manipulation of riparian buffer width along forest streams. *Journal of Applied Ecology* 40: 1060–1076.

Kim, Y. and N. Kim. 1997. Cold hardiness in *Spodoptera exigua* (Lepidoptera: Noctuidae). *Environmental Entomology* 26: 1117–1123.

Klowden, M. 2008. *Physiological Systems in Insects*. San Diego: Elsevier/Academic.

Knight, T. M. and R. D. Holt. 2005. Fire generates spatial gradients in herbivory: An example from a Florida sandhill ecosystem. *Ecology* 86: 587–593.

Kölsch, G., K. Jakobi, G. Wegener, and H. J. Braune. 2002. Energy metabolism and metabolic rate of the alder leaf beetle, *Agelastica alni* (L.) (Coleoptera: Chrysomelidae) under aerobic and anaerobic conditions: A microcalorimetric study. *Journal of Insect Physiology* 48: 143–151.

Koptur, S., M. C. Rodriguez, S. F. Oberbauer, C. Weekley, and A. Herndon. 2002. Herbivore-free time? Damage to new leaves of woody plants after Hurricane Andrew. *Biotropica* 34: 547–554.

Koricheva, J., S. Larsson, and E. Haukioja. 1998. Insect performance on experimentally stressed woody plants: A meta-analysis. *Annual Review of Entomology* 43: 195–216.

Kranthi, K. R., S. Naidu, C. S. Dhawad, A. Tatwawadi, K. Mate, E. Patil, A. A. Bharose, G. T. Behare, R. M. Wadaskar, and S. Kranthi. 2005. Temporal and intra-plant variability in Cry1Ac expression in Bt cotton and its influence on the survival of the cotton bollworm, *Helicoverpa armigera* (Hübner) (Noctuidae: Lepidoptera). *Current Science* 89: 291–298.

Krantz, G. W. and J. L. Mellott. 1972. Studies on phoretic specificity in *Macrocheles mycotrupetes* and *M. peltotrupetes* Krantz and Mellott (Acari: Macrochelidae), associates of geotrupine Scarabaeidae. *Acarologia* 14: 317–344.

Kundzewicz, Z. W., Y. Hirabayashi, and S. Kanae. 1998. River floods in the changing climate—observations and projections. *Water Resources Management* 24: 2633–2646.

Kurz, W. A., C. C. Dymond, G. Stinson, G. J. Rampley, E. T. Neilson, A. L. Carroll, T. Ebata, and L. Safranyik. 2008. Mountain pine beetle and forest carbon feedback to climate change. *Nature* 452: 987–990.

Lambeets, K., M. L. Vendegehuchte, J-P. Maelfait, and D. Bonte. 2008. Understanding the impact of flooding on trait-displacements and shifts in assemblage structure of predatory arthropods on river banks. *Journal of Animal Ecology* 77: 1162–1174.

Lancaster, J. 1996. Scaling the effects of predation and disturbance in a patchy environment. *Oecologia* 107: 321–331.

Little, D., C. Gouhier-Darimont, F. Bruessow, and P. Reymond. 2007. Oviposition by pierid butterflies triggers defense responses in *Arabidopsis*. *Plant Physiology* 143: 784–800.

MacLean, D. A. 2004. Predicting forest insect disturbance regimes for use in emulating natural disturbance. In *Emulating Natural Forest Landscape Disturbances: Concepts and Applications*, A. H. Perera, L. J. Buse, and M. G. Weber, eds. 69–82. New York: Columbia University Press.

Maeno, K. and S. Tanaka. 2008. Phase-specific developmental and reproductive strategies in the desert locust. *Bulletin of Entomological Research* 98: 527–534.

Maeno, K., S. Tanaka, and K. Harano. 2011. Tactile stimuli perceived by the antennae cause the isolated females to produce gregarious offspring in the desert locust, *Schistocerca gregaria*. *Journal of Insect Physiology* 57: 74–82.

Mahon, R. J. and K. M. Olsen. 2009. Limited survival of a Cry2Ab-resistant strain of *Helicoverpa armigera* (Lepidoptera: Noctuidae) on Bollgard II. *Journal of Economic Entomology* 102: 708–716.

Marske, K. A., M. A. Ivie, and G. M. Hilton. 2007. Effects of volcanic ash on the forest canopy insects of Montserrat, West Indies. *Environmental Entomology* 36: 817–825.

Martin-R., M., J. R. Cox, F. Ibarra-F., D. G. Alston, R. E. Banner, and J. C. Malecheck. 1999. Spittlebug and buffelgrass responses to summer fires in Mexico. *Journal of Range Management* 52: 621–625.

Mathias, D., L. Jacky, W. E. Bradshaw, and C. M. Holzapfel. 2007. Quantitative trait loci associated with photoperiodic response and stage of diapause in the pitcher-plant mosquito, *Wyeomyia smithii*. *Genetics* 176: 391–402.

Matthews, R. W. and J. R. Matthews. 2010. *Insect Behavior*, 2nd ed. Dordrecht, The Netherlands: Springer.

Mattson, W. J. 1980. Herbivory in relation to plant nitrogen content. *Annual Review of Ecology and Systematics* 11: 119–161.

Mattson, W. J. and R. A. Haack. 1987. The role of drought in outbreaks of plant-eating insects. *BioScience* 37: 110–118.

McClure, M. S. 1990. Role of wind, birds, deer, and humans in the dispersal of hemlock woolly adelgid (Homoptera: Adelgidae). *Environmental Entomology* 19: 36–43.

McCormick, M. P., L. W. Thomason, and C. R. Trepte. 1995. Atmospheric effects of the Mt. Pinatubo eruption. *Nature* 373: 399–404.

McCullough, D. G., R. A. Werner, and D. Neumann. 1998. Fire and insects in northern and boreal forest ecosystems of North America. *Annual Review of Entomology* 43: 107–127.

Meisel, J. E. 2006. Thermal ecology of the neotropical army ant, *Eciton burchellii*. *Ecological Applications* 16: 913–922.

Merlin, C., R. J. Gegear, and S. M. Reppert. 2009. Antennal circadian clocks coordinate sun compass orientation in migratory monarch butterflies. *Science* 325: 1700–1704.

Mertl, A. L., K. T. R. Wilkie, and J. F. A. Traniello. 2009. Impact of flooding on the species richness, density, and composition of Amazonian litter-nesting ants. *Biotropica* 41: 633–641.

Miller, K. K. and M. R. Wagner. 1984. Factors influencing pupal distribution of the pandora moth (Lepidoptera: Saturniidae) and their relationship to prescribed burning. *Environmental Entomology* 13: 430–431.

Mlot, M. J., C. A. Tovey, and D. L. Hu. 2011. Fire ants self-assemble into waterproof rafts to survive floods. *Nature* 108: 7669–7673.

Moretti, M. and C. Legg. 2009. Combining plant and animal traits to assess community functional responses to disturbance. *Ecography* 32: 299–309.

Ottea, J. and R. Leonard. 2006. Insecticide/acaricide resistance and management strategies. In *Use and Management of Insecticides, Acaricides, and Transgenic Crops*, J. N. All and M. F. Treacy, eds. Lanham, MD: Entomological Society of America.

Paquin, P. and D. Coderre. 1997. Deforestation and fire impact on edaphic insect larvae and other macroarthropods. *Environmental Entomology* 26: 21–30.

Parker, T. J., K. M. Clancy, and R. L. Mathiasen. 2006. Interactions among fire, insects, and pathogens in coniferous forests of the interior western United States and Canada. *Agricultural and Forest Entomology* 8: 167–189.

Pfadt, R. E. and D. M. Hardy. 1987. A historical look at rangeland grasshoppers and the value of grasshopper control programs. In *Integrated Pest Management on Rangeland*, J. L. Capinera, ed. 183–195, Boulder, CO: Westview Press.

Pollard, A. J. and A. J. M. Baker. 1997. Deterrence of herbivory by zinc hyperaccumulation in *Thlaspi caerulescens* (Brassicaceae). *New Phytologist* 135: 655–658.

Pope, D. N., R. N. Coulson, W. S. Fargo, J. A. Gagne, and C. W. Kelly. 1980. The allocation process and between-tree survival probabilities in *Dendroctonus frontalis* infestations. *Researches in Population Ecology* 22: 197–210.

Price, P. W. 1991. The plant vigor hypothesis and herbivore attack. *Oikos* 62: 244–251.

Price, P. W. 1997. *Insect Ecology*, 3rd ed. New York: John Wiley & Sons.

Raffa, K. F., B. H. Aukema, B. J. Bentz, A. L. Carroll, J. A. Hicke, M. G. Turner, and W. H. Romme. 2008. Cross-scale drivers of natural disturbances prone to anthropogenic amplification: The dynamics of bark beetle eruptions. *BioScience* 58: 501–517.

Raffa, K. F., T. W. Phillips, and S. M. Salom. 1993. Strategies and mechanisms of host colonization by bark beetles. In *Beetle-Pathogen Interactions in Conifer Forests*, T. D. Schowalter and G. M. Filip, eds. 103–128. London: Academic.

Ralph, S. G., H. Yueh, M. Friedmann, D. Aeschliman, J. A. Zeznik, C. C. Nelson, Y. S. N. Butterfield, R. Kirkpatrick, J. Liu, S. J. M. Jones, et al. 2006. Conifer defence against insects: Microarray gene expression profiling of Sitka spruce (*Picea sitchensis*) induced by mechanical wounding or feeding by spruce budworms (*Choristoneura occidentalis*) or white pine weevils (*Pissodes strobi*) reveals large-scale changes of the host transcriptome. *Plant, Cell and Environment* 29: 1545–1570.

Rankin, M. A. and J. C. A. Burchsted. 1992. The cost of migration in insects. *Annual Review of Entomology* 37: 533–559.

Rasmussen, E. M. and J. M. Wallace. 1983. Meterological aspects of the El Niño/ southern oscillation. *Science* 222: 1195–1202.

Raubenheimer, D. and S. J. Simpson. 2003. Nutrient balancing in grasshoppers: Behavioural and physiological correlates of diet breadth. *Journal of Experimental Biology* 206: 1669–1681.

Renwick, J. A. A. 2002. The chemical world of crucivores: Lures, treats, and traps. *Entomologia Experimentalis et Applicata* 104: 35–42.

Riley, C. V. 1885. *Fourth Report of the United States Entomological Commission, Being a Revised Edition of Bulletin No. 3, and the Final Report on the Cotton Worm, Together with a Chapter on the Boll Worm.* Washington, DC: U.S. Department of Agriculture.

Rodrigues, A., L. Cunha, A. Amaral, J. Medeiros, and P. Garcia. 2008. Bioavailability of heavy metals and their effects on the midgut cells of a phytophagous insect inhabiting volcanic environments. *Science of the Total Environment* 406: 116–122.

Rodriguez, J. M., ed. 1972. *Insect and Mite Nutrition: Significance and Implications in Ecology and Pest Management.* Amsterdam: North-Holland Publishing.

Roland, J. and W. J. Kaupp. 1995. Reduced transmission of forest tent caterpillar (Lepidoptera: Lasiocampidae) nuclear polyhedrosis virus at the forest edge. *Environmental Entomology* 24: 1175–1178.

Roussel, J. S. and D. F. Clower 1957. Resistance to the chlorinated hydrocarbon insecticides in the boll weevil. *Journal of Economic Entomology* 50: 463–468.

Rykken, J. J., A. R. Moldenke, and D. H. Olson. 2007. Headwater riparian forest-floor invertebrate communities associated with alternative forest management practices. *Ecological Applications* 17: 1168–1183.

St. Pierre, M. J. and S. D. Hendrix. 2003. Movement patterns of *Rhyssomatus lineaticollis* Say (Coleoptera: Curculionidae) within and among *Asclepias syriaca* (Asclepiadaceae) patches in a fragmented landscape. *Ecological Entomology* 28: 579–586.

Sartwell, C. and R. E. Stevens. 1975. Mountain pine beetle in ponderosa pine: Prospects for silvicultural control in second-growth stands. *Journal of Forestry* 73: 136–140.

Schmelz, E. A., M. J. Carroll, S. LeClere, S. M. Phipps, J. Meredith, P. S. Chourey, H. T. Alborn, and P. E. A. Teal. 2006. Fragments of ATP synthase mediate plant perception of insect attack. *Proceedings of the National Academy of Sciences USA* 103: 8894–8899.

Schmelz, E. A., S. LeClere, M. J. Carroll, H. T. Alborn, and P. E. A. Teal. 2007. Cowpea chloroplastic ATP synthase is the source of multiple plant defense elicitors during insect herbivory. *Plant Physiology* 144: 793–805.

Schmitz, H. 1997. Infrared detection in a beetle. *Nature* 386: 773–774.

Schowalter, T. D. 1989. Canopy arthropod community structure and herbivory in old-growth and regenerating forests in western Oregon. *Canadian Journal of Forest Research* 19: 318–322.

Schowalter, T. D. 1995. Canopy arthropod communities in relation to forest age and alternative harvest practices in western Oregon. *Forest Ecology and Management* 78: 115–125.

Schowalter, T. D. 2011. *Insect Ecology: An Ecosystem Approach*, 3rd ed. San Diego: Elsevier / Academic.

Schowalter, T. D. 2012. Insect responses to major landscape-level disturbance. *Annual Review of Entomology* 57: 1–20.

Schowalter, T. D. and L. M. Ganio. 2003. Diel, seasonal and disturbance-induced variation in invertebrate assemblages. In *Arthropods of Tropical Forests*, Y. Basset, V. Novotny, S. E. Miller, and R. L. Kitching, eds. 315–328. Cambridge, UK: Cambridge University.

Schowalter, T. D., D. C. Lightfoot, and W. G. Whitford. 1999. Diversity of arthropod responses to host-plant water stress in a desert ecosystem in southern New Mexico. *American Midland Naturalist* 142: 281–290.

Schowalter, T. D., D. N. Pope, R. N. Coulson, and W. S. Fargo. 1981a. Patterns of southern pine beetle (*Dendroctonus frontalis* Zimm.) infestation enlargement. *Forest Science* 27: 837–849.

Schowalter, T. D. and P. Turchin. 1993. Southern pine beetle infestation development: Interaction between pine and hardwood basal areas. *Forest Science* 39: 201–210.

Schowalter, T. D., J. W. Webb, and D. A. Crossley, Jr. 1981b. Community structure and nutrient content of canopy arthropods in clearcut and uncut forest ecosystems. *Ecology* 62: 1010–1019.

Schowalter, T. D., Y. L. Zhang, and J. J. Rykken. 2003. Litter invertebrate responses to variable density thinning in western Washington forest. *Ecological Applications* 13: 1204–1211.

Schütz, S., B. Weissbecker, H. E. Hummel, K.-H. Apel, H. Schmitz, and H. Bleckmann. 1999. Insect antenna as a smoke detector. *Nature* 398: 298–299.

Seastedt, T. R. and D. A. Crossley, Jr. 1981. Microarthropod response following cable logging and clear-cutting in the southern Appalachians. *Ecology* 62: 126–135.

Showalter, A. M., S. Heuberger, B. E. Tabashnik, and Y. Carrière. 2009. A primer for using transgenic insecticidal cotton in developing countries. *Journal of Insect Science* 9: 1–39.

Shure, D. J. and D. L. Phillips. 1991. Patch size of forest openings and arthropod populations. *Oecologia* 86: 325–334.

Shure, D. J. and L. A. Wilson. 1993. Patch-size effects on plant phenolics in successional openings of the southern Appalachians. *Ecology* 74: 55–67.

Soderlund, D. M. and J. R. Bloomquist. 1990. Molecular mechanisms of insecticide resistance. In *Pesticide Resistance in Arthropods*, R. T. Roush and B. E. Tabashnik, eds. 58–96. New York: Chapmann & Hall.

Stapp, P., M. F. Antolin, and M. Ball. 2004. Patterns of extinction in prairie dog metapopulations: Plague outbreaks follow El Niño events. *Frontiers in Ecology and the Environment* 2: 235–240.

Stephen, F. M., C. W. Berisford, D. L. Dahlsten, P. Fenn, and J. C. Moser. 1993. Invertebrate and microbial associates. In *Beetle-Pathogen Interaction in Conifer Forests*, T. D. Schowalter and G. M. Filip, eds. 129–153. London: Academic.

Sterner, R. W. and J. J. Elser. 2002. *Ecological Stoichiometry: The Biology of Elements from Molecules to the Biosphere*. Princeton, NJ: Princeton University Press.

Stige, L. C., K.-S. Chan, Z. Zhang, D. Frank, and N. C. Stenseth. 2007. Thousand-year-long Chinese time series reveals climatic forcing of decadal locust dynamics. *Proceedings of the National Academy of Sciences USA* 104: 16188–16193.

Stiles, J. H. and R. H. Jones. 1998. Distribution of the red imported fire ant, *Solenopsis invicta*, in road and powerline habitats. *Landscape Ecology* 13: 335–346.

Strauss, S. Y., R. E. Irwin, and V. M. Lambrix. 2004. Optimal defence theory and flower petal colour predict variation in the secondary chemistry of wild radish. *Journal of Ecology* 92: 132–141.

Strom, B. L., L. M. Roton, R. A. Goyer, and J. R. Meeker. 1999. Visual and semiochemical disruption of host finding in the southern pine beetle. *Ecological Applications* 9: 1028–1038.

Tabashnik, B. E., F. R. Groeters, N. Finson, Y. B. Liu, M. W. Johnson, D. G. Heckel, K. Luo, and M. L. Adang. 1996. Resistance to *Bacillus thuringiensis* in *Plutella xylostella*: The moth heard round the world. In *Molecular Genetics and Evolution of Pesticide Resistance*, American Chemical Society Symposium Series 645, 130–140. Washington, DC: American Chemical Society.

Tabashnik, B. E., Y. B. Liu, N. Finson, L. Masson, and D. G. Heckel. 1997. One gene in diamondback moth confers resistance to four *Bacillus thuringiensis* toxins. *Proceedings of the National Academy of Sciences USA* 94: 1640–1644.

Tanaka, S. 2012. Do desert locust hoppers develop gregarious characteristics by watching a video? *Journal of Insect Physiology* 58: 1060–1071.

Taylor, S. L. and D. A. MacLean. 2009. Legacy of insect defoliators: Increased wind-related mortality two decades after a spruce budworm outbeak. *Forest Science* 55: 256–267.

Thaler, J. S., M. J. Stout, R. Karban, and S. S. Duffey. 2001. Jasmonate-mediated induced plant resistance affects a community of herbivores. *Ecological Entomology* 26: 312–324.

Thies, C., I. Steffan-Dewenter, and T. Tscharntke. 2003. Effects of landscape context on herbivory and parasitism at different spatial scales. *Oikos* 101: 18–25.

Tisdale, R. A. and M. R. Wagner. 1990. Effects of photoperiod, temperature, and humidity on oviposition and egg development of *Neodiprion fulviceps* (Hymenoptera: Diprionidae) on cut branches of ponderosa pine. *Environmental Entomology* 19: 456–458.

Torres, J. A. 1988. Tropical cyclone effects on insect colonization and abundance in Puerto Rico. *Acta Científica* 2: 40–44.

Torres, J. A. 1992. Lepidoptera outbreaks in response to successional changes after the passage of Hurricane Hugo in Puerto Rico. *Journal of Tropical Ecology* 8: 285–298.

Trumble, J. and M. Sorensen. 2008. Selenium and the elemental defense hypothesis. *New Phytologist* 177: 569–572.

Ulyshen, M. D., S. Horn, B. Barnes, and K. J. K. Gandhi. 2010. Impacts of prescribed fire on saproxylic beetles in loblolly pine logs. *Insect Conservation and Diversity* 3: 247–251.

Van Bael, S. A., A. Aiello, A. Valderrama, E. Medianero, M. Samaniego, and S. J. Wright. 2004. General herbivore outbreak following an El Niño-related drought in a lowland Panamanian forest. *Journal of Tropical Ecology* 20: 625–633.

Van Bael, S. A., M. C. Valencia, E. I. Rojas, N. Gómez, D. M. Windsor, and E. A. Herre. 2009. Effects of foliar endophytic fungi on the preference and performance of the leaf beetle *Chelymorpha alternans* in Panama. *Biotropica* 41: 221–225.

Van Straalen, N. M. and H. A. Verhoef. 1997. The development of a bioindicator system for soil acidity based on arthropod pH preferences. *Journal of Applied Ecology* 34: 217–232.

Vasconcelos, H. L., E. H. M. Vieira-Neto, and F. M. Mundim. 2006. Roads alter the colonization dynamics of a keystone herbivore in neotropical savannas. *Biotropica* 38: 661–665.

Veblen, T. T., K. S. Hadley, E. M. Nel, T. Kitzberger, M. Reid, and R. Villalba. 1994. Disturbance regime and disturbance interactions in a Rocky Mountain subalpine forest. *Journal of Ecology* 82: 125–135.

Vickerman, D. B., J. K. Young, and J. T. Trumble. 2002. Effect of selenium-treated alfalfa on development, survival, feeding, and oviposition preferences of *Spodoptera exigua* (Lepidoptera: Noctuidae). *Environmental Entomology* 31: 953–959.

Voelckel, C., U. Schittko, and I. T. Baldwin. 2001. Herbivore-induced ethylene burst reduces fitness costs of jasmonate- and oral secretion-induced defenses in *Nicotiana attenuata*. *Oecologia* 127: 274–280.

Wagner, T. L., R. M. Feldman, J. A. Gagne, J. D. Cover, R. N. Coulson, and R. M. Schoolfield. 1981. Factors affecting gallery construction, oviposition, and reemergence of *Dendroctonus frontalis* in the laboratory. *Annals of the Entomological Society of America* 74: 255–273.

Waring, G. L. and P. W. Price. 1990. Plant water stress and gall formation (Cecidomyiidae: *Asphondylia* spp.) on creosote bush (*Larrea tridentata*). *Ecological Entomology* 15: 87–95.

Webb, M. R. and A. S. Pullin. 1998. Effects of submergence by winter floods on diapausing caterpillars of a wetland butterfly, *Lycaena dispar batavus*. *Ecological Entomology* 23: 96–99.

Wellington, W. G. 1980. Dispersal and population change. In *Dispersal of Forest Insects: Evaluation, Theory and Management Implications*. Proceedings of the International Union of Forest Research Organizations Conference, A. A. Berryman and L. Safranyik, eds. 11–24. Pullman, WA: Washington State University Cooperative Extension Service.

White, P. S. and S. T. A. Pickett. 1985. Natural disturbance and patch dynamics: An introduction. In *Ecology of Natural Disturbance and Patch Dynamics*, S. T. A. Pickett and P. S. White, eds. New York: Academic.

White, T. C. R. 1969. An index to measure weather-induced stress of trees associated with outbreaks of psyllids in Australia. *Ecology* 50: 905–909.

Whitford, W. G. 1992. Effects of climate change on soil biotic communities and soil processes. In *Global Warming and Biological Diversity*, R. L. Peters and T. E. Lovejoy, eds. 124–136. New Haven, CT: Yale University Press.

Whittaker, R. H. and P. P. Feeny. 1971. Allelochemics: Chemical interactions between species. *Science* 171: 757–770.

Wickman, B. E. 1964. Attack habits of *Melanophila consputa* on fire-killed pines. *Pan-Pacific Entomologist* 40: 183–186.

Wikars, L. O. 2002. Dependence on fire in wood-living insects: An experiment with burned and unburned spruce and birch logs. *Journal of Insect Conservation* 6: 1–12.

Wikars, L. O. and J. Schimmel. 2001. Immediate effects of fire-severity on soil invertebrates in cut and uncut pine forests. *Forest Ecology and Management* 141: 189–200.

Willig, M. R. and G. R. Camilo. 1991. The effect of Hurricane Hugo on six invertebrate species in the Luquillo Experimental Forest of Puerto Rico. *Biotropica* 23: 455–461.

Willig, M. R. and L. R. Walker. 1999. Disturbance in terrestrial ecosystems: Salient themes, synthesis, and future directions. In *Ecosystems of the World: Ecosystems of Disturbed Ground*, L. R. Walker, ed. 747–767. Amsterdam, The Netherlands: Elsevier Science.

Wilson, E. O. 1986. The organization of flood evacuation in the ant genus *Pheidole* (Hymenoptera: Formicidae). *Insectes Sociaux* 33: 458–469.

Windsor, D. M. 1990. *Climate and Moisture Variability in a Tropical Forest: Long-term Records from Barro Colorado Island, Panamá*. Washington, DC: Smithsonian Institution Press.

Witcosky, J. J., T. D. Schowalter, and E. M. Hansen. 1986. The influence of time of precommercial thinning on the colonization of Douglas-fir by three species of root-colonizing insects. *Canadian Journal of Forest Research* 16: 745–749.

Zalucki, M. P., L. P. Brower, and A. Alonso-M. 2001. Detrimental effects of latex and cardiac glycosides on survival and growth of first-instar monarch butterfly larvae *Danaus plexippus* feeding on the sandhill milkweed *Asclepias humistrata*. *Ecological Entomology* 26: 212–224.

Zehnder, C. B. and M. D. Hunter. 2009. More is not necessarily better: The impact of limiting and excessive nutrients on herbivore population growth rates. *Ecological Entomology* 34: 535–543.

Zenk, M. H. and M. Juenger. 2007. Evolution and current status of the phytochemistry of nitrogenous compounds. *Phytochemistry* 68: 2757–2772.

Zerm, M. and J. Adis. 2003. Exceptional anoxia resistance in larval tiger beetle, *Phaeoxantha klugii* (Coleoptera: Cicindelidae). *Physiological Entomology* 28: 150–153.

Zhou, J., W. K.-M. Lau, P. M. Masuoka, R. G. Andre, J. Chamberlin, P. Lawyer, and L. W. Laughlin. 2002. El Niño helps spread Bartonellosis epidemics in Peru. *EOS, Transactions, American Geophysical Union* 83: 157, 160–161.

4

Changes in Insect Abundance and Distribution

> First, by changing or manipulating the environment, man has created conditions that permit certain species to increase their population densities.
>
> **Stern et al. (1959)**

Although the frequent appearance of insects en masse with no apparent increase in abundance supported early perceptions of "spontaneous generation" and "acts of God," insect population dynamics are governed by relatively predictable responses to changing environmental conditions. The small size, sensitivity to temperatures and moisture conditions, and high reproductive and dispersal capacities of insects allow them to respond rapidly, and often dramatically, to environmental change.

Under favorable environmental conditions, some species have the capacity to increase population size by several orders of magnitude in a few generations, thereby appearing en masse with little warning. The largest congregation of animals ever recorded was a swarm of Rocky Mountain locusts, *Melanoplus spretus*, flying over Lincoln, Nebraska, on June 16, 1875, that was estimated to include at least 125 billion insects covering more than 23,000 km^2 (Riley 1878; Lockwood 2001). Under such conditions, insects can interfere with human interests in a variety of ways. During the locust plagues of the 1870s, locusts were so numerous that their crushed bodies on railroad tracks reduced traction sufficiently to stop trains, especially on an upgrade, but benefits to crop production also were noted (Riley 1878).

Under adverse conditions, populations can virtually disappear for long time periods or become extinct. For example, two major agricultural pests of the 1800s, the Rocky Mountain locust and cotton leafworm, *Alabama argillacea*, both became extinct as a result of agricultural intensification and loss of core habitat (Riley 1885; Lockwood and DeBrey 1990; Lockwood 2001; Wagner 2009).

Devastating outbreaks have brought insects into direct conflict with humans but also supported focused entomological research that has expanded our understanding of factors affecting insect population dynamics and, more recently, insect effects on ecosystem conditions and services. Consequently, methods and models for describing population change are best developed for economically important insects.

Predicting the effects of global change has become a major goal of research on population dynamics. Insect populations respond in various ways to changes in habitat conditions and resource quality, and their responses to current and historic environmental changes help us to anticipate responses to future environmental changes. Disturbances, in particular, influence population systems abruptly (see Chapter 3). Factors that normally regulate population size, such as resource availability and predation, also are affected by disturbances (see Chapter 3). This chapter describes factors controlling changes in insect abundance and distribution over time and space.

4.1 Population Fluctuation through Time

BOX 4.1   PREDICTING POPULATION DYNAMICS

The simplest way to describe population growth for any species is initial population size multiplied by the per capita rate of increase (Figure B4.1) (Berryman 1981, 1997; Price 1997). Per capita natality (birthrate), mortality (death rate), immigration, and emigration per unit time are integrated as the instantaneous or intrinsic rate of increase, r:

$$r = (N + I) - (M + E), \qquad (4.1)$$

where N = natality, I = immigration, M = mortality, and E = emigration— all instantaneous rates.

For insects, with relatively discrete, nonoverlapping generations, life table data for individual cohorts (group of individuals born at about the same time) are easier to use than are time-specific data used by actuaries for human demographics. When cohort data are available, r can be estimated as

$$r = \frac{\log_e R_0}{T}, \qquad (4.2)$$

where R_0 is replacement rate (i.e., the number of female offspring per female per generation) and T is generation time. Population increase is conveniently expressed in terms of females because males are rare or absent for many insect species (e.g., aphids) at least for many generations.

The rate of population change for species with overlapping generations is a function of per capita rate of change (r) and current population size (N_t). The simplest model to describe exponential population growth is

$$N_{t+1} = N_t + rN_t. \qquad (4.3)$$

This equation also can be written as

$$N_{t+1} = N_t e^{rt}. \qquad (4.4)$$

For insect species with nonoverlapping generations, the replacement rate, R_0, represents the per capita rate of increase from one generation to the next. This parameter can be used in place of r for such insects. The resulting expression for geometric population growth is

$$N_t = R_0^t N_0, \qquad (4.5)$$

where N_t is the population size after t generations.

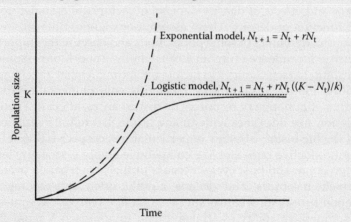

FIGURE B4.1
Exponential and logistic models of population growth. The exponential model describes an indefinitely increasing population, whereas the logistic model describes a population reaching an asymptote at the carrying capacity of the environment (K).

Equations 4.3–4.5 describe density-independent population growth that continues to grow exponentially without limit (Figure B4.1). However, density-dependent competition, predation, and other factors interact to limit population growth, typically as a population exceeds the carrying capacity of current ecosystem conditions (Schowalter 2011).

A logistic model (Figure B4.1), often called the Pearl-Verhulst equation (Pearl and Reed 1920; Berryman 1981; Price 1997), was developed
to reflect limitation of population growth at the carrying capacity of the
environment (K):

$$N_{t+1} = \frac{N_t + rN_t(K - N_t)}{K}. \qquad (4.6)$$

This model describes a sigmoid (S-shaped) curve that reaches equilibrium at K. If $N < K$, then the population will increase up to $N = K$. If the
ecosystem is disturbed in a way that $N > K$, then the population will
decline to $N = K$.

General models such as the Pearl-Verhulst model do not predict
the dynamics of real populations accurately because the logistic growth
model is limited by several assumptions. First, individuals are assumed
to be equal in their reproductive potential. Clearly, immature insects and
males do not produce offspring, and females vary in their productivity,
depending on nutrition, access to oviposition sites, and so on. Second,
population adjustment to changing density is assumed to be instantaneous, and effects of density-dependent factors are assumed to be a
linear function of density. These assumptions ignore time lags, which
may control dynamics of some populations and obscure the importance
of density dependence (Turchin 1990). Finally, r and K are assumed to
be constant. In fact, changes in factors (including K) that affect natality,
mortality, and dispersal affect r. Changing environmental conditions,
including elimination of resources by disturbances, affect K. Therefore,
population size fluctuates with an amplitude that reflects variations in
K and the life history strategy of particular insect species. Species with
high reproductive rates and low competitive ability (r strategy) tend to
undergo boom-and-bust cycles because of their tendency to overshoot
K, deplete resources, and decline rapidly, often approaching their
extinction threshold, whereas species with low reproductive rates and
high competitive ability (K strategy) tend to approach K more slowly
and maintain relatively stable population sizes near K (Boyce 1984).
Modeling real populations of interest, then, requires development of
complex models with additional parameters that correct these short-
comings. A few examples include the following.

Nonlinear density dependent processes and delayed feedback can be
addressed by allowing r to vary as

$$r = r_{max} - sN_t - T, \qquad (4.7)$$

where r_{max} is the maximum per capita rate of increase, s represents the
strength of interaction between individuals in the population, and T is

the time delay in the feedback response (Berryman 1981). The sign and magnitude of s also can vary, depending on the relative dominance of competitive and cooperative interactions:

$$s = s_p - s_m N_t,\qquad(4.8)$$

where s_p is the maximum benefit from cooperative interactions and s_m is the competitive effect, assuming that s is a linear function of population density at time t (Berryman 1981). An extinction threshold, E, representing the minimum population size that can remain viable, can be incorporated by adding a term forcing population change to be negative below this threshold:

$$N_{t+1} = N_t + rN_t((K - N_t)/K)((N_t - E)/E).\qquad(4.9)$$

Similarly, the effect of factors influencing natality, mortality, and dispersal can be incorporated into the model to improve representation of r.

The effect of other species interacting with a population was addressed first by Lotka (1925) and Volterra (1926). The Lotka-Volterra equation for the effect of a species competing for the same resources includes a term that reflects the degree to which the competing species reduces carrying capacity:

$$N_{1(t+1)} = N_{1t} + r_1 N_{1t}((K_1 - N_{1t} - \alpha N_{2t})/K_1),\qquad(4.10)$$

where N_1 and N_2 are populations of two competing species, and α is a competition coefficient that measures the per capita inhibitive effect of species 2 on species 1.

Similarly, the effects of a predator on a prey population can be incorporated into the logistic model (Lotka 1925; Volterra 1926) as

$$N_{1(t+1)} = N_{1t} + r_1 N_{1t} - \rho N_{1t} N_{2t},\qquad(4.11)$$

where N_1 is prey population density, N_2 is predator population density, and ρ is a predation constant. This equation assumes random movement of prey and predator, prey capture and consumption for each encounter with a predator, and no self-limiting density effects for either population (Pianka 1974; Price 1997).

May (1981) and Dean (1983) modified the logistic model to include effects of mutualists on population growth. Gutierrez (1996) and Royama (1992) discussed additional population modeling approaches, including incorporation of age and mass structure and population refuges from predation. Clearly, the increasing complexity of these models, as more parameters are included, requires computerization for prediction of population trends.

The utility of models often is limited by a number of problems. The effects of multiple interacting factors typically must be modeled as the direct effects of individual factors, in the absence of multifactorial experiments to assess interactive effects. Effects of host condition often are particularly difficult to quantify for modeling purposes because factors affecting host biochemistry remain poorly understood for most species. Moreover, models must be initialized with adequate data on current population parameters and environmental conditions. Traditional approaches to modeling have emphasized deterministic (based on known relationships between population growth and a particular environmental factor) or stochastic (allowed to vary randomly, based on typical ranges of population responses to an environmental factor) variables.

Chaos theory has proven to be useful in developing models of outbreak behavior (Hassell et al. 1991; Logan and Allen 1992; Cavalieri and Koçak 1994, 1995; Cushing et al. 2003). "Chaos" does not imply the popular concept of total unpredictability, rather that outbreaks often reflect responses to unique combinations of variables at a particular point in time, such as those resulting from disturbances, which are difficult to predict. Small changes in the values of these variables can trigger exponential population growth, the "butterfly effect."

Insect population dynamics models typically are developed to address "pest" effects on commodity values. Few population dynamics models explicitly incorporate effects of population change on ecosystem processes or services. However, a growing number of studies are providing data on effects of insect herbivore or detritivore abundance on primary productivity, water and nutrient fluxes, and/or diversity and abundances of other organisms (Klock and Wickman 1978; Leuschner 1980; Seastedt et al. 1983; Seastedt and Crossley 1984; Schowalter et al. 1991; Christenson et al. 2002; Chapman et al. 2003; Hunter et al. 2003; Frost and Hunter 2004; Classen et al. 2005; Fonte and Schowalter 2005; Whitham et al. 2006; see also Chapter 7).

Despite limitations, population dynamics models are valuable tools for synthesizing a vast and complex body of data, for identifying critical gaps in our understanding of factors affecting populations, and for predicting or simulating responses to environmental changes. Such predictions represent hypotheses that can be tested over time. Therefore, models represent our state-of-the-art understanding of population dynamics, can be used to focus future research on key questions, and can contribute to improved efficiency of management or manipulation of important processes. Population dynamics models are the most rigorous tools available for projecting survival or recovery of endangered species and outbreaks of potential pests and their effects on ecosystem services.

We have long records of population change for some insects, including 1,173 years for the larch budmoth, *Zeiraphera diniana*, in the European Alps (Esper et al. 2007), 1,000 years for the oriental migratory locust, *Locusta migratoria manilensis*, in China (Figure 4.1) (Ma 1958; Konishi and Itô 1973; Stige et al. 2007), and 622 years for the Pandora moth, *Coloradia pandora*, in western North America (Speer et al. 2001). Such long records aid greatly in identifying environmental factors responsible for population change. If environmental conditions change in a way that favors insect population growth, the population will increase until regulatory factors (see section 4.2) reduce and finally stop population growth rate.

Some populations can vary in density by several orders of magnitude (Mason and Luck 1978; Royama 1984; Mason 1996; Schell and Lockwood 1997), but most populations vary less than this (Berryman 1981; Strong et al. 1984). The amplitude and frequency of population fluctuations can be used to describe three general patterns. Stable populations fluctuate relatively little over time whereas irruptive and cyclic populations show wide fluctuations.

Irruptive populations increase sporadically to peak numbers followed by a decline. Certain combinations of life history traits may be conducive to irruptive fluctuation. Larsson et al. (1993), Nothnagle and Schultz (1987), and Koricheva et al. (2012) identified differences in life history attributes between irruptive and nonirruptive species of sawflies and Lepidoptera from European and North American forests. Irruptive species generally are

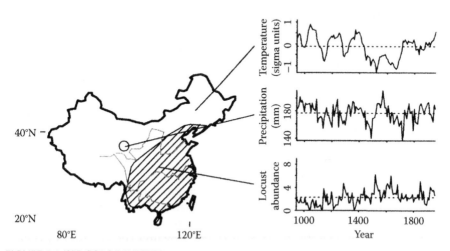

FIGURE 4.1 (SEE COLOR INSERT.)
One thousand years of locust abundance and climate data from China. The present distribution of *Locusta migratoria manilensis* is shown by hatched lines on the map. Large rivers (blue lines) from north to south are the Yellow, Yangtze, and Pearl. Temperature data are a composite index for all of China. Precipitation data are from northeastern Qinghai (circle). Locust data are for all of China. (From Stige, L. C., K.-S. Chan, Z. Zhang, D. Frank, and N. C. Stenseth, *Proceedings of the National Academy of Sciences USA*, 104, 16188–16193, 2007. With permission.)

controlled by only one or a few factors, especially drought (Konishi and Itô 1973; Mattson and Haack 1987; Priesser and Strong 2004), whereas populations of nonirruptive species are controlled by many factors. In addition, irruptive Lepidoptera and sawfly species tend to be gregarious, have a single generation per year, feed in the spring, and are sensitive to changes in quality or availability of their particular resources, whereas nonirruptive species do not share this combination of traits. Bark beetles do not show any apparent differences in life history traits between irruptive and nonirruptive species (Koricheva et al. 2012).

Cyclic populations oscillate at relatively regular intervals, generating the greatest interest among ecologists. Cyclic patterns can be seen over different time scales and reflect a variety of interacting factors. Several forest Lepidoptera exhibit cycles with periods of about 10 years, 20 years, 30 years, or 40 years (Mason and Luck 1978; Berryman 1981; Royama 1992; Swetnam and Lynch 1993; Price 1997; Esper et al. 2007), or combinations of cycles (Speer et al. 2001). For example, the larch budmoth has shown outbreaks every 9.3 years, on average, over a 1,173-yr period in the European Alps, and spruce budworm, *Choristoneura fumiferana*, populations have peaked at approximately 25- to 30-year intervals over a 250-year period in eastern North America (Royama 1984), whereas Pandora moth populations have shown a combination of 20- and 40-year cycles over a 622-year period in western North America (Speer et al. 2001).

Explanations for cyclic population dynamics include climatic cycles and changes in insect gene frequencies or behavior, food quality, or susceptibility to disease that occur during large changes in insect abundance (Myers 1988). Climatic cycles may trigger insect population cycles directly through changes in mortality or indirectly through changes in host condition or susceptibility to pathogens. However, regular irruptions of *Z. diniana* have occurred for more than 1,000 years in the European Alps independently of long-term warming and cooling trends (Esper et al. 2007). Depletion of food resources during an outbreak may impose a time lag for recovery of depleted resources to levels capable of sustaining renewed population growth (Clark 1979). Epizootics of entomopathogens may occur only above threshold densities. High genetic heterogeneity for disease resistance can lead to intense selection for resistance during disease-driven population collapse, followed by increased infection risk during population growth due to the cost of resistance (Elderd et al. 2008). Sparse populations near their extinction threshold (see section 4.3) may require several years to recover sufficient numbers for rapid population growth. Berryman (1996), Royama (1992), Turchin (1990), and Turchin et al. (1999) have demonstrated the importance of delayed effects (time lags) of regulatory factors (especially predation or parasitism) to the generation of cyclic pattern.

For some insects, especially Lepidoptera, population cycles are synchronized over large areas, up to 1,200 km apart (i.e., at a continental scale), suggesting the influence of a common widespread trigger such as climate, sunspot, lunar, or ozone cycles (Clark 1979; Royama 1984; 1992; Price 1997; Speer et al. 2001; Liebhold et al. 2004; Johnson et al. 2005; Økland et al. 2005). Moran (1953) suggested, and Royama (1992) demonstrated (using models), that synchronized cycles could result from correlations among controlling factors. Liebhold et al. (2004) found that synchrony arises from three primary mechanisms: (a) dispersal among demes (local populations of a species that may be more or less isolated from other local populations) transfers individuals from growing demes to smaller demes, (b) congruent dependence of multiple demes on exogenous synchronizing factors, such as temperature or precipitation (the "Moran" effect), and (c) trophic interactions with other species that integrate demes via mortality (see Chapter 9). Raimondo et al. (2004) reported that generalist predators could explain the observed synchronous population dynamics of multiple prey species. On the other hand, synchronous emergence of periodical cicadas, *Magicicada* spp., was found to (in turn) synchronize abundances of 15 of 37 predaceous bird species evaluated by Koenig and Liebhold (2005). Bird populations sharing the same cicada brood showed greater intraspecific spatial synchrony than did bird populations in the ranges of different cicada broods. Liebhold et al. (2006) used simulation modeling to demonstrate that geographic variation in direct and delayed density dependence diminished synchrony resulting from stochastic forcing by geographic variation but not synchrony resulting from dispersal processes. Hence, the cause of synchrony can be independent of the cause of the cyclic pattern of fluctuation.

Generally, peak abundances are maintained only for a few (2–3) years, followed by relatively precipitous declines. Changes in population size can be described by four distinct phases (Mason and Luck 1978). The endemic phase is the low population level maintained between outbreaks. The beginning of an outbreak cycle is triggered by a disturbance, release from predation, or other environmental change that allows the population to increase in size above a release threshold. This threshold represents a population size at which reproductive momentum results in escape of at least a portion of the population from normal regulatory factors, such as crowding or predation. Despite the importance of this threshold to population outbreaks, few studies have established its size for any insect species. Schowalter et al. (1981) reported that local outbreaks of southern pine beetles, *Dendroctonus frontalis*, occurred when demes reached a critical size of about 100,000 beetles by early June. Above the release threshold, survival is relatively high and population growth continues during the release phase. During this period, emigration peaks and the population spreads to other suitable habitat patches. Resources eventually become limiting, as a result of depletion by the growing population, and predators and pathogens respond to increased prey/ host density and stress. Population growth slows and abundance reaches a peak. Competition, predation, and pathogen epizootics initiate and accelerate population decline. Intraspecific competition and predation rates then decline as the population reenters the endemic phase.

Populations of many species fluctuate at amplitudes that are insufficient to cause economic damage and, therefore, do not attract attention. Some of these species may experience more conspicuous outbreaks under changing environmental conditions, for example, climate change, introduction into new habitats, or large-scale conversion of natural ecosystems to managed ecosystems (Mattson and Haack 1987; Williams and Liebhold 2002; Van Bael et al. 2004).

Outbreaks of some insect populations have become more frequent and intense in crop systems or natural monocultures where food resources are relatively unlimited, where crop cultivars have been developed to favor resource allocation to growth at the expense of defense, or where manipulation of disturbance frequency has created favorable conditions (Kareiva 1983; Wickman 1992; Schowalter and Turchin 1993; Heiermann and Schütz 2008; Raffa et al. 2008; see Chapter 5). In other cases, the frequency of recent outbreaks has remained within ranges for frequencies of historic outbreaks, but the extent or severity has increased as a result of anthropogenic changes in vegetation structure or disturbance regime (Speer et al. 2001).

In some cases, insect populations have grown and spread widely when human settlement exposed them to suitable crop monocultures. The Colorado potato beetle, *Leptinotarsa decemlineata*, subsisted on wild solanaceous hosts in western North America until westward movement of settlers brought it into contact with cultivated potatoes in the Midwest during the late 1800s (Riley 1883; Stern et al. 1959; Hitchner et al. 2008), allowing it to spread eastward and, eventually, to Europe. Similarly, the cotton boll weevil, *Anthonomus grandis*, coevolved with scattered wild *Gossypium* spp., including *G. hirsutum*, in tropical Mesoamerica until citrus cultivation in the 1890s provided overwintering food resources that allowed the insect to spread into subtropical cotton-growing regions of south Texas and northern Argentina (Showler 2009). Subsequently, rapid reproduction in the spring by overwintering adults permitted spread throughout the U.S. Cotton Belt (Showler 2009).

4.2 Factors Affecting Population Size

A number of abiotic and biotic factors affect insect population size. Factors that increase reproduction or reduce mortality lead to population growth, whereas factors that reduce reproduction or increase mortality lead to population decline. Abiotic factors include extreme temperature, moisture, and atmospheric and water chemistry (see Chapter 3). Biotic factors include suitability and availability of food resources and predation. Abiotic factors generally operate independently of insect population size (density-independent factors), whereas biotic factors, especially competition and predation, often have increasing effect as insect abundance increases (density-dependent factors). Density-dependent factors are of particular interest because the negative feedback they provide tends to prevent insect population growth above an abundance that can be sustained under given environmental conditions (carrying capacity), as described in Section 4.2.3.

4.2.1 Abiotic Factors

Insect populations are highly sensitive to changes in abiotic conditions, such as temperature, water availability, and so on, which affect insect growth and survival (see Chapter 2). Changes in population size of some insects have been related directly to changes in climate or to disturbances (Greenbank 1963; Reice 1985; Kozár 1991; Porter and Redak 1996). Paleontological data show increased herbivory during warmer periods. Currano et al. (2008) reported that the amount and diversity of leaf damage by herbivorous insects was significantly greater (mean 57%) during the Paleocene-Eocene thermal maximum (PETM) 56 million years ago than during either the preceding Paleocene (< 38%) or subsequent Eocene (33%) epochs (Figure 4.2). The amount and diversity of herbivore damage on angiosperm leaves were correlated positively with temperature change 55–59 million years ago.

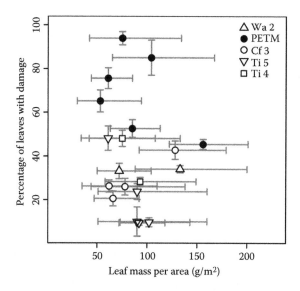

FIGURE 4.2
Estimated leaf mass per area (LMA) and damage frequency for individual plant species from sites representing time periods before (Tiffanian 4a and 5b [Ti 4 and 5, 57.5–58.9 Mya] and Clarkforkian 3 [Cf 3, 55.9 Mya]), during (PETM, 55.8 Mya), and after (Wasatchian 2 [Wa 2, 55.2 Mya]) the Paleocene-Eocene Thermal Maximum (PETM). LMA values are species means, and error bars represent 95% confidence intervals. Error bars for herbivory represent one standard deviation based on a binomial sampling distribution. (From Currano, E. D., P. Wilf, S. L. Wing, C. C. Labandeira, E. C. Lovelock, and D. L. Royer, *Proceedings of the National Academy of Sciences USA*, 105, 1960–1964, 2008. With permission.)

Changes in climate may disrupt synchrony between herbivores and host plants or between predator and prey populations when interacting species' phenologies do not respond equally to changing temperature (Lawrence et al. 1997; Hunter and Elkinton 2000; Visser and Holleman 2001; Bale et al. 2002; Watt and McFarlane 2002; Visser and Both 2005; Logan et al. 2006; Hance et al. 2007; Klapwijk et al. 2010). For example, Lawrence et al. (1997) and Hunter and Elkinton (2000) found that cohorts of lepidopteran larvae placed on foliage at increasing times before or after budbreak showed reduced survival and increased development times, relative to larvae placed on foliage near the time of budbreak. Hunter and Elkinton (2000) found that predation had the opposite effect, indicating that disruption of phenological synchrony of budbreak, herbivore egghatch, and predation as a result of climate change could greatly affect herbivore population dynamics.

Climate change or disturbance affects resource values for insects. Concentrations of plant defensive chemicals change seasonally and annually as a result of environmental changes (Cronin et al. 2001; Mopper et al. 2004) and disturbance (Nebeker et al. 1993; Hunter and Forkner 1999). Cronin et al. (2001) monitored preferences of a stem-galling fly, *Eurosta solidaginis*, among the same 20 clones of goldenrod, *Solidago altissima*, over a 12-year period and found that preference for, and performance on, the different clones was uncorrelated among years. These data indicated that genotype x environmental interaction affected nutritional quality of clones for this herbivore. Increased exposure to UVB radiation reduced concentration of gallic acid and increased concentration of flavenoid aglycone in southern beech, *Nothofagus antarctica* (Rousseaux et al. 2004). Cipollini (1997) found that wind increased concentrations of peroxidase, cinnamyl alcohol-dehydrogenase, and lignin in bean, *Phaseolus vulgaris*, and reduced oviposition and population growth of two-spotted spider mites, *Tetranychus urticae*. Loss of riparian habitat, primarily as a result of agricultural intensification in western North America, may have led to extinction of the historically important Rocky Mountain grasshopper, *M. spretus* (Lockwood and DeBrey 1990; Lockwood 2001).

Many environmental changes occur relatively slowly and affect insect populations gradually as a result of subtle shifts in genetic structure and individual fitness. Other environmental changes occur more abruptly and may trigger rapid change in population size because of a sudden alteration of natality, mortality, or dispersal. Esper et al. (2007) found that larch budmoth, *Z. diniana*, showed regular population outbreaks at nine-year intervals for 1,200 years, during warming and cooling climate periods, but have failed to increase since 1981, a period during which temperatures have increased above the historic range. Disturbances are particularly important triggers for inducing population change because of their acute disruption of population structure and of resource, substrate, and other ecosystem conditions. Drought has been identified as an important trigger for population irruptions of many species, including locusts, moths, and bark beetles, promoted by host stress and crowding (Konishi and Itô 1973; Mattson and

Haack 1987; Schowalter et al. 1999; Priesser and Strong 2004; van Bael et al. 2004; Breshears et al. 2005), but can reduce populations of other species (see Figure 3.5) (Schowalter et al. 1999). Storm damage also increases resources for some insects (Schowalter and Ganio 2003; Hanewinkel et al. 2008).

Population responses to disturbance vary, depending on scale (see Chapter 3). Few natural experiments have addressed the effects of scale. Clearly, a larger scale event should affect environmental conditions and populations within the disturbed area more than would a smaller scale event. Shure and Phillips (1991) compared arthropod abundances in clear-cuts of different sizes in the southeastern United States. They suggested that the greater differences in arthropod densities in larger clear-cuts reflected the steepness of environmental gradients from the clear-cut into the surrounding forest. The surrounding forest has a greater effect on environmental conditions within a small canopy opening than within a larger opening.

4.2.2 Biotic Factors

Intra- and interspecific competition, for limited resources, and predation are the primary biotic factors limiting population growth. Malthus (1789) wrote the first theoretical treatise describing the increasing competition for limited resources by growing populations. As competition for finite resources becomes intense, fewer individuals obtain sufficient resources to survive, reproduce, or disperse. Similarly, a rich literature on predator-prey interactions generally, and biocontrol agents in particular, has shown the important density-dependent effects of predators, parasitoids, and parasites on prey populations (Tinbergen 1960; van den Bosch et al. 1982; Carpenter et al. 1985; Marquis and Whelan 1994; Van Driesche and Bellows 1996; Parry et al. 1997; Price 1997). Predation rates typically increase as prey abundance increases, up to a point at which predators become satiated. Predators respond behaviorally and numerically to changes in prey density. Predators can be attracted to an area of high prey abundance, a behavioral response, and increase production of offspring as food supply increases, a numerical response (Koenig and Liebhold 2005). Parasites are not subject to satiation, and natural epizootics commonly terminate outbreaks (Brookes et al. 1978).

4.2.2.1 Resource Suitability

Most plants do not provide adequate amounts or balances of all essential nutrients. This forces herbivorous insects to make choices among available hosts to maximize ingestion of limiting nutrients without ingesting toxic amounts of nonlimiting nutrients. Generalists that feed on a variety of plant species may optimize nutrient balance by feeding sequentially on multiple plant species that differ in their nutrient balances, but specialists that feed on a particular plant species must face the trade-off between overeating nutrients that occur in excess and undereating nutrients that occur in insufficient amounts (Joern and Behmer 1998; Lee et al. 2002, 2003; Simpson et al. 2002; Sterner and Elser 2002; Raubenheimer and Simpson 2003; Behmer 2009). In contrast, animal tissues have relatively balanced nutrient values for predators.

Furthermore, most, if not all, plants produce various defenses that prevent feeding, are toxic, or that interfere with digestion (Harborne 1994; see Box 3.1). These include spines and lignin-toughened tissues (Ausmus 1977; Scriber and Slansky 1981); resins and latex that prevent penetration of plant tissues (Rhoades 1977; Nebeker et al. 1993; Agrawal and Kono 2009); complex multicyclic phenols that bind proteins and destroy gut tissues (Feeny 1969; Barbehenn et al. 2008); terpenoids, including cardiac glycosides produced by milkweed and sequestered by monarch butterflies, *Danaus plexippus*, for defense against bird predators (Brower et al. 1968); photooxidants, such as quinones and furanocoumarins, that cause severe sunburn and epidermal necrosis on exposure to sunlight (Berenbaum 1987); alkaloids, such as strychnine and epinephrine, that cause serious cardiovascular and nervous system failure (Shonle and Bergelson 2000; Jackson et al. 2002); cyanogenic glucosides that are hydrolyzed by gut enzymes into glucose and hydrogen cyanide (Zenk and Juenger 2007); and even insect growth hormone analogues and alarm pheromones that interfere with insect development or deter host-seeking individuals (Gibson and Pickett 1983; Schmelz et al. 2002; see Box 3.1). These defenses require energy and nutrient expenditure by insects to detoxify plant compounds or to continue searching for more suitable food resources (Tanaka and Suzuki 1998; Zera and Zhao 2006).

Each plant species has a characteristic chemical composition ("fingerprint") that limits herbivory to those insect species adapted to that chemical fingerprint (see Box 3.1). For such species, these host chemicals may serve to identify a suitable host and therefore function as an attractant. Because insects identify hosts primarily by chemical cues, they will tend to feed on plants providing attractive cues, regardless of phylogenetic relationship.

Some of these compounds are normally present in plant tissues (constitutive), whereas other, more targeted compounds (especially proteinase inhibitors), can be induced by insect feeding or other injury (Karban and Baldwin 1997; Kessler and Baldwin 2002). In some cases, feeding injury releases volatile signalers, especially jamonic acid and ethylene, that communicate this injury and elicit production of defenses among neighboring plants, even of different species, in advance of herbivore spread (Baldwin and Schultz 1983; Farmer and Ryan 1990; Stout and Bostock 1999; Dolch and Tscharntke 2000; Thaler et al. 2001; Kessler and Baldwin 2002; Kessler et al. 2006). However, production of these defenses also requires expenditure of energy and limiting nutrients (especially nitrogen, in the case of alkaloids and cyanogenic glucosides) and may be suspended when plants are stressed by adverse environmental conditions. As a result, stressed plants become suitable for feeding by a wider variety of generalist, as well as specialist, herbivores (Kessler et al. 2004).

4.2.2.2 Resource Availability

The abundance, distribution, and apparency of acceptable resources determine their availability in space and time to searching insects (Courtney 1985, 1986; Bozer et al. 1996; Eggert and Wallace 2003). Resources are most available when distributed evenly at nonlimiting concentrations or densities. Organisms living under such conditions need not move widely to locate new resources and tend to be relatively sedentary. For example, aphids and scale insects that capture resources from flowing solutions in plant vessels may enjoy relatively nonlimiting resources for many generations.

Necessary resources typically are less concentrated, available at suboptimal ratios with other resources, or are unevenly distributed in space and time. This requires that organisms occur at times when resources will be most available and/or select habitats where required resources are most concentrated or in most efficient balance and search for new sources as current resources become depleted.

Life histories of many insect species are synchronized with periods when host resources and nutritional value are most available (Feeny 1970; Varley and Gradwell 1970; Lawrence et al. 1997). Filip et al. (1995) reported that the foliage of many tropical trees has higher nitrogen and water content earlier in the wet season than later in the wet season. Lawrence et al. (1997) caged successive cohorts of spruce budworm, *C. fumiferana*, larvae on white spruce at different phenological stages of the host. Cohorts that began feeding three to four weeks before budbreak and completed larval development prior to the end of shoot elongation developed significantly faster and showed significantly greater survival rate and adult mass than did cohorts caged later. These results indicate that the phenological window of opportunity for this insect was sharply defined by the period of shoot elongation, during which foliar nitrogen, phosphorus, potassium, copper, sugars, and water were higher than in mature needles.

In some cases, insects that depend on resources that occur unpredictably, such as insects feeding on seeds of trees that produce seed crops irregularly (masting), are capable of remaining in diapause for prolonged periods. Emergence of some portion of each year's cohort increases the probability that some individuals will emerge during a period of favorable resource availability. Turgeon et al. (1994) reported that 70 species of Diptera, Lepidoptera, and Hymenoptera that feed on conifer cones or seeds can remain in diapause for as long as seven years.

Most insects must search for suitable food resources that are unevenly distributed across landscapes. This requires that insects optimize the trade-off between the suitability of discovered resources and the risk of predation or failure to find more suitable resources during extended search (Townsend and Hughes 1981; Schultz 1983; Stephens and Krebs 1986; Kamil et al. 1987; Behmer 2009). Suitable resources represent net gain, but nonnutritive or toxic resources represent costs in terms of time, energy, or nutrient resources expended in detoxification or continued search. Defensive chemicals reduce nutritional value of a food resource, but defended plants often are eaten when more suitable hosts are unavailable or cannot be found (Courtney 1985, 1986). Continued search also increases exposure to predators or other mortality agents. Foraging can be optimized by searching for more nutritive food and risking attention of predators, accepting less nutritive food, or defending against predation (Schultz 1983). Natural selection can favor a reduction in cost among any of these three options, within the constraints of

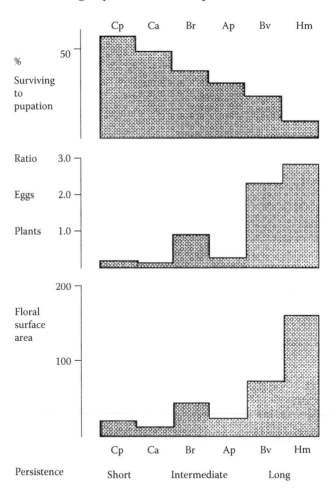

FIGURE 4.3
Trade-off between plant suitability for larval survival (top) and efficiency of oviposition site selection by adult pierids, *Anthocharis cardamines*, as indicated by the ratio of eggs per host species (middle) and plant apparency, that is, floral surface area and longevity (bottom). Searching females preferentially oviposit on the most conspicuous plants although these are not the most suitable food plants for their larvae. Cp = *Cardamine pratensis*; Ca = *C. amara*; Br = *Brassica rapa*; Ap = *Allaria petiolata*; Bv = *Barbarea vulgaris*; and Hm = *Hesperis matronalis*. (From Courtney, S. P., *Oikos*, 44, 91–98, 1985. With permission.)

the other two. Insects can maximize foraging efficiency by spending more time where hosts are concentrated (the Resource Concentration Hypothesis), and ignore patches where hosts are less available, until their resource value declines below the average for the landscape (Kareiva 1983; Bell 1990).

Because insects are small and have relatively poor long-distance vision, they rely primarily on host odors to find suitable hosts. However, odors from nonhosts can interfere with detection of host odors, making discovery of hosts difficult in diverse vegetation or landscapes (Courtney 1985, 1986; Visser 1986; Hambäck et al. 2003). Alternatively, the same volatile chemical emitted by unrelated species may provide the basis for host switching, such as attraction of *Culex* mosquitoes to nonanal produced by both birds and humans (Syed and Leal 2009). Consequently, host-seeking insects often must accept less suitable host resources that limit survival and reproduction (Figure 4.3) (Courtney 1985, 1986). Such trade-offs limit the degree to which any single environmental factor can impose directional selection for adaptation.

4.2.2.3 Predation

Predation and parasitism reduce survival and reproduction of prey or host species. Predation typically is concentrated on the most abundant potential prey species (Figure 4.4). A variety of insects and other arthropods (e.g., predaceous beetles, dragonflies, spiders, mites), and vertebrates (particularly insectivorous fish, amphibians, reptiles, birds and mammals) feed largely or exclusively on insects (Tinbergen 1960; Dial and Roughgarden 1995; Gardner and Thompson 1998; Kawaguchi and Nakano 2001; Allan et al. 2003; Beard et al. 2003; Baxter et al. 2005). Aquatic and terrestrial insects support major freshwater fisheries, including salmonids (Cloe and Garman 1996; Wipfli 1997). Terrestrial insects constitute more than half the diets for salmonids and other insectivorous fish species, primarily during summer months

FIGURE 4.4 (SEE COLOR INSERT.)
Predation tends to concentrate on the most abundant prey, including conspecifics under crowded conditions, as in the case of this dragonfly preying on another dragonfly.

(Kawaguchi and Nakano 2001; Allan et al. 2003; Baxter et al. 2005). Reduced availability of insect prey reduces productivity and abundance of these fish (Baxter et al. 2007). Stewart and Woolbright (1996) calculated, from gut contents, that tree frog, *Eleutherodactylus coqui*, adults at densities of about 3300 ha^{-1} in Puerto Rican rainforest, consumed 10,000 insects ha^{-1} per night; 17,000 pre-adult frogs ha^{-1} ate an additional 100,000 insects ha^{-1} per night. Gut contents consisted primarily of ants, crickets, and cockroaches, three of the most abundant canopy taxa on foliage at this site.

4.3 Regulation of Population Size

When population size exceeds the number of individuals that can be supported by existing resources, competition and other factors reduce population size until it reaches levels in balance with resource supply. This equilibrium population size, which can be sustained indefinitely by available resources, is termed the carrying capacity of the environment. Carrying capacity is not constant, but depends on factors that affect the abundance and suitability of necessary resources, including the intensity of competition with other species that also use those particular resources.

Density-independent factors modify population size, but only density-dependent factors can regulate population size, in the sense of stabilizing abundance near carrying capacity. Regulation requires environmental feedback, such as through density-dependent mechanisms that reduce population growth at high densities but allow population growth at low densities (Schowalter 2011).

Regulation can be accomplished through the dependence of populations on resource supply ("bottom-up"). Suitable food is most often invoked as the limiting resource, but suitable shelter and oviposition sites also may become limiting as populations grow. Increased quality or availability of resources promote population growth. As populations grow, resources become increasingly limited and objects of intense competition, reducing reproduction and increasing mortality and dispersal, thereby reducing population growth. As population size declines, resources become relatively more available and support population growth. Hence, a population should tend to fluctuate around the size (carrying capacity) that can be sustained by resource supply. For example, Schowalter and Turchin (1993) demonstrated that growth of southern pine beetle populations, measured as number of host trees killed, was significant only under conditions of high host density and low nonhost density (Figure 4.5). Jactel and Brockerhoff (2007) conducted a meta-analysis of 119 studies that compared insect herbivory in monoculture and mixed-species forests. Overall, herbivory was significantly higher in monocultures than in more diverse forests, and the trend was more significant for

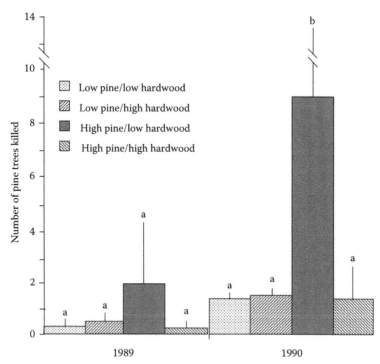

FIGURE 4.5
Effect of host (pine) and nonhost (hardwood) densities on population growth of the southern pine beetle, measured as pine mortality in 1989 (Mississippi) and 1990 (Louisiana). Low pine = 11–14 m^2/ha basal area; high pine = 23–29 m^2/ha basal area; low hardwood = 0–4 m^2/ha basal area; high hardwood = 9–14 m^2/ha basal area. Vertical lines indicate standard error of the mean. Bars under the same letter did not differ at an experiment-wise error rate of $P < 0.05$ for data combined for the two years. (From Schowalter, T. D. and P. Turchin, *Forest Science*, 39, 201–210, 1993. With permission.)

generalists than for specialists but depended on the particular composition of tree species.

Regulation also can be accomplished through the response of predators and parasites to increasing host availability ("top-down"). As prey abundance increases, predators and parasites encounter more prey. Predators respond functionally to increased abundance of a prey species by learning to acquire prey more efficiently and respond numerically by increasing population size as food supply increases. Increased intensity of predation reduces prey numbers. Reduced prey availability limits food supply for predators and reduces the intensity of predation. Hence a prey population should fluctuate around the size determined by intensity of predation. Predation has been widely recognized as reducing prey population density. Appreciation for this lies at the heart of predator control policies designed to increase abundances of commercial or game species by alleviating population control by predators. However, mass starvation and declining genetic quality of populations protected from nonhuman predators have demonstrated the importance of predation to maintenance of viable prey populations through selective predation on old, injured, or diseased individuals (Peterson 1999; Wilmers et al. 2006).

As a result of these changing perceptions, predator reintroduction programs are being implemented in some regions. At the same time, recognition of the important role of insectivorous species in controlling insect populations has justified augmentation of predator abundances, often through introduction of exotic species, for biological control purposes (van den Bosch et al. 1982; Van Driesche and Bellows 1996). Population regulation by predators or parasites has been supported by population growth following predator or parasite removal (Oksanan 1983; Carpenter and Kitchell 1987, 1988; Marquis and Whelan 1994; Dial and Roughgarden 1995; Turchin et al. 1999; Beard et al. 2003) or population decline following predator or parasite augmentation (Priesser and Strong 2004). Manipulative experiments have shown that increased abundance at one predator trophic level causes reduced abundance of the next lower trophic level and increased abundance at the second trophic level down, a "trophic cascade" (Carpenter and Kitchell 1987, 1988; Letourneau and Dyer 1998; Mooney 2007). However, in many cases, predators appear to respond to prey abundance without regulating prey populations (Parry et al. 1997), and the effect of predation and parasitism often is delayed (Turchin et al. 1999) and hence less obvious than the effects of resource supply.

Which factors are most important in maintaining insect population size? Most populations are controlled by combinations of factors (Hunter and Price 1992; Power 1992; Harrison and Cappuccino 1995; Polis and Strong 1996). Denno et al. (2002, 2003) manipulated both bottom-up (host plant biomass and nutrition) and top-down (predation by spiders) factors and concluded that bottom-up factors predominated in regulating populations of six sap-feeding insect species. The impact of predation was significant only for two species of planthoppers, *Prokelisia* spp., and was mediated by vegetation biomass and complexity. Stiling and Moon (2005) simultaneously manipulated host plant, *Borrichia frutescens*, nutritional (nitrogen) quality and density, and parasitoid abundance and measured effects on abundances of a meristem-galling fly, *Asphondylia borrichiae*, and a planthopper, *Pissonotus quadripustulatus*. Abundances of both herbivores were positively related to plant nitrogen, but not plant density at the scale of the study, and to parasitoid removal. Density-dependent competition and dispersal, as well as increased predation, eventually reduce a population to a size at which these regulatory factors become less operative.

Whereas density dependence acts in a regulatory (stabilizing) manner through negative feedback (i.e., acting to slow or stop continued growth), inverse density dependence has been thought to act in a destabilizing manner. Allee (1931) first proposed that positive feedback creates unstable thresholds (i.e., an extinction threshold below which a population inevitably

declines to extinction) and the release threshold above which the population grows uncontrollably until resource depletion or epizootics decimate
the population (Begon and Mortimer 1981; Berryman 1996, 1997). Between
these thresholds, density dependent factors should maintain stable populations near carrying capacity, a property known as the Allee effect. However,
positive feedback may ensure population persistence at low densities and is
counteracted, in most species, by the effects of crowding, resource depletion,
and predation at higher densities.

Clearly, conditions that bring populations near release or extinction
thresholds are of particular interest to resource managers, for preventing
outbreaks of potential pests and for preventing extinction of valued species.
Host plant density and stress have been identified as factors promoting population irruptions (Mattson and Haack 1987; Schowalter and Turchin 1993;
Koricheva et al. 1998, 2000; Priesser and Strong 2004). Low-density populations are particularly vulnerable to the failure of potential mates to find
each other (Gascoigne et al. 2009; Kramer et al. 2009; Yamanaka and Liebhold
2009).

4.4 Population Fluctuation in Space

As populations change in size, they also change in spatial distribution of
individuals. Population movement (epidemiology) across landscapes and
watersheds (stream continuum) reflects integration of physiological and
behavioral attributes with landscape or watershed structure. Growing populations tend to spread across the landscape as dispersal leads to colonization
of new habitats, whereas declining populations tend to constrict into more
or less isolated refuges. Isolated populations of irruptive or cyclic species can
coalesce during outbreaks, facilitating genetic exchange.

Insect populations show considerable spatial variation in densities in
response to geographic variation in habitat conditions and resource quality.
The spatial representation of populations can be described across a range of
scales from microscopic to global. The pattern of population distribution can
change over time as population size and environmental conditions change.
Two general types of spatial variation are represented by the expansion of
growing populations and by the discontinuous pattern of fragmented populations, or metapopulations.

4.4.1 Population Expansion

Growing populations tend to spread geographically as density-dependent
dispersal leads to colonization of nearby resources. This spread occurs in
two ways. First, diffusion from the origin, as density increases, produces a

gradient of decreasing density toward the fringe of the expanding population (Grilli and Gorla 1997). Second, long distance dispersal leads to colonization of vacant patches and "proliferation" or "jump dispersal" of the population (Hanski and Simberloff 1997; Suarez et al. 2001; Tobin et al. 2007). Subsequent growth and expansion of these new demes can lead to population coalescence, with local "hot spots" of superabundance that eventually disappear as resources in these sites are depleted and individuals disperse.

Spread of demes from population refuges can be synchronous over landscapes, as described earlier. Alternatively, population expansion can occur as traveling waves, typified by partial synchrony with a gradient in the degree of population change as a function of distance (Johnson et al. 2004). For example, Johnson et al. (2004) showed that spatial dynamics of larch budmoths in the European Alps from 1961 to 1998 were described as a series of traveling waves from multiple epicenters in favorable habitats and that landscape heterogeneity (gradients and connectivity of habitat suitability) alone was capable of inducing waves. Furthermore, population spread often occurs as pulses of range expansion under favorable environmental conditions (e.g., contact with patches of suitable habitat) interspersed with periods of relative stasis (Johnson et al. 2006).

The speed at which a population expands likely affects the efficiency of density-dependent regulatory factors. Populations that expand slowly may experience immediate density-dependent negative feedback in zones of high density, whereas induction of negative feedback may be delayed in rapidly expanding populations, as dispersal slows increase in density. Therefore, density-dependent factors should operate with a longer time lag in populations capable of rapid dispersal during irruptive population growth.

The speed, extent, and duration of population spread are determined by abiotic and biotic factors. Insect species with annual life cycles often show incremental colonization and population expansion. Kozár (1991) reported that several insect species showed rapid range expansion northward in Europe during the 1970s, likely reflecting warming temperatures during this period. Similarly, Jepson et al. (2008) reported northward expansion of two cyclic geometrid moths, *Operophtera brumata* and *Epirrita autumnata*, associated with continued warming during the past 20 years. Population expansion of bark beetles, spruce budworm, western harvester ants, and grasshoppers during outbreaks is associated with warmer, drier periods (Greenbank 1963; Capinera 1987; Mattson and Haack 1987; DeMers 1993). Environmental change or disturbances can terminate the spread of sensitive populations. Frequently disturbed systems, such as crop systems or streams subject to annual scouring, limit population spread to the intervals between recolonization and subsequent disturbance (Reice 1985; Matthaei and Townsend 2000).

Populations of species with relatively slow dispersal may expand only to the limits of a suitable patch during the favorable period. Spread beyond the patch also depends on the suitability of neighboring patches (Figure 4.6)

(Liebhold and Elkinton 1989; Haynes and Cronin 2003; Baum et al. 2004; Johnson et al. 2004). Populations can spread more rapidly and extensively across landscapes dominated by host species, such as agricultural and silvicultural systems, than in more heterogeneous systems in which unsuitable patches limit spread (Schowalter and Turchin 1993; Haynes and Cronin 2003; Onstad et al. 2003; Johnson et al. 2004).

Finally, the status of competitors, predators, or parasites with which the expanding population comes in contact can limit further expansion. Lounibos et al. (2003) examined factors responsible for the higher abundance of invasive container mosquitoes, *Aedes albopictus*, in areas where two species

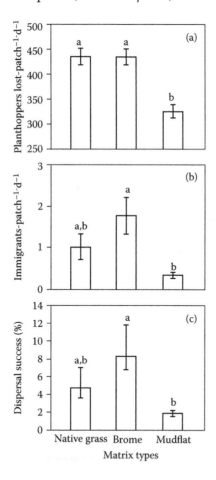

FIGURE 4.6
Effect of surrounding matrix on rate of planthopper movement among patches. (a) Planthopper loss from cordgrass patch in which released; (b) rate of planthopper immigration into satellite patches; and (c) percentage of planthoppers lost from the central release patch that successfully immigrated into any of the eight surrounding patches. Vertical lines represent 1 SE. Bars with different letters are significantly different at $P < 0.05$. (From Haynes, K. J. and J. T. Cronin, *Ecology*, 84, 2856 2866, 2003. With permission.)

of native *Wyeomyia* mosquitoes are absent. Tests in experimental containers revealed that *Wyeomyia* larvae did not deter oviposition by *A. albopictus*. However, fourth instar *Wyeomyia* larvae (but not first instar larvae) significantly reduced growth and survival of *A. albopictus* larvae, indicating that interspecific larval interaction determined the geographic distribution of *A. albopictus*.

The direction of population expansion is constrained by environmental gradients, by wind or water flow, and by unsuitable patches. Gradients in temperature, moisture, or chemical concentrations restrict the directions in which insect populations can spread, based on tolerance ranges to these factors (Heliövaara and Väisänen 1986, 1993; see Chapter 3). Even relatively homogeneous environments, such as enclosed stored grain, are subject to gradients in internal temperatures that affect spatial change in granivore populations (Flinn et al. 1992). Furthermore, direction and flow rate of wind or water have considerable influence on insect movement. Insects with limited capability to move against air or water currents move primarily downwind or downstream, whereas insects capable of movement toward attractive cues move primarily upwind or upstream. Insects that are sensitive to stream temperature, flow rate, or chemistry may be restricted to spread along linear stretches of the stream. Jepson and Thacker (1990) reported that recolonization of agricultural fields by carabid beetles dispersing from population centers was delayed by extensive use of pesticides in neighboring fields.

Schowalter et al. (1981) described the progressive colonization of individual trees or groups of trees by southern pine beetle, *D. frontalis*, populations in east Texas. A centroid was calculated for each day as the center of beetle mass (weighted abundance of beetles among the x,y coordinates of colonized trees). The distances between centroids on successive days was a measure of the rate of population movement. Populations moved at a rate of 0.9 m/day, primarily in the direction of the nearest group of available trees. The probability that a tree would be colonized depended on its distance from currently occupied trees. Trees within 6 m of sources of dispersing beetles had a 14–17% probability of being colonized, compared with less than 4% probability for trees further than 6 m from sources of dispersing beetles. Population spread in most cases ended at canopy gaps where no trees were available within 6 m.

Research on related bark beetles has confirmed the importance of host tree density for population spread of these insects (Sartwell and Stevens 1975; Brown et al. 1987; Amman et al. 1988; Mitchell and Preisler 1992). Schowalter and Turchin (1993) demonstrated that patches of relatively dense pure pine forest are essential to growth and spread of southern pine beetle populations from experimental refuge trees (Figure 4.5). Experimentally established founding populations spread from initially colonized trees surrounded by dense pure pine forest but not from trees surrounded by sparse pines or pine/hardwood mixtures.

Under suitable conditions, insect populations can spread rapidly. Reay-Jones et al. (2007) used pheromone-baited traps to measure the spread of invasive Mexican rice borer, *Eoreuma loftini*, through the Texas rice belt from 2000 to 2005. These data and the date of first appearance of this species by county since 1980 indicated an average rate of spread of 23 km yr^{-1}. Henne et al. (2007) measured rate of spread of a phorid fly, *Pseudacteon tricuspis*, introduced as a biological control agent for the red imported fire ant, *Solenopsis invicta*, in Louisiana during the first six years after release. Annual rate of spread was slow during the first two years, as the fly population became established, increased rapidly during years three to four and reached an asymptote of 15–25 km yr^{-1} during years five to six.

A critical aspect of population spread is the degree of continuity of hospitable resources or patches on the landscape. As described earlier for the southern pine beetle, unsuitable patches can interrupt population spread unless population density or growth is sufficient to maintain high dispersal rates across inhospitable patches. Similarly, Meisel (2006) reported that army ants, *Eciton burchellii*, in Costa Rica were restricted to forest fragments and consistently avoided entering pastures, where midday temperatures reached more than 50°C; ants survived less than three minutes at this temperature and only 18 minutes at 43°C. Heterogeneous landscapes composed of a variety of patch types force insects to expend their acquired resources detoxifying less acceptable resources or searching for more acceptable resources. Therefore, heterogeneous landscapes should tend to limit population growth and spread, whereas more homogeneous landscapes, such as large areas devoted to plantation forestry, pasture grasses, or major crops, provide conditions more conducive to sustained population growth and spread. The particular composition of landscape mosaics may be as important as patch size and isolation to insect movement and population distribution (Haynes and Cronin 2003). Furthermore, herbivores and predators may respond differently to landscape structure. Herbivores were more likely to be absent from small patches than large patches, whereas predators were more likely to be absent from more isolated patches than from less isolated patches in agricultural landscapes in Germany (Zabel and Tscharntke 1998).

Corridors or stepping-stones (small intermediate patches) can facilitate population spread among suitable patches across otherwise unsuitable patches. Haddad et al. (2003) found that ten taxa—representing plants, insects, and mammals—consistently showed greater movement toward patches connected by corridors or stepping-stones than toward unconnected patches. Populations of the western harvester ant, *Pogonomyrmex occidentalis*, do not expand across patches subject to frequent anthropogenic disturbance (specifically, soil disruption through agricultural activities) but are able to expand along well-drained, sheltered roadside ditches (DeMers 1993). Roads often provide a disturbed habitat with conditions suitable for dispersal of weedy vegetation and associated insects among disturbed patches. Roadside conditions also may increase plant suitability for herbivorous insects and facilitate movement across landscapes fragmented by roads (Spencer and Port 1988; Spencer et al. 1988). However, for some insects, the effect of corridors and stepping-stones may depend on the composition of the surrounding matrix. Baum et al. (2004) reported that experimental corridors and stepping-stones significantly increased colonization of prairie cordgrass, *Spartina pectinata*, patches by planthoppers, *Prokelisia crocea*, in a low resistance matrix composed of exotic, nonhost brome, *Bromus inermis*, that is conducive to planthopper dispersal but not in a high resistance matrix composed of mudflat that interferes with planthopper dispersal, relative to control matrices without corridors or stepping-stones.

An important consequence of rapid population growth and dispersal is the colonization of marginally suitable resources or patches where populations could not persist in the absence of continuous influx. Whereas small populations of herbivores, such as locusts or bark beetles, may show considerable selectivity in acceptance of potential hosts, rapidly growing populations often eat all potential hosts in their path. Dense populations of the range caterpillar, *Hemileuca oliviae*, disperse away from population centers, as grasses are depleted, and form an expanding ring, leaving denuded grassland in their wake. Landscapes that are conducive to population growth and spread because of widespread homogeneity of resources facilitate colonization of surrounding patches and more isolated resources because of the large numbers of dispersing insects. Epidemic populations of southern pine beetles, generated in the homogenous pine forests of the southern Coastal Plain during the drought years of the mid-1980s, produced sufficient numbers of dispersing insects to discover and kill most, otherwise resistant, pitch pines, *Pinus rigida*, in the southern Appalachian Mountains (Schowalter 2011).

4.4.2 Metapopulation Dynamics

A metapopulation is a population composed of relatively isolated demes maintained by some degree of dispersal among suitable patches (Hanski and Simberloff 1997; Harrison and Taylor 1997). Metapopulation structure can be identified at various scales (Massonnet et al. 2002), depending on the scale of distribution and the dispersal ability of the population (see Figure 3.3). For example, metapopulations of some sessile, host-specific insects, such as scale insects, can be distinguished among host plants at a local scale, although the insect occurs commonly over a wide geographic range (Edmunds and Alstad 1978). Local populations of black flies (Simuliidae) can be distinguished at the scale of isolated stream sections characterized by particular substrate, water velocity, temperature, proximity to lake outlets, and so on, whereas many species occur over a broad geographic area (Hirai et al. 1994; Adler and McCreadie 1997). Many litter-feeding species occur throughout patches of a particular vegetation type, but that particular vegetation type and associated populations are fragmented at the landscape scale (Grove 2002).

Metapopulation structure is most apparent where patches of suitable habitat or food resources are distinct and isolated due to natural environmental heterogeneity (e.g., desert or montane landscapes) or to anthropogenic fragmentation. The spatial pattern of metapopulations reflects a number of interacting factors, including patch size, isolation, and quality (e.g., resource availability and disturbance frequency), and insect dispersal ability (Fleishman et al. 2002; Summerville et al. 2002; Frouz and Kindlmann 2006) and largely determines gene flow, species viability and, perhaps, evolution of life history strategies (Colegrave 1997). Hence, attention to spatially structured populations has increased rapidly in recent years.

Metapopulation structure can develop in a number of ways. One is through the colonization of distant resources, and subsequent population development, that occurs during expansion of the source population (see earlier). A second is through the isolation of population remnants during population decline. A third represents a stable population structure in a heterogeneous environment, in which vacant patches are colonized as local extinction occurs in other patches.

The colonization of new patches as dispersal increases during population growth is an important mechanism for initiating new demes and facilitating population persistence on the landscape. The large number of dispersants generated during rapid population growth maximizes the probability that suitable resources will be colonized over a considerable area and that more founders will infuse the new demes with greater genetic heterogeneity (Hedrick and Gilpin 1997). Species with ruderal life histories generally exhibit considerable dispersal capacity and often arrive at sites quite remote from their population sources (Edwards and Sugg 1990). Such species quickly find and colonize disturbed sites and represent a widely occurring "weedy" fauna. By contrast, species with more competitive strategies show much slower rates of dispersal and may travel shorter distances consistent with their more stable population sizes and adaptation to more stable habitats (St. Pierre and Hendrix 2003). Such species can be threatened by rapid changes in environmental conditions that exterminate demes more rapidly than new demes are established (Hanski 1997; Hedrick and Gilpin 1997).

If conditions for population growth continue, the outlying demes may grow and coalesce with the expanding source population. This process contributes to more rapid expansion of growing populations than would occur only as diffusive spread at the fringes of the source population. A well-known example of this is the pattern of gypsy moth population expansion during outbreaks in eastern North America. New demes appear first on ridgetops in the direction of the prevailing wind because of the wind-driven dispersal of ballooning larvae. These demes grow and spread downslope, merging in the valleys. Similarly, swarms of locusts may move great distances to initiate new demes beyond the current range of the population (Riley 1878; Lockwood and DeBrey 1990).

As a population retreats during decline, local demes typically persist in isolated refuges, establishing the post-outbreak metapopulation structure. Refuges are characterized by relatively lower population densities that escape the density-dependent decline of the surrounding population. These surviving demes may remain relatively isolated until the next episode of population growth and represent the sources of the next population expansion. The existence and distribution of refuges is extremely important to population persistence. For example, bark beetle populations typically persist as scattered demes in isolated lightning-struck, diseased, or injured trees that can be colonized by small numbers of beetles (Flamm et al. 1993). Such trees appear on the landscape with sufficient frequency and proximity to beetle refuges that endemic populations are maintained (Coulson et al. 1983). Croft and Slone (1997) and Strong et al. (1997) reported that predaceous mites quickly find colonies of spider mites. New leaves on expanding shoots provide important refuges for spider mite colonists by increasing their distance from predators associated with source colonies.

If suitable refuges are unavailable, too isolated, or of limited persistence, a population may decline to extinction. Under these conditions, the numbers and low heterozygosity of dispersants generated by remnant demes are insufficient to ensure viable colonization of available habitats. For most species, life history strategies represent successful adaptations that balance population processes with natural rates of patch dynamics (i.e., the rates of appearance and disappearance of suitable patches across the landscape). For example, Leisnham and Jamieson (2002) reported that immigration and emigration rates of the mountain stone weta, *Hemideina maori*, were equivalent (0.023 per capita). However, anthropogenic activities have dramatically altered natural rates and landscape pattern of patch turnover and put many species at risk of extinction (Lockwood and DeBrey 1990; Fielding and Brusven 1995; Vitousek et al. 1997).

Even very abundant species can become vulnerable to extinction. Lockwood and DeBrey (1990) suggested that loss of critical refuges as a result of anthropogenically altered landscape structure led to the extinction of a previously widespread and periodically irruptive grasshopper species. The Rocky Mountain grasshopper, *M. spretus*, occurred primarily in permanent breeding grounds in valleys of the northern Rocky Mountains but was considered to be one of the most serious agricultural pests in western North America prior to 1900 (Riley 1878, 1883). Large swarms periodically migrated throughout the western United States and Canada during the mid-1800s, destroying crops over areas as large as 330,000 km² before declining precipitously (Riley 1883). The frequency and severity of outbreaks declined during the 1880s, and the last living specimen was collected in 1902. Macroscale changes during this period (e.g., climate changes, disappearance of Native Americans and bison, and introduction of livestock) do not seem adequate by themselves to explain this extinction. However, the population refuges (breeding ground) for this species during the late 1800s were riparian habitats (Riley 1883) where agricultural activity (e.g., tillage, irrigation, trampling by cattle, introduction of nonnative plants and birds) was concentrated. Hence, competition between humans and grasshoppers for refugia with suitable oviposition and nymphal development sites may have been the factor leading to extinction of *M. spretus* (Lockwood and DeBrey 1990).

Stable metapopulation structures are maintained by balances between source and sink habitats on the landscape. Frouz and Kindlmann (2006) described patterns of colonization and extinction for a soil-dwelling chironomid fly, *Smittia atterima*, in the Czech Republic. Larvae were most abundant in open, disturbed habitats. However, source habitats were vulnerable to desiccation and local extinction of larvae during summer. Smaller populations produced in surrounding, more densely vegetated areas tended to be more stable. Dispersing individuals from these sink habitats subsequently recolonized the source habitats, maintaining a stable population distribution.

4.4.3 Habitat Connectivity

As described earlier, habitat homogeneity facilitates population spread over landscapes. However, habitats often are heterogeneous over landscapes, and unsuitable patches can interrupt population spread (Onstad et al. 2003). In such cases, availability of corridors connecting otherwise isolated habitat patches are critical to population growth and spread (Haddad et al. 2003). For example, roads and other disturbed corridors facilitate movement of species associated with disturbed habitats (Spencer and Port 1988; Spencer et al. 1988; DeMers 1993; Haddad 1999, 2000); corridors of undisturbed habitat connecting undisturbed patches are necessary to ensure adequate dispersal of species characterizing undisturbed habitats (Collinge 2000; Várkonyi et al. 2003).

Várkonyi et al. (2003) used mark-recapture techniques to track movement of two species of noctuid moths, *Xestia speciosa*, a habitat generalist that can be found in natural and managed spruce forests and also in pine-dominated forest throughout Finland, and *X. fennica*, a species more restricted to natural spruce forests in northern Finland. They found that both species preferred to move along spruce forest corridors and avoid entering clear-cuts and regenerating forest. Movement of *X. speciosa* generally covered longer distances, whereas movement of *X. fennica* was characterized by shorter distances confined within corridors. However, *X. fennica* was capable of longer-distance dispersal across more open areas.

Haddad (1999, 2000) demonstrated that corridors between patches of open-habitat, embedded in pine, *Pinus* spp., forest, significantly increased interpatch dispersal of buckeye, *Junonia coenia*, and variegated fritillary, *Euptoieta claudia*, butterflies. Haddad and Baum (1999) found that three butterfly species (*J. coenia*, *E. claudia*, and cloudless sulphur: *Phoebis sennae*) characterizing open habitat reached higher population densities in patches connected by corridors than in isolated patches; a fourth species, the spicebush swallowtail, *Papilio troilus*, did not show any preference for open versus pine habitat and did not differ in density between connected or isolated patches. Collinge (2000) also reported variable effects of corridors on grassland insect movement. Corridors slightly increased the probability of colonization by less vagile species but did not affect recolonization by rare species. One of three focus species significantly preferred corridors, whereas the other two moved independently of corridors. These studies indicated that corridors may facilitate movement of organisms among patches, but their effect depends on species characteristics, landscape context, patch size, corridor length, and environmental variation.

Riparian habitats provide unique conditions for specialized terrestrial assemblages and facilitate movement of some terrestrial species through fragmented landscapes (Sabo et al. 2005; Rykken et al. 2007a, b). Riparian habitat widths of at least 30 m on either side of streams appear necessary to provide an adequate corridor effect (Rykken et al. 2007a, b). However, the distinct habitat conditions characterizing riparian corridors may not be suitable for dispersing upland species in areas with steep elevational gradients.

Riparian corridors also may be necessary to maintain habitat conditions for populations of some stream invertebrates. Reduction in riparian canopy cover significantly increases water temperature, especially in the summer (Kiffney et al. 2003; Rykken et al. 2007a). Davies and Nelson (1994) found that mayfly (Ephemeroptera) and stonefly (Plecoptera) densities in streams were significantly and positively correlated with the width of adjacent riparian forest buffers in Tasmania, mirroring effects of buffer width on stream temperature. Changes in riparian composition (e.g., deciduous vs. evergreen) also influence seasonal gradients in temperature; stonefly densities were significantly higher in streams bordered by young deciduous forest compared with streams through old-growth coniferous forest (Frady et al. 2007).

4.5 Summary

Insect populations can change dramatically in size over short periods of time and spread quickly over large areas during population growth or virtually disappear into small isolated refuges during population decline. Environmental changes, especially disturbances, can trigger outbreaks of species that respond to altered abundance and distribution of resources or to relaxed predation. However, resources are not unlimited for any species, and insect populations, like others, are normally regulated near the carrying capacity of the environment by a combination of the availability of suitable food resources and negative feedback resulting from competition and predation.

Abiotic factors that exceed the tolerance ranges of exposed insects reduce population size. Disturbances are particularly important events that cause abrupt changes in temperature, moisture, or chemical conditions. However, these factors are only weakly related to population density and cannot function predictably to reduce population size during outbreaks. Predation and competition for limited resources are directly related to population density, that is, intensity of competition and predation increases as population density increases. Therefore, predation and competition typically function to slow, and finally reverse, population growth. Resource availability and predation can be manipulated to some extent to prevent outbreaks, although agricultural and silvicultural practices more often maximize resources for pests and minimize habitat for predators, thereby inducing pest outbreaks.

As populations grow, individuals typically move from crowded to less crowded habitats, thereby initiating population spread over the landscape. The mosaic of suitable and unsuitable patches, and degree of connectivity among suitable patches, across the landscape can either promote continued population growth and spread among suitable patches or restrict population spread and facilitate density-dependent feedback that slows and reverses population growth. However, large populations generate sufficient numbers of dispersing individuals to ensure that remote or isolated habitats can be colonized. Large areas dominated by particular host plant species or planted to particular agricultural crops result in elevated populations of associated insects over large areas. Clearly, as long as humans homogenize agricultural and forest ecosystems and landscapes, we will continue to be liable for the exorbitant costs of controlling outbreaks. Such costs could be reduced greatly by managing agricultural and forest systems for greater plant diversity and retention of natural habitats that harbor natural regulatory agents.

References

Adler, P. H. and J. W. McCreadie. 1997. The hidden ecology of black flies: Sibling species and ecological scale. *American Entomologist* 43: 153–161.

Agrawal, A. A. and K. Konno. 2009. Latex: A model for understanding mechanisms, ecology, and evolution of plant defense against herbivory. *Annual Review of Ecology, Evolution and Systematics* 40: 311–331.

Allan, J. D., M. S. Wipfli, J. P. Caouette, A. Prussian, and J. Rodgers. 2003. Influence of streamside vegetation on inputs of terrestrial invertebrates to salmonid food webs. *Canadian Journal of Fisheries and Aquatic Sciences* 60: 309–320.

Allee, W. C. 1931. *Animal Aggregations: A Study in General Sociology*. Chicago, IL: University of Chicago Press.

Amman, G. D., M. D. McGregor, R. F. Schmitz, and R. D. Oakes. 1988. Susceptibility of lodgepole pine to infestation by mountain pine beetles following partial cutting of stands. *Canadian Journal of Forest Research* 18: 688–695.

Ausmus, B. S. 1977. Regulation of wood decomposition rates by arthropod and annelid populations. *Ecological Bulletin (Stockholm)* 25: 180–192.

Baldwin, I. T. and J. C. Schultz. 1983. Rapid changes in tree leaf chemistry induced by damage: Evidence for communication between plants. *Science* 221: 277–279.

Bale, J. S., G. J. Masters, I. D. Hodkinson, C. Awmack, T. M. Bezemer, V. K. Brown, J. Butterfield, A. Buse, J. C. Coulson, J. Farrar, et al. 2002. Herbivory in global climate change research: Direct effects of rising temperature on insect herbivores. *Global Change Biology* 8: 1–16.

Barbehenn, R., Q. Weir, and J.-P. Salminen. 2008. Oxidation of ingested phenolics in the tree-feeding caterpillar *Orgyia leucostigma* depends on foliar chemical composition. *Journal of Chemical Ecology* 34: 748–756.

Baum, K. A., K. J. Haynes, F. P. Dillemuth, and J. T. Cronin. 2004. The matrix enhances the effectiveness of corridors and stepping stones. *Ecology* 85: 2671–2676.

Baxter, C. V., K. D. Fausch, M. Murakami, and P. L. Chapman. 2007. Invading rainbow trout usurp a terrestrial prey subsidy from native charr and reduce their growth and abundance. *Oecologia* 153: 461–470.

Baxter C. V., K. D. Fausch, and W. C. Saunders. 2005. Tangled webs: Reciprocal flows of invertebrate prey link streams and riparian zones. *Freshwater Biology* 50: 201–220.

Beard, K. H., A. K. Eschtruth, K. A. Vogt, D. J. Vogt, and F. N. Scatena. 2003. The effects of the frog *Eleuthrodactylus coqui* on invertebrates and ecosystem processes at two scales in the Luquillo Experimental Forest, Puerto Rico. *Journal of Tropical Ecology* 19: 607–617.

Begon, M. and M. Mortimer. 1981. *Population Ecology: A Unified Study of Animals and Plants*. Oxford, UK: Blackwell Scientific.

Behmer, S. T. 2009. Insect herbivore nutrient regulation. *Annual Review of Entomology* 54: 165–187.

Bell, W. J. 1990. Searching behavior patterns in insects. *Annual Review of Entomology* 35: 447–467.

Berenbaum, M. R. 1987. Charge of the light brigade: Phototoxicity as a defense against insects. In *Light-Activated Pesticides*, J. R. Heitz and K. R. Downum, eds. 206–216. Washington, DC: American Chemical Society.

Berryman, A. A. 1981. *Population Systems: A General Introduction*. New York: Plenum.

Berryman, A. A. 1996. What causes population cycles of forest Lepidoptera? *Trends in Ecology and Evolution* 11: 28–32.

Berryman, A. A. 1997. On the principles of population dynamics and theoretical models. *American Entomologist* 43: 147–151.

Boyce, M. S. 1984. Restitution of r- and K-selection as a model of density-dependent natural selection. *Annual Review of Ecology and Systematics* 15: 427–447.

Bozer, S. F., M. S. Traugott, and N. E. Stamp. 1996. Combined effects of allelochemical-fed and scarce prey of the generalist insect predator *Podisus maculiventris*. *Ecological Entomology* 21: 328–334.

Breshears, D. D., N. S. Cobb, P. M. Rich, K. P. Price, C. D. Allen, R. G. Balice, W. H. Romme, J. H. Kastens, M. L. Floyd, J. Belnap, et al. 2005. Regional vegetation die-off in response to global-change-type drought. *Proceedings of the National Academy of Sciences USA* 102: 15144–15148.

Brookes, M.. H., R. W. Stark, and R. W. Campbell, eds. 1978. *The Douglas-Fir Tussock Moth: A Synthesis*. USDA Forest Service Technical Bulletin 1585. Washington, DC: USDA.

Brower, L. P., W. N. Ryerson, L. L. Coppinger, and S. C. Glazier. 1968. Ecological chemistry and the palatability spectrum. *Science* 161: 1349–1351.

Brown, M. V., T. E. Nebeker, and C. R. Honea. 1987. Thinning increases loblolly pine vigor and resistance to bark beetles. *Southern Journal of Applied Forestry* 11: 28–31.

Capinera, J. L. 1987. Population ecology of rangeland grasshoppers. In *Integrated Pest Management on Rangeland: A Shortgrass Prairie Perspective*, J. L. Capinera, ed. 162–182. Boulder, CO: Westview.

Carpenter, S. R. and J. F. Kitchell. 1987. The temporal scale of variance in lake productivity. *American Naturalist* 129: 417–433.

Carpenter, S. R. and J. F. Kitchell. 1988. Consumer control of lake productivity. *BioScience* 38: 764–769.

Carpenter, S. R., J. F. Kitchell, and J. R. Hodgson. 1985. Cascading trophic interactions and lake productivity. *BioScience* 35: 634–639.

Cavalieri, L. F. and H. Koçak. 1994. Chaos in biological control systems. *Journal of Theoretical Biology* 169: 179–187.

Cavalieri, L. F. and H. Koçak. 1995. Intermittent transition between order and chaos in an insect pest population. *Journal of Theoretical Biology* 175: 231–234.

Chapman, S. K., S. C. Hart, N. S. Cobb, T. G. Whitham, and G. W. Koch. 2003. Insect herbivory increases litter quality and decomposition: An extension of the acceleration hypothesis. *Ecology* 84: 2867–2876.

Christenson, L. M., G. M. Lovett, M. J. Mitchell, and P. M. Groffman. 2002. The fate of nitrogen in gypsy moth frass deposited to an oak forest floor. *Oecologia* 131: 444–452.

Cipollini, D. F., Jr. 1997. Wind-induced mechanical stimulation increases pest resistance in common bean. *Oecologia* 111: 84–90.

Clark, W. C. 1979. Spatial structure relationship in a forest insect system: Simulation models and analysis. *Mitteilungen der Schweizerischen Entomologischen Gesellschaft* 52: 235–257.

Classen, A. T., S. C. Hart, T. G. Whitham, N. S. Cobb, and G. W. Koch. 2005. Insect infestations linked to changes in microclimate: Important climate change implications. *Soil Science Society of America Journal* 69: 2049–2057.

Cloe, W. W., III and G. C. Garman. 1996. The energetic importance of terrestrial arthropod inputs to three warm-water streams. *Freshwater Biology* 36: 105–114.

Colegrave, N. 1997. Can a patchy population structure affect the evolution of competition strategies? *Evolution* 51: 483–492.

Collinge, S. K. 2000. Effects of grassland fragmentation on insect species loss, colonization, and movement patters. *Ecology* 81: 2211–2226.

Coulson, R. N., P. B. Hennier, R. O. Flamm, E. J. Rykiel, L. C. Hu, and T. L. Payne. 1983. The role of lightning in the epidemiology of the southern pine beetle. *Zeitschrift für angewandte Entomologie* 96: 182–193.

Courtney, S. P. 1985. Apparency in coevolving relationships. *Oikos* 44: 91–98.

Courtney, S. P. 1986. The ecology of pierid butterflies: Dynamics and interactions. *Advances in Ecological Research* 15: 51–131.

Croft, B. A. and D. H. Slone. 1997. Equilibrium densities of European red mite (Acari: Tetranychidae) after exposure to three levels of predaceous mite diversity on apple. *Environmental Entomology* 26: 391–399.

Cronin, J. T., W. G. Abrahamson, and T. P. Craig. 2001. Temporal variation in herbivore host-plant preference and performance: Constraints on host-plant adaptation. *Oikos* 93: 312–320.

Currano, E. D., P. Wilf, S. L. Wing, C. C. Labandeira, E. C. Lovelock, and D. L. Royer. 2008. Sharply increased insect herbivory during the Paleocene-Eocene Thermal Maximum. *Proceedings of the National Academy of Sciences USA* 105: 1960–1964.

Cushing, J. M., R. F. Costantino, B. Dennis, R. A. Desharnais, and S. M. Henson. 2003. *Chaos in Ecology: Experimental Nonlinear Dynamics*. San Diego: Academic/Elsevier.

Davies, P. E. and M. Nelson. 1994. Relationship between riparian buffer widths and the effect of logging in stream habitat, invertebrate community response, and fish abundance. *Australian Journal of Marine and Freshwater Research* 45: 1289–1305.

Dean, A. M. 1983. A simple model of mutualism. *American Naturalist* 121: 409–417.

DeMers, M. N. 1993. Roadside ditches as corridors for range expansion of the western harvester ant (*Pogonomyrmex occidentalis* Cresson). *Landscape Ecology* 8: 93–102.

Denno, R. F., C. Gratton, H. Döbel, and D. L. Finke. 2003. Predation risk affects relative strength of top-down and bottom-up impacts on insect herbivores. *Ecology* 84: 1032–1044.

Denno, R. F., C. Gratton, M. A. Peterson, G. A. Langellotto, D. L. Finke, and A. F. Huberty. 2002. Bottom-up forces mediate natural-enemy impact in a phytophagous insect community. *Ecology* 83: 1443–1458.

Dial, R. and J. Roughgarden. 1995. Experimental removal of insectivores from rain forest canopy: Direct and indirect effects. *Ecology* 76: 1821–1834.

Dolch, R. and T. Tscharntke. 2000. Defoliation of alders (*Alnus glutinosa*) affects herbivory by leaf beetles on undamaged neighbors. *Oecologia* 125: 504–511.

Edmunds, G. F., Jr. and D. N. Alstad. 1978. Coevolution in insect herbivores and conifers. *Science* 199: 941–945.

Edwards, J. S. and P. Sugg. 1990. Arthropod fallout as a resource in the recolonization of Mt. St. Helens. *Ecology* 74: 954–958.

Eggert, S. L. and J. B. Wallace. 2003. Reduced detrital resources limit *Pycnopsyche gentilis* (Trichoptera: Limnephilidae) production and growth. *Journal of the North American Benthological Society* 22: 388–400.

Elderd, B. D., J. Dushoff, and G. Dwyer. 2008. Host-pathogen interactions, insect outbreaks, and natural selection for disease resistance. *American Naturalist* 172: 829–842.

Esper, J., U. Büntgen, D. C. Frank, D. Nievergelt, and A. Liebhold. 2007. 1200 years of regular outbreaks in alpine insects. *Proceedings of the Royal Society B* 274: 671–679.

Farmer, E. E. and C. A. Ryan. 1990. Interplant communication: Airborne methyl jasmonate induces synthesis of proteinase inhibitors in plant leaves. *Proceedings of the National Academy of Sciences USA* 87: 7713–7716.

Feeny, P. P. 1969. Inhibitory effect of oak leaf tannins on the hydrolysis of proteins by trypsin. *Phytochemistry* 8: 2119–2126.

Feeny, P. P. 1970. Seasonal changes in oak leaf tannins and nutrients as a cause of spring feeding by winter moth caterpillars. *Ecology* 51: 565–581.

Fielding, D. J. and M. A. Brusven. 1995. Ecological correlates between rangeland grasshopper (Orthoptera: Acrididae) and plant communities of southern Idaho. *Environmental Entomology* 24: 1432–1441.

Filip, V., R. Dirzo, J. M. Maass, and J. Sarukhán. 1995. Within- and among-year variation in the levels of herbivory on the foliage of trees from a Mexican tropical deciduous forest. *Biotropica* 27: 78–86.

Flamm, R. O., P. E. Pulley, and R. N. Coulson. 1993. Colonization of disturbed trees by the southern pine beetle guild (Coleoptera: Scolytidae). *Environmental Entomology* 22: 62–70.

Fleishman, E., C. Ray, P. Sjögren-Gulve, C. L. Boggs, and D. D. Murphy. 2002. Assessing the roles of patch quality, area, and isolation in predicting metapopulation dynamics. *Conservation Biology* 16: 706–716.

Flinn, P. W., D. W. Hagstrum, W. E. Muir, and K. Sudayappa. 1992. Spatial model for simulating changes in temperature and insect population dynamics in stored grain. *Environmental Entomology* 21: 1351–1356.

Fonte, S. J. and T. D. Schowalter 2005. The influence of a neotropical herbivore (*Lamponius portoricensis*) on nutrient cycling and soil processes. *Oecologia* 146: 423–431.

Frady, C., S. Johnson, and J. Li. 2007. Stream macroinvertebrate community responses as legacies of forest harvest at the H. J. Andrews Experimental Forest, Oregon. *Forest Science* 53: 281–293.

Frost, C. J. and M. D. Hunter. 2004. Insect canopy herbivory and frass deposition affect soil nutrient dynamics and export in oak mesocosms. *Ecology* 85: 3335–3347.

Frouz, J. and P. Kindlmann. 2006. The role of sink to source re-colonisation in the population dynamics of insects living in unstable habitats: An example of terrestrial chironomids. *Oikos* 93: 50–58.

Gardner, K. T. and D. C. Thompson. 1998. Influence of avian predation on a grasshopper (Orthoptera: Acrididae) assemblage that feeds on threadleaf snakeweed. *Environmental Entomology* 27: 110–116.

Gascoigne, J., L. Berec, S. Gregory, and F. Courchamp. 2009. Dangerously few liaisons: A review of mate-finding Allee effects. *Population Ecology* 51: 355–372.

Gibson, R. W. and J. A. Pickett. 1983. Wild potato repels aphids by release of aphid alarm pheromone. *Nature* 302: 608–609.

Greenbank, D. O. 1963. The development of the outbreak. In *The Dynamics of Epidemic Spruce Budworm Populations*, R. F. Morris, ed. *Memoirs of the Entomological Society of Canada* 31: 19–23.

Grilli, M. P. and D. E. Gorla. 1997. The spatio-temporal pattern of *Delphacodes kuscheli* (Homoptera: Delphacidae) abundance in central Argentina. *Bulletin of Entomological Research* 87: 45–53.

Grove, S. J. 2002. Saproxylic insect ecology and the sustainable management of forests. *Annual Review of Ecology and Systematics* 33: 1–23.

Gutierrez, A. P. 1996. *Applied Population Ecology: A Supply-Demand Approach*. New York: John Wiley & Sons.

Haddad, N. 2000. Corridor length and patch colonization by a butterfly, *Junonia coenia*. *Conservation Biology* 14: 738–745.

Haddad, N. M. 1999. Corridor and distance effects on interpatch movements: A landscape experiment with butterflies. *Ecological Applications* 9: 612–622.

Haddad, N. M. and K. A. Baum. 1999. An experimental test of corridor effects on butterfly densities. *Ecological Applications* 9: 623–633.

Haddad, N. M., D. R. Browne, A. Cunningham, B. J. Danielson, D. J. Levey, S. Sargent, and T. Spira. 2003. Corridor use by diverse taxa. *Ecology* 84: 609–615.

Hambäck, P. A., J. Pettersson, and L. Ericson. 2003. Are associational refuges species-specific? *Functional Ecology* 17: 87–93.

Hance, T., J. van Baaren, P. Vernon, and G. Boivin. 2007. Impact of extreme temperatures on parasitoids in a climate change perspective. *Annual Review of Entomology* 52: 107–126.

Hanewinkel, M., J. Breidenbach, T. Neeff, and E. Kublin. 2008. Seventy-seven years of natural disturbances in a mountain forest area—the influence of storm, snow, and insect damage analysed with a long-term time series. *Canadian Journal of Forest Research* 38: 2249–2261.

Hanski, I. 1997. Metapopulation dynamics: From concepts and observations to predictive models. In *Metapopulation Biology: Ecology, Genetics and Evolution*, I. A. Hanski and M. E. Gilpin, eds. 69–91. San Diego: Academic.

Hanski, I. and D. Simberloff. 1997. The metapopulation approach, its history, conceptual domain, and application to conservation. In *Metapopulation Biology: Ecology, Genetics, and Evolution*, I. A. Hanski and M. E. Gilpin, eds. 5–26. San Diego: Academic.

Harborne, J. B. 1994. *Introduction to Ecological Biochemistry*, 4th ed. London: Academic.

Harrison, S. and N. Cappuccino. 1995. Using density-manipulation experiments to study population regulation. In *Population Dynamics: New Approaches and Synthesis*, N. Cappuccino and P. W. Price, eds. 131–147. San Diego: Academic.

Harrison, S. and A. D. Taylor. 1997. Empirical evidence for metapopulation dynamics. In *Metapopulation Biology: Ecology, Genetics, and Evolution*, I. A. Hanski and M. E. Gilpin, eds. 27–42. San Diego: Academic.

Hassell, M. P., H. N. Comins, and R. M. May. 1991. Spatial structure and chaos in insect population dynamics. *Nature* 353: 255–258.

Haynes, K. J. and J. T. Cronin. 2003. Matrix composition affects the spatial ecology of a prairie planthopper. *Ecology* 84: 2856–2866.

Hedrick, P. W. and M. E. Gilpin. 1997. Genetic effective size of a metapopulation. In *Metapopulation Biology: Ecology, Genetics, and Evolution*, I. A. Hanski and M. E. Gilpin, eds. 165–181. San Diego: Academic.

Heiermann, J. and S. Schütz. 2008. The effect of the tree species ratio of European beech (*Fagus sylvatica* L.) and Norway spruce (*Picea abies* (L.) Karst.) on polyphagous and monophagous pest species—*Lymantria monacha* L. and *Calliteara pudibunda* L. (Lepidoptera: Lymantriidae) as an example. *Forest Ecology and Management* 255: 1161–1166.

Heliövaara, K. and R. Väisänen. 1986. Industrial air pollution and the pine bark bug, *Aradus cinnamomeus* Panz. (Het., Aradidae). *Zeitschrift für angewandte Entomologie* 101: 469–478.

Heliövaara, K. and R. Väisänen. 1993. *Insects and Pollution*. Boca Raton, FL: CRC Press.

Henne, D. C., S. J. Johnson, and J. T. Cronin. 2007. Population spread of the introduced red imported fire ant parasitoid, *Pseudacteon tricuspis* Borgmeier (Diptera: Phoridae), in Louisiana. *Biological Control* 42: 97–104.

Hirai, H., W. S. Procunier, J. O. Ochoa, and K. Uemoto. 1994. A cytogenetic analysis of the *Simulium ochraceum* species complex (Diptera: Simuliidae) in Central America. *Genome* 37: 36–53.

Hitchner, E. M., T. P. Kuhar, J. C. Dickens, R. R. Youngman, P. B. Schultz, and D. G. Pfeiffer. 2008. Host plant choice experiments of Colorado potato beetle (Coleoptera: Chrysomelidae) in Virginia. *Journal of Economic Entomology* 101: 859–865.

Hunter, A. F. and J. S. Elkinton. 2000. Effects of synchrony with host plant on populations of a spring-feeding lepidopteran. *Ecology* 81: 1248–1261.

Hunter, M. D., S. Adl, C. M. Pringle, and D. C. Coleman. 2003. Relative effects of macroinvertebrates and habitat on the chemistry of litter during decomposition. *Pedobiologia* 47: 101–115.

Hunter, M. D. and P. W. Price. 1992. Playing chutes and ladders: Heterogeneity and the relative roles of bottom-up and top-down forces in natural communities. *Ecology* 73: 724–732.

Hunter, M. D. and R. E. Forkner. 1999. Hurricane damage influences foliar polyphenolics and subsequent herbivory on surviving trees. *Ecology* 80: 2676–2682.

Jackson, D. M., A. W. Johnson, and M. G. Stephenson. 2002. Survival and development of *Heliothis virescens* (Lepidoptera: Noctuidae) larvae on isogenic tobacco lines with different levels of alkaloids. *Journal of Economic Entomology* 95: 1294–1302.

Jactel, H. and E. G. Brockerhoff. 2007. Tree diversity reduces herbivory by forest insects. *Ecology Letters* 10: 835–848.

Jepsen, J. U., S. B. Hagen, R. A. Ims, and N. G. Yaccoz. 2008. Climate change and outbreaks of the geometrids *Operophthera brumata* and *Epirrita autumnata* in subarctic birch forest: Evidence of a recent outbreak range expansion. *Journal of Animal Ecology* 77: 257–264.

Jepson, P. C. and J. R. M. Thacker. 1990. Analysis of the spatial component of pesticide side-effects on non-target invertebrate populations and its relevance to hazard analysis. *Functional Ecology* 4: 349–355.

Joern, A. and S. T. Behmer. 1998. Impact of diet quality on demographic attributes in adult grasshoppers and the nitrogen limitation hypothesis. *Ecological Entomology* 23: 174–184.

Johnson, D. M., O. N. Bjørnstad, and A. M. Liebhold,. 2004. Landscape geometry and traveling waves in the larch budmoth. *Ecology Letters* 7: 967–974.

Johnson, D. M., A. M. Liebhold, O. N. Bjørnstad, and M. L. McManus. 2005. Circumpolar variation in periodicity and synchrony among gypsy moth populations. *Journal of Animal Ecology* 74: 882–892.

Johnson, D. M., A. M. Liebhold, P. C. Tobin, and O. N. Bjørnstad. 2006. Allee effects and pulsed invasion by the gypsy moth. *Nature* 444: 361–363.

Kamil, A. C., J. R. Krebs, and H. R. Pulliam, eds. 1987. *Foraging Behavior*. New York: Plenum.

Karban, R. and I. T. Baldwin. 1997. *Induced Responses to Herbivory*. Chicago, IL: University of Chicago Press.

Kareiva, P. 1983. Influence of vegetation texture on herbivore populations: Resource concentration and herbivore movement. In *Variable Plants and Herbivores in Natural and Managed Systems*, R. F. Denno and M. S. McClure, eds. 259–289. New York: Academic.

Kawaguchi Y. and S. Nakano. 2001. Contribution of terrestrial invertebrates to the annual resource budget for salmonids in forest and grassland reaches of a headwater stream. *Freshwater Biology* 46: 303–316.

Kessler, A. and I. T. Baldwin. 2002. Plant responses to herbivory: The emerging molecular analysis. *Annual Review of Plant Biology* 53: 299–328.

Kessler, A., R. Halitschke, and I. T. Baldwin. 2004. Silencing the jasmonate cascade: Induced plant defenses and insect populations. *Science* 305: 665–668.

Kessler, A., R. Halitschke, C. Diezel, and I. T. Baldwin. 2006. Priming of plant defense responses in nature by airborne signaling between *Artemisia tridentata* and *Nicotiana attenuata*. *Oecologia* 148: 280–292.

Kiffney, P. M., J. S. Richardson, and J. P. Bull. 2003. Responses of periphyton and insects to experimental manipulation of riparian buffer width along forest streams. *Journal of Applied Ecology* 40: 1060–1076.

Klapwijk, M. J., B. C. Gröbler, K. Ward, D. Wheeler, and O. T. Lewis. 2010. Influence of experimental warming and shading on host-parasitoid synchrony. *Global Change Biology* 16: 102–112.

Klock, G. O. and B. E. Wickman. 1978. Ecosystem effects. In *The Douglas-Fir Tussock Moth: A Synthesis*, M. H. Brookes, R. W. Stark, and R. W. Campbell, eds. 90–95. USDA Forest Service Technical Bulletin 1585. Washington, DC: USDA Forest Service.

Koenig, W. D. and A. M. Liebhold. 2005. Effects of periodical cicada emergences on abundances and synchrony of avian populations. *Ecology* 86: 1873–1882.

Konishi, M. and Y. Itô. 1973. Early entomology in East Asia. In *History of Entomology*, R. F. Smith, T. E. Mittler, and C. N. Smith, eds. 1–20. Palo Alto, CA: Annual Reviews.

Koricheva, J., M. J. Klapwijk, and C. Björkman. 2012. Life history traits and host plant use in defoliators and bark beetles: Implications for population dynamics. In *Insect Outbreaks Revisited*, P. Barbosa, K. Letourneau, and A. A. Agrawal, eds. 177–196. Chichester: Wiley/Blackwell.

Koricheva, J., S. Larsson, and E. Haukioja. 1998. Insect performance on experimentally stressed woody plants: A meta-analysis. *Annual Review of Entomology* 43: 195–216.

Koricheva, J., C. P. H. Mulder, B. Schmid, J. Joshi, and K. Huss-Danell. 2000. Numerical responses of different trophic groups of invertebrates to manipulations of plant diversity in grasslands. *Oecologia* 125: 271–282.

Kozár, F. 1991. Recent changes in the distribution of insects and the global warming. In *Proceedings of the 4th European Congress of Entomology and the 13th International Symposium for the Entomofauna of Central Europe*. Gödöllö, Hungary.

Kramer, A. M., B. Dennis, A. M. Liebhold, and J. M. Drake. 2009. The evidence for Allee effects. *Population Ecology* 51: 341–354.

Larsson, S., C. Bjorkman, and N. A. C. Kidd. 1993. Outbreaks in diprionid sawflies: Why some species and not others? In *Sawfly Life History Adaptations to Woody Plants*, M. R. Wagner and K. F. Raffa, eds. 453–483. San Diego: Academic.

Lawrence, R. K., W. J. Mattson, and R. A. Haack. 1997. White spruce and the spruce budworm: Defining the phenological window of susceptibility. *Canadian Entomologist* 129: 291–318.

Lee, K. P., S. T. Behmer, S. J. Simpson, and D. Raubenheimer. 2002. A geometric analysis of nutrient regulation in the generalist caterpillar *Spodoptera littoralis* (Boisduval). *Journal of Insect Physiology* 48: 655–665.

Lee, K. P., D. Raubenheimer, S. T. Behmer, and S. J. Simpson. 2003. A correlation between macronutrient balancing and insect host-plant range: Evidence from the specialist caterpillar *Spodoptera exempta* (Walker). *Journal of Insect Physiology* 49: 1161–1171.

Leisnham, P. T. and I. G. Jamieson. 2002. Metapopulation dynamics of a flightless alpine insect *Hemideina maori* in a naturally fragmented habitat. *Ecological Entomology* 27: 574–580.

Letourneau, D. K. and L. A. Dyer. 1998. Density patterns of *Piper* ant-plants and associated arthropods: Top-predator trophic cascades in a terrestrial system? *Biotropica* 30: 162–169.

Leuschner, W. A. 1980. Impacts of the southern pine beetle. In *The Southern Pine Beetle*, R. C. Thatcher, J. L. Searcy, J. E. Coster, and G. D. Hertel, eds. 137–151. USDA Forest Service Technical Bulletin 1631. Washington, DC: USDA Forest Service.

Liebhold, A., D. M. Johnson, and O. N. Bjørnstad. 2006. Geographic variation in density-dependent dynamics impacts the synchronizing effect of dispersal and regional stochaisticity. *Population Ecology* 48: 131–138.

Liebhold, A. M. and J. S. Elkinton. 1989. Characterizing spatial patterns of gypsy moth regional defoliation. *Forest Science* 35: 557–568.

Liebhold, A., W. D. Koenig, and O. N. Bjørnstad. 2004. Spatial synchrony in population dynamics. *Annual Review of Ecology, Evolution and Systematics* 35: 467–490.

Lockwood, J. A. 2001. Voices from the past: What we can learn from the Rocky Mountain locust. *American Entomologist* 47: 208–215.

Lockwood, J. A. and L. D. DeBrey. 1990. A solution for the sudden and unexplained extinction of the Rocky Mountain grasshopper (Orthoptera: Acrididae). *Environmental Entomology* 19: 1194–1205.

Logan, J. A. and J. C. Allen. 1992. Nonlinear dynamics and chaos in insect populations. *Annual Review of Entomology* 37: 455–477.

Logan, J. D., W. Wolesensky, and A. Joern. 2006. Temperature-dependent phenology and predation in arthropod systems. *Ecological Modelling* 196: 471–482.

Lotka, A. J. 1925. *Elements of Physical Biology*. Baltimore, MD: Williams and Wilkins.

Lounibos, L. P., G. F. O'Meara, N. Nishimura, and R. L. Escher. 2003. Interactions with native mosquito larvae regulate the production of *Aedes albopictus* from bromeliads in Florida. *Ecological Entomology* 28: 551–558.

Ma, S.-C. 1958. The population dynamics of the oriental migratory locust (*Locusta migratoria manilensis* Mayen) in China. *Acta Entomologica Sinica* 8: 1–40.

Malthus, T. R. 1789. *An Essay on the Principle of Population as it Affects the Future Improvement of Society*. London: Johnson.

Marquis, R. J. and C. J. Whelan. 1994. Insectivorous birds increase growth of white oak through consumption of leaf-chewing insects. *Ecology* 75: 2007–2014.

Mason, R. R. 1996. Dynamic behavior of Douglas-fir tussock moth populations in the Pacific Northwest. *Forest Science* 42: 182–191.

Mason, R. R. and R. F. Luck. 1978. Population growth and regulation. In *The Douglas-Fir Tussock Moth: A Synthesis*, M. H. Brookes, R. W. Stark, and R. W. Campbell, eds. 41–47. USDA Forest Service Technical Bulletin 1585. Washington, DC: USDA Forest Service.

Massonnet, B., J.-C. Simon, and W. G. Weisser. 2002. Metapopulation structure of the specialized herbivore *Macrosiphoniella tanacetaria* (Homoptera, Aphididae). *Molecular Ecology* 11: 2511–2521.

Matthaei, C. D. and C. R. Townsend. 2000. Long-term effects of local disturbance history on mobile stream invertebrates. *Oecologia* 125: 119–126.

Mattson, W. J. and R. A. Haack. 1987. The role of drought in outbreaks of plant-eating insects. *BioScience* 37: 110–118.

May, R. M. 1981. Models for two interacting populations. In *Theoretical Ecology: Principles and Applications*, R. M. May, ed. 78–104. Oxford, UK: Blackwell Scientific.

Meisel, J. E. 2006. Thermal ecology of the neotropical army ant, *Eciton burchellii*. *Ecological Applications* 16: 913–922.

Mitchell, R. G. and H. Preisler. 1992. Analysis of spatial patterns of lodgepole pine attacked by outbreak populations of mountain pine beetle. *Forest Science* 29: 204–211.

Mooney, K. A. 2007. Tritrophic effects of birds and ants on a canopy food web, tree growth, and phytochemistry. *Ecology* 88: 2005–2014.

Mopper, S., Y. Wang, C. Criner, and K. Hasenstein. 2004. *Iris hexagona* hormonal responses to salinity stress, leafminer herbivory, and phenology. *Ecology* 85: 38–47.

Moran, P. A. P. 1953. The statistical analysis of the Canadian lynx cycle. II. Synchronization and meteorology. *Australian Journal of Zoology* 1: 291–298.

Myers, J. H. 1988. Can a general hypothesis explain population cycles of forest Lepidoptera? *Advances in Ecological Research* 18: 179–242.

Nebeker, T. E., J. D. Hodges, and C. A. Blanche. 1993. Host response to bark beetle and pathogen colonization. In *Beetle-Pathogen Interactions in Conifer Forests*, T. D. Schowalter and G. M. Filip, eds. 157–173. London: Academic.

Nothnagle, P. J. and J. C. Schultz. 1987. What is a forest pest? In _Insect Outbreaks_, P. Barbosa and J. C. Schultz, eds. 59–80. San Diego: Academic.

Økland, B., A. M. Liebhold, A. Bjørnstad, N. Erbilgin, and P. Krokene. 2005. Are bark beetle outbreaks less synchronous than forest Lepidoptera outbreaks? _Oecologia_ 146: 365–372.

Oksanen, L. 1983. Trophic exploitation and arctic phytomass patterns. _American Naturalist_ 122: 45–52.

Onstad, D. W., D. W. Crowder, S. A. Isard, E. Levine, J. L. Spencer, M. E. O'Neal, S. T. Ratcliffe, M. E. Gray, L. W. Bledsoe, C. D. Di Fonzo, et al. 2003. Does landscape diversity slow the spread of rotation-resistant western corn rootworm (Coleoptera: Chrysomelidae)? _Environmental Entomology_ 32: 992–1001.

Parry, D., J. R. Spence, and W. J. A. Volney. 1997. Responses of natural enemies to experimentally increased populations of the forest tent caterpillar, _Malacosoma disstria_. _Ecological Entomology_ 22: 97–108.

Pearl, R. and L. J. Reed. 1920. On the rate of growth of the population of the United States since 1790 and its mathematical representation. _Proceedings of the National Academy of Sciences USA_ 6: 275–288.

Peterson, R. O. 1999. Wolf–moose interaction on Isle Royale: The end of natural regulation? _Ecological Applications_ 9: 10–16.

Pianka, E. R. 1974. _Evolutionary Ecology_. New York: Harper & Row.

Polis, G.A. and D. R. Strong. 1996. Food web complexity and community dynamics. _American Naturalist_ 147: 813–846.

Porter, E. E. and R. A. Redak. 1996. Short-term recovery of grasshopper communities (Orthoptera: Acrididae) of a California native grassland after prescribed burning. _Environmental Entomology_ 25: 987–992.

Power, M. E. 1992. Top-down and bottom-up forces in food webs: Do plants have primacy? _Ecology_ 73: 733–746.

Price, P. W. 1997. _Insect Ecology_, 3rd ed. New York: John Wiley & Sons.

Priesser, E. L. and D. R. Strong. 2004. Climate affects predator control of an herbivore outbreak. _American Naturalist_ 163: 754–762.

Raffa, K. F., B. H. Aukema, B. J. Bentz, A. L. Carroll, J. A. Hicke, M. G. Turner, and W. H. Romme. 2008. Cross-scale drivers of natural disturbances prone to anthropogenic amplification: The dynamics of bark beetle eruptions. _BioScience_ 58: 501–517.

Raimondo, S., M. Turcáni, J. Patoèka, and A. M. Liebhold. 2004. Interspecific synchrony among foliage-feeding forest Lepidoptera species and the potential role of generalist predators as synchronizing agents. _Oikos_ 107: 462–470.

Raubenheimer, D. and S. J. Simpson. 2003. Nutrient balancing in grasshoppers: Behavioural and physiological correlates of diet breadth. _Journal of Experimental Biology_ 206: 1669–1681.

Reay-Jones, F. P. F., L. T. Wilson, M. O. Way, T. E. Reagan, and C. E. Carlton. 2007. Movement of Mexican rice borer (Lepidoptera: Crambidae) through the Texas rice belt. _Journal of Economic Entomology_ 100: 54–60.

Reice, S. R. 1985. Experimental disturbance and the maintenance of species diversity in a stream community. _Oecologia_ 67: 90–97.

Rhoades, D. F. 1977. The antiherbivore chemistry of _Larrea_. In _Creosote Bush: Biology and Chemistry of Larrea in New World Deserts_, T. J. Mabry, J. H. Hunziker, and D. R. DiFeo, Jr., eds. 135–175. Stroudsburg, PA: Dowden, Hutchinson & Ross.

Riley, C. V. 1878. *First Annual Report of the United States Entomological Commission for the Year 1877 Relating to the Rocky Mountain Locust and the Best Methods of Preventing Its Injuries and of Guarding Against Its Invasions, in Pursuance of an Appropriation Made by Congress for This Purpose.* Washington, DC: U.S. Department of Agriculture.

Riley, C. V. 1883. *Third Report of the United States Entomological Commission, Relating to the Rocky Mountain Locust, the Western Cricket, the Army-Worm, Canker Worms, and the Hessian Fly, Together with Descriptions of Larvae of Injurious Forest Insects, Studies on the Embryological Development of the Locust and of Other Insects, and on the Systematic Position of the Orthoptera in Relation to Other Orders of Insects.* Washington, DC: U.S. Department of Agriculture.

Riley, C. V. 1885. *Fourth Report of the United States Entomological Commission, Being a Revised Edition of Bulletin No. 3, and the Final Report on the Cotton Worm, Together with a Chapter on the Boll Worm.* Washington, DC: U.S. Department of Agriculture.

Rousseaux, M. C., R. Julkunen-Tiitto, P. S. Searles, A. L. Scopel, P. J. Aphalo, and C. L. Ballaré. 2004. Solar UV-B radiation affects leaf quality and insect herbivory in the southern beech tree *Nothofagus antarctica*. *Oecologia* 138: 505–512.

Royama, T. 1984. Population dynamics of the spruce budworm *Choristoneura fumiferana*. *Ecological Monographs* 54: 429–462.

Royama, T. 1992. *Analytical Population Dynamics*. London: Chapman & Hall.

Rykken, J. J., A. R. Moldenke, and D. H. Olson. 2007a. Headwater riparian forest-floor invertebrate communities associated with alternative forest management practices. *Ecological Applications* 17: 1168–1183.

Rykken, J. J., S. S. Chan, and A. R. Moldenke. 2007b. Headwater riparian microclimate patterns under alternative forest management treatments. *Forest Science* 53: 270–280.

Sabo, J. L., R. Sponseller, M. Dixon, K. Gade, T. Harms, J. Heffernan, A. Jani, G. Katz, C. Soykan, J. Watts, and J. Welter. 2005. Riparian zones increase regional species richness by harboring different, not more, species. *Ecology* 86: 56–62.

St. Pierre, M. J. and S. D. Hendrix. 2003. Movement patterns of *Rhyssomatus lineaticollis* Say (Coleoptera: Curculionidae) within and among *Asclepias syriaca* (Asclepiadaceae) patches in a fragmented landscape. *Ecological Entomology* 28: 579–586.

Sartwell, C. and R. E. Stevens. 1975. Mountain pine beetle in ponderosa pine: Prospects for silvicultural control in second-growth stands. *Journal of Forestry* 73: 136–140.

Schell, S. P. and J. A. Lockwood. 1997. Spatial characteristics of rangeland grasshopper (Orthoptera: Acrididae) population dynamics in Wyoming: Implications for pest management. *Environmental Entomology* 26: 1056–1065.

Schmelz, E. A., R. J. Brebeno, T. E. Ohnmeiss, and W. S. Bowers. 2002. Interactions between *Spinacia oleracea* and *Bradysia impatiens*: A role for phytoecdysteroids. *Archives of Insect Biochemistry and Physiology* 51: 204–221.

Schowalter, T. D. 2011. *Insect Ecology: An Ecosystem Approach*, 3rd ed. San Diego: Elsevier/Academic.

Schowalter, T. D., S. J. Fonte, J. Geagan, and J. Wang. 2011. Effects of manipulated herbivore inputs on nutrient flux and decomposition in a tropical rainforest in Puerto Rico. *Oecologia* 167: 1141–1149.

Schowalter, T. D. and L. M. Ganio. 2003. Diel, seasonal, and disturbance-induced variation in invertebrate assemblages. In *Arthropods of Tropical Forests*, Y. Basset, V. Novotny, S. E. Miller, and R. L. Kitching, eds. 315–328. Cambridge, UK: Cambridge University Press.

Schowalter, T. D., D. C. Lightfoot, and W. G. Whitford. 1999. Diversity of arthropod responses to host-plant water stress in a desert ecosystem in southern New Mexico. *American Midland Naturalist* 142: 281–290.

Schowalter, T. D., D. N. Pope, R. N. Coulson, and W. S. Fargo. 1981. Patterns of southern pine beetle (*Dendroctonus frontalis* Zimm.) infestation enlargement. *Forest Science* 27: 837–849.

Schowalter, T. D., T. E. Sabin, S. G. Stafford, and J. M. Sexton. 1991. Phytophage effects on primary production, nutrient turnover, and litter decomposition of young Douglas-fir in western Oregon. *Forest Ecology and Management* 42: 229–243.

Schowalter, T. D. and P. Turchin. 1993. Southern pine beetle infestation development: Interaction between pine and hardwood basal areas. *Forest Science* 39: 201–210.

Schultz, J. C. 1983. Habitat selection and foraging tactics of caterpillars in heterogeneous trees. In *Variable Plants and Herbivores in Natural and Managed Systems*, R. F. Denno and M. S. McClure, eds. 61–90. New York: Academic.

Scriber, J. M. and F. Slansky, Jr. 1981. The nutritional ecology of immature insects. *Annual Review of Entomology* 26: 183–211.

Seastedt, T. R. and D. A. Crossley, Jr. 1984. The influence of arthropods on ecosystems. *BioScience* 34: 157–161.

Seastedt, T. R., D. A. Crossley, Jr., and W. W. Hargrove. 1983. The effects of low-level consumption by canopy arthropods on the growth and nutrient dynamics of black locust and red maple trees in the southern Appalachians. *Ecology* 64: 1040–1048.

Shonle, I. and J. Bergelson. 2000. Evolutionary ecology of the tropane alkaloids of *Datura stramonium* L. (Solanaceae). *Evolution* 54: 778–788.

Showler, A. T. 2009. Roles of host plants in boll weevil range expansion beyond tropical Mesoamerica. *American Entomologist* 55: 234–242.

Shure, D. J. and D. L. Phillips. 1991. Patch size of forest openings and arthropod populations. *Oecologia* 86: 325–334.

Simpson, S. J., D. Raubenheimer, S. T. Behmer, A. Whitworth, and G. A. Wright. 2002. A comparison of nutritional regulation in solitarious and gregarious phase nymphs of the desert locust *Schistocerca gregaria*. *Journal of Experimental Biology* 205: 121–129.

Speer, J. H., T. W. Swetnam, B. E. Wickman, and A. Youngblood. 2001. Changes in Pandora moth outbreak dynamics during the past 622 years. *Ecology* 82: 679–697.

Spencer, H. J. and G. R. Port. 1988. Effects of roadside conditions on plants and insects. II. Soil conditions. *Journal of Applied Ecology* 25: 709–715.

Spencer, H. J., N. E. Scott, G. R. Port, and A. W. Davison. 1988. Effects of roadside conditions on plants and insects. I. Atmospheric conditions. *Journal of Applied Ecology* 25: 699–707.

Stephens, D. W. and J. R. Krebs. 1986. *Foraging Theory*. Princeton, NJ: Princeton University Press.

Stern, V. M., R. F. Smith, R. van den Bosch, and K. S. Hagen. 1959. The integration of chemical and biological control of the spotted alfalfa aphid. Part 1. The integrated control concept. *Hilgardia* 29: 81–101.

Sterner, R. W. and J. J. Elser. 2002. *Ecological Stoichiometry: The Biology of Elements from Molecules to the Biosphere*. Princeton, NJ: Princeton University Press.

Stewart, M. M. and L. L. Woolbright. 1996. Amphibians. In *The Food Web of a Tropical Rain Forest*, D. P. Reagan and R. B. Waide, eds. 273–320. Chicago, IL: University of Chicago Press.

Stige, L. C., K.-S. Chan, Z. Zhang, D. Frank, and N. C. Stenseth. 2007. Thousand-year-long Chinese time series reveals climatic forcing of decadal locust dynamics. *Proceedings of the National Academy of Sciences USA* 104: 16188–16193.

Stiling, P. and D. C. Moon. 2005. Quality or quantity: The direct and indirect effects of host plants on herbivores and their natural enemies. *Oecologia* 142: 413–420.

Stout, M. and R. M. Bostock. 1999. Specificity of induced responses to arthropods and pathogens. In *Induced Plant Defenses Against Pathogens and Herbivores: Biochemistry, Ecology, and Agriculture*, A. A. Agrawal, S. Tuzun, and E. Bent, eds. 183–209. St. Paul, MN: American Phytopathological Society.

Strong, D. R., J. H. Lawton, and T. R. E. Southwood. 1984. *Insects on Plants: Community Patterns and Mechanisms*. Cambridge, MA: Harvard University Press.

Strong, W. B., B. A. Croft, and D. H. Slone. 1997. Spatial aggregation and refugia of the mites *Tetranychus urticae* and *Neoseiulus fallacis* (Acari: Tetranychidae, Phytoseiidae) on hop. *Environmental Entomology* 26: 859–865.

Suarez, A. V., D. A. Holway, and T. J. Case. 2001. Patterns of spread in biological invasions dominated by long-distance jump dispersal: Insights from Argentine ants. *Proceedings of the National Academy of Sciences USA* 98: 1095–1100.

Summerville, K. S., J. A. Veech, and T. O. Crist. 2002. Does variation in patch use among butterfly species contribute to nestedness at fine spatial scales? *Oikos* 97: 195–204.

Swetnam, T. W. and A. M. Lynch. 1993. Multicentury, regional-scale patterns of western spruce budworm outbreaks. *Ecological Monographs* 63: 399–424.

Syed, Z. and W. S. Leal. 2009. Acute olfactory response of *Culex* mosquitoes to a human- and bird-derived attractant. *Proceedings of the National Academy of Sciences USA* 106: 18803–18808.

Tanaka, S. and Y. Suzuki. 1998. Physiological trade-offs between reproduction, flight capability, and longevity in a wing-dimorphic cricket, *Modicogryllus confirmatus*. *Journal of Insect Physiology* 44: 121–129.

Thaler, J. S., M. J. Stout, R. Karban, and S. S. Duffey. 2001. Jasmonate-mediated induced plant resistance affects a community of herbivores. *Ecological Entomology* 26: 312–324.

Tinbergen, L. 1960. The natural control of insects in pinewoods. I. Factors influencing the intensity of predation by songbirds. *Archives Neerlandaises de Zoologie* 13: 265–343.

Tobin, P. C., A. M. Liebhold, and E. A. Roberts. 2007. Comparison of methods for estimating the spread of a non-indigenous species. *Journal of Biogeography* 34: 305–312.

Townsend, C. R. and R. N. Hughes. 1981. Maximizing net energy returns from foraging. In *Physiological Ecology: An Evolutionary Approach to Resource Use*, C. R. Townsend and P. Calow, eds. 86–108. Oxford, UK: Blackwell Scientific.

Turchin, P. 1990. Rarity of density dependence or population regulation with lags? *Nature* 344: 660–663.

Turchin, P., A. D. Taylor, and J. D. Reeve. 1999. Dynamical role of predators in population cycles of a forest insect. *Science* 285: 1068–1071.

Turgeon, J. J., A. Roques, and P. de Groot. 1994. Insect fauna of coniferous seed cones: Diversity, host plant interactions, and management. *Annual Review of Entomology* 39: 179–212.

Van Bael, S. A., A. Aiello, A. Valderrama, E. Medianero, M. Samaniego, and S. J. Wright. 2004. General herbivore outbreak following an El Niño-related drought in a lowland Panamanian forest. *Journal of Tropical Ecology* 20: 625–633.

van den Bosch, R., P. S. Messenger, and A. P. Gutierrez. 1982. *An Introduction to Biological Control*. New York: Plenum.

Van Driesche, R. G. and T. Bellows. 1996. *Biological Control*. New York: Chapman & Hall.

Várkonyi, G., M. Kuussaari, and H. Lappalainen. 2003. Use of forest corridors by boreal *Xestia* moths. *Oecologia* 137: 466–474.

Varley, G. C. and G. R. Gradwell. 1970. Recent advances in insect population dynamics. *Annual Review of Entomology* 15: 1–24.

Visser, J. H. 1986. Host odor perception in phytophagous insects. *Annual Review of Entomology* 31: 121–144.

Visser, M. E. and C. Both. 2005. Shifts in phenology due to global climate change: The need for a yardstick. *Proceedings of the Royal Society B* 272: 2561–2569.

Visser, M. E. and L. J. M. Holleman. 2001. Warmer springs disrupt the synchrony of oak and winter moth phenology. *Proceedings of the Royal Society B* 268: 289–294.

Vitousek, P. M., H. A. Mooney, J. Lubchenco, and J. M. Melillo. 1997. Human domination of earth's ecosystems. *Science* 277: 494–499.

Volterra, V. 1926. Fluctuations in the abundance of a species considered mathematically. *Nature* 118: 558–560.

Wagner, D. L. 2009. Ode to Alabama: The meteoric fall of a once extraordinarily abundant moth. *American Entomologist* 55: 170–173.

Watt, A. D. and A. M. McFarlane. 2002. Will climate change have a different impact on different trophic levels? Phenological development of winter moth *Opherophtera brumata* and its host plants. *Ecological Entomology* 27: 254–256.

Whitham, T. G., J. K. Bailey, J. A. Schweitzer, S. M. Shuster, R. K. Bangert, C. J. LeRoy, E. V. Lonsdorf, G. J. Allan, S. P. DiFazio, B. M. Potts, D. G. Fischer, et al. 2006. A framework for community and ecosystem genetics: From genes to ecosystems. *Nature Reviews Genetics* 7: 510–523.

Wickman, B. E. 1992. *Forest Health in the Blue Mountains: The Influence of Insects and Diseases*. USDA Forest Service General Technical Report PNW-GTR-295. Portland, OR: USDA Forest Service, Pacific Northwest Research Station.

Williams, D. W. and A. M. Liebhold. 2002. Climate change and the outbreak ranges of two North American bark beetles. *Agricultural and Forest Entomology* 4: 87–99.

Wilmers, C. C., E. Post, R. O. Peterson, and J. A. Vucetich. 2006. Predator disease out-break modulates top-down, bottom-up, and climatic effects on herbivore population dynamics. *Ecology Letters* 9: 383–389.

Wipfli, M. S. 1997. Terrestrial invertebrates as salmonid prey and nitrogen sources in streams: Contrasting old-growth and young-growth riparian forests in southeastern Alaska, U.S.A. *Canadian Journal of Fisheries and Aquatic Science* 54: 1259–1269.

Yamanaka, T. and A. M. Liebhold. 2009. Spatially implicit approaches to understand the manipulation of mating success for insect invasion management. *Population Ecology* 51: 427–444.

Zabel, J. and T. Tscharntke. 1998. Does fragmentation of *Urtica* habitats affect phytophagous and predatory insects differentially? *Oecologia* 116: 419–425.

Zenk, M. H. and M. Juenger. 2007. Evolution and current status of the phytochemistry of nitrogenous compounds. *Phytochemistry* 68: 2757–2772.

Zera, A. J. and Z. Zhao. 2006. Intermediary metabolism and life-history trade-offs: Differential metabolism of amino acids underlies the dispersal-reproduction trade-off in a wing-polymorphic cricket. *American Naturalist* 167: 889–900.

5

How Do Ecosystems Provide Services?

[The butterfly] fell to the floor, an exquisite thing, a small thing that could upset balances and knock down a line of small dominoes and then big dominoes and then gigantic dominoes. . . . Killing one butterfly couldn't be that important! Could it?

Ray Bradbury (1952)

Most modern urbanites tend to believe that they are divorced from the natural world and subject only to social principles. Most people have little appreciation for ecosystems or how ecosystem services are provided. They simply take these services for granted. As long as food is available and inexpensive at the local grocery, people do not even think about how it is produced, transported, and marketed before it reaches the grocery. As long as freshwater is available from the faucet, they do not consider what ecosystem processes control the quality and supply of their water. Awareness of threats to these services certainly rises on occasions when famine or drought limits supply.

Furthermore, most people, including many scientists, believe that ecosystems are random collections of plants and animals that interact in interesting ways. This view of ecosystems is convenient because it absolves us from any responsibility for the effects of species' disappearance or replacement by different species. In fact, general public apathy and lack of understanding of the species interactions that support ecosystem services has driven harvest practices and environmental policies that overexploit, undermine, and alter these interactions and the ecological processes that support them—to our detriment.

Even ecosystems altered as severely as those in urban areas function in much the same way as natural ecosystems, albeit with some important differences. Natural ecosystems exchange inputs and outputs with neighboring ecosystems, primarily as soluble nutrients or detritus, whereas urban ecosystems receive inputs from distant ecosystems, often from overseas, with considerable cost of labor and nonsustainable fossil fuel consumption. Urban outputs include toxic compounds or highly concentrated nutrients that disrupt neighboring and downstream ecosystems. Natural ecosystems are not exposed (except as pollutants) to the variety of industrial and household chemicals that pervade urban ecosystems. Furthermore, whereas vegetation in natural ecosystems shades the ground, reduces surface temperatures, and facilitates infiltration of water into the ground, urban ecosystems replace most vegetation with impervious materials that absorb solar energy, create heat islands, and increase runoff and flooding downstream (Arnfield 2003). Finally, most natural ecosystems are not overwhelmingly dominated by a single species (i.e., humans) that largely directs transportation pathways of resources and wastes over large areas.

Ironically, insects are the only other animals on the planet that are capable of engineering ecosystems and landscapes to an extent and rate that rival anthropogenic changes, but (unlike humans) insect outbreaks rarely last more than a few years. Their effects include increased biodiversity, increased rates of nutrient flux, and long-term ecosystem productivity, effects that can contribute to the sustainability of ecosystem services.

It is important to note that ecosystems do not provide services, on which we depend, solely for our benefit, nor can we force ecosystems to provide services once they have been degraded. Ecosystem services are provided as a result of complicated networks of interactions among species, unique to each ecosystem. These interactions harness energy from the sun and water and nutrients from the atmosphere and ground, channel water, and nutrients and produce biomass in its infinite variety of forms that we, along with all the other species, use for our survival. The ecosystem processes that control the supply of services are regulated by positive and negative feedbacks among species that stabilize these processes and their products much as a thermostat controls temperature in a home or office. Some ecosystems, such as forests, can buffer effects of environmental changes that would alter these interactions, whereas other ecosystems are relatively fragile and are easily degraded by overuse to states that cannot recover, at least on the time scale of human lifetimes. Ecosystems with less organic structure, such as grasslands, are relatively more resilient (capable of recovering to predisturbance conditions) than are ecosystems with greater organic structure, such as forests (Schowalter 2011).

Whether or not we recognize or appreciate the importance of ecosystem integrity, including the diversity of organisms that define it, our alteration of ecosystem conditions inevitably jeopardizes future ecosystem conditions and services. The bottom line is that we do not know how the loss of any species, no matter how seemingly unimportant, will cascade through ecosystem processes to affect the supply of ecosystem services. Overuse of ecosystem resources or disruption of critical processes by humans has undermined ecosystem ability to provide services in the past (Xue et al. 1990; Zheng and Eltahir 1998; Janssen et al. 2008; Lentz and Hockaday 2009). Interruptions in ecosystem services have led to social unrest, war, and cultural collapse in the past and likely will do so in the future (Zhang et al. 2007). Accordingly, the goal of sustainable use should be protection of ecosystem integrity and processes that ensure a consistent rate of production or supply of ecosystem services that meet human needs, not maximum use that undermines ecosystem integrity and services (Maser 2009).

This chapter provides a description of ecosystem processes that underlie ecosystem services. Insects are dominant components of terrestrial and freshwater ecosystems and have the capacity to control structure and processes. Therefore, understanding how ecosystems function in changing environments, and the roles and responses of insects, is critical to managing ecosystems in ways that minimize future conflicts with insects that affect ecosystem services or human health.

5.1 Ecosystem Structure and Function

Ecosystems are characterized by structure and function, both of which support ecosystem services. Structure reflects species composition and the distribution of energy and matter (biomass). Biomass is the basis of harvestable resources, including food, medical and industrial products, fiber and timber (Figure 5.1). Sequestration of carbon in biomass also offsets to some extent the release of carbon into the atmosphere from fossil fuel combustion, but this ecosystem service is undermined by vegetation removal, especially deforestation. The amount of biomass also determines the ability of ecosystems to withstand changes in resource supply or modify climate and protect functions that are necessary to maintain structure. Forests, for example, have

FIGURE 5.1 (SEE COLOR INSERT.)
Ecosystem structure varies widely among terrestrial ecosystem types. (a) Deserts have harsh environmental conditions with relatively small biomass and structure, but dominant plants, such as organ pipe (*Stenocereus therberi*) and saguaro (*Carnegiea gigantea*) cacti, often are hundreds of years old. (b) Grasslands and savannas have moderate environmental conditions, but frequent fire prevents most tree development. Intermediate plant biomass is able to reproduce aboveground plant parts quickly following loss, a characteristic necessary to support the large biomass of herbivores, such as the white rhinoceros (*Ceratotherium simum*) and nearly equal biomass of less conspicuous invertebrates. Rich soils are particularly valued for agriculture. (c) Forests have the largest biomass and most complex structure. Note the heavily fire-scarred base of the giant sequoia, *Sequoiadendron giganteum*, which depends on infrequent fire to maintain optimal soil conditions and minimal competition for water and nutrients.

FIGURE 5.1 (CONTINUED)
Ecosystem structure varies widely among terrestrial ecosystem types. (a) Deserts have harsh environmental conditions with relatively small biomass and structure, but dominant plants, such as organ pipe (*Stenocereus therberi*) and saguaro (*Carnegiea gigantea*) cacti, often are hundreds of years old. (b) Grasslands and savannas have moderate environmental conditions, but frequent fire prevents most tree development. Intermediate plant biomass is able to reproduce aboveground plant parts quickly following loss, a characteristic necessary to support the large biomass of herbivores, such as the white rhinoceros (*Ceratotherium simum*) and nearly equal biomass of less conspicuous invertebrates. Rich soils are particularly valued for agriculture. (c) Forests have the largest biomass and most complex structure. Note the heavily fire-scarred base of the giant sequoia, *Sequoiadendron giganteum*, which depends on infrequent fire to maintain optimal soil conditions and minimal competition for water and nutrients.

considerable capacity to reduce local temperature and increase precipitation (see section 5.3). This capacity is lost when forests are removed from the landscape, exacerbating global warming (Schlesinger et al. 1990; Foley et al. 2003a, b; Juang et al. 2007; Janssen et al. 2008).

Function reflects ways in which energy and matter are exchanged among individuals through feeding relationships and between the community of organisms and abiotic pools (i.e., atmosphere, sediments, and oceans). Many functions are critical to the maintenance of structure. For example, photosynthesis is the engine by which ecosystems capture and store energy as biomass. Pollination is critical to reproduction and the survival of many plant species. Decomposition releases energy and nutrients from detritus for plant uptake and production of new plant tissues. Nutrients are tightly cycled and conserved in many (especially nutrient-limited) ecosystems. In most cases, specific species are responsible for particular fluxes. Loss of any species can jeopardize rates or directions of fluxes that make nutrients available for other species. If nutrient capital is removed through harvest or erosion, ecosystems can become resource limited and subject to degradation (e.g., desertification) as nutrient-limited species disappear (Wood et al. 2009).

The accumulation of living biomass, the basis for all provisioning services, depends on balances among energy and nutrient transfers. Energy is the ecological currency with which individual organisms acquire water and nutrients for growth and reproduction (Schowalter 2011). The energy or nutrient budget of any organism can be expressed by the equation $I = P + R + E$, where I = ingestion, P = production, R = respiration, E = egestion, and $I - E = P + R =$ assimilation (A). Energy is required to fuel metabolism, so only part of the energy assimilated by an organism is available for growth and reproduction. The remainder is lost through respiration. Fitness depends on the extent to which an organism acquires sufficient amounts of suitable food resources to satisfy the requirements for growth and the costs of escaping from predators and of searching for food and mates, all of which are necessary to reproduce.

Plants store (or "fix") solar energy, water, and atmospheric carbon dioxide in carbohydrates through the process of photosynthesis. The total rate of carbon fixation through photosynthesis is termed gross primary productivity (GPP); the rate after subtraction of respiration is net primary productivity (NPP), which becomes the food resource for all consumers, including herbivores, predators, and detritivores. However, most plants require the services of nitrogen-fixing or nitrifying bacteria to provide organic nitrogen, which is available in limited supply in most ecosystems, for production of nucleic acids, amino acids, and other compounds. Primary productivity varies among ecosystem types, with the highest rates in forests, marshes, and estuaries, where resources are abundant, and the lowest rates in deserts, tundra, and open ocean, where resources are most limiting (Figure 5.2) (Whittaker 1970; Waring and Running 2007). However, when adjusted for global area of each ecosystem type, open ocean and tropical forests are revealed to make the greatest contributions to global primary productivity, and hence, to carbon sequestration (Figure 5.2).

Insects and other heterotherms (organisms that do not regulate body temperature internally) require little energy to maintain thermal homeostasis, generally respiring only 60–90% of assimilated energy, compared with more than 97% for homeotherms (birds and mammals that regulate body temperature internally) (Golley 1968; Phillipson 1981; Wiegert and Petersen 1983; Fitzgerald 1995). Because of this difference in respiration requirements,

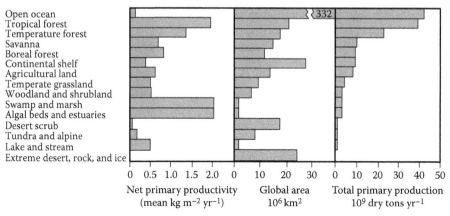

FIGURE 5.2

Net primary production, total area, and contribution to global net primary production of the major biomes. Note that although open ocean has relatively low rates of net primary production, its large area makes it the major contributor globally. The second most important contributor to global net primary production is tropical forests that currently are being converted at a rapid rate to less productive grassland and agricultural land (data from Whittaker 1970). (From Schowalter, T. D., *Insect Ecology: An Ecosystem Approach*, Elsevier, San Diego, 2011. With permission.)

insects and other heterotherms convert acquired resources into animal biomass far more efficiently than do homeotherms, making them a more efficient source of food for an increasingly hungry human population. In fact, insect biomass in terrestrial and freshwater ecosystems frequently exceeds vertebrate biomass, even in grassland ecosystems where large herds of grazing vertebrates are most obvious (Table 5.1) (Odum, 1957, 1970; Gosz et al. 1978; Watts et al. 1982; Fisher 1986; Coupland 1993; Gardiner et al. 2005). Insects become even more prominent during outbreaks when populations may increase up to 1,000-fold. When organisms die and decompose, the energy stored in their biomass is lost from the system as heat.

Nutrients, on the other hand, are conserved and cycled with varying degrees of efficiency among atmospheric, oceanic, geologic, and biotic "pools" (storage locations, including species biomass) (Figure 5.3). The cumulative movement of energy and nutrients through plants, herbivores, predators, and detritivores determines the development of community structure and distribution of biomass (physical structure) in the ecosystem.

Structure and function are interrelated in that structure affects rates and directions of energy and matter fluxes and, in turn, these fluxes affect the development and maintenance of structure. Some functions are common to all ecosystems (e.g., consumption and decomposition of organic materials). Photosynthesis and nitrogen fixation, the two processes that support all life on earth by capturing and storing carbon and nitrogen in organic material, require particular environmental conditions. Headwater streams have inadequate access to solar energy, so they have few, if any, organisms capable of

TABLE 5.1

Biomass of Plants, Insects, and Vertebrates in Ecosystems for Which These Have
All Been Measured

	Plants (g m^{-2})	Insects (g m^{-2})	Vertebrates (g m^{-2})
Tropical lowland forest: Peru[1]	39,000	5.4	0.15
Tropical rainforest: Puerto Rico[2]	26,000	4.0	0.69
Tropical grassland: Serengeti[3]	3,000	0.76	2.3
Temperate deciduous forest: various[4]	20,000	5.0	0.11
Temperate coniferous forest: various[5]	30,000	2.4	0.08
Temperate grassland: Colorado[6]	2,300	0.62	1.1
Cropland: Poland[7]	1,260	5.8	0.20
Spring: Florida[8]	809	22	31
Stream: Arizona[9]	350	3.0	50

Plant biomasses are dry mass, whereas vertebrate biomasses are fresh mass; insects are dry
mass, except where noted.

[1] DeWalt and Chave (2004); Endo et al. (2010); Lavelle and Pashanasi (1989); insect mass is fresh
mass.
[2] Odum (1970).
[3] Lamotte and Bourliére (1983); data include native ungulates.
[4] Monk and Day (1988); Perry et al. (2008); Turcek (1971).
[5] Perry et al. (2008); Turcek (1971).
[6] Coupland and Van Dyne (1979); data for vertebrates include stocked cattle (1.09 g m^2).
[7] Ryszkowski (1979).
[8] Odum (1957).
[9] Fisher (1986).

photosynthesis; energy for support of headwater stream ecosystems comes
from organic detritus washing into streams. This resource base supports a
diverse assemblage of aquatic detritivores (insects and other invertebrates)
that support important fisheries (e.g., salmonids) and can be destroyed by
terrestrial activities that increase sedimentation and consequent burial of
critical resources. Woody ecosystems require specialized consumers and
detritivores to utilize and decompose complex lignin-cellulose resources
(Progar et al. 2000). Decomposition is necessary to release energy and nutrients from organic detritus and make these nutrients available for uptake into
new plant growth (Coleman et al. 2004; Wood et al. 2009).

Ecosystems differ in rates of processes and accumulation of biomass that
affect services. Deserts are relatively impoverished in biomass and species
diversity and provide limited services. Periodic flash floods cause massive erosion that requires water filtration prior to human consumption.
Grasslands are characterized by short, fire- and drought-tolerant vegetation,
that supports characteristic herds of large grazers (that supported generations
of hunter-gatherers) and deep, organic-rich soils. Grasslands are particularly
valued for crop production and livestock grazing and have become threatened
worldwide as a result of conversion to agricultural uses. Forests are capable
of accumulating biomass in wood because environmental conditions and

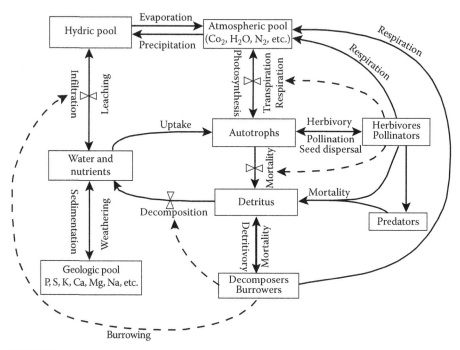

FIGURE 5.3

Conceptual diagram of ecosystem water and nutrient cycling. Boxes represent storage compartments, lines represent fluxes between storage compartments, and hourglasses represent regulatory factors. Solid lines are direct transfers of matter and dashed lines are informational or regulatory pathways. (From Schowalter, T. D., *Insect Ecology: An Ecosystem Approach*, Elsevier, San Diego, 2011. With permission.)

disturbance frequency are suitable for primary production in excess of consumption over long time periods. Once developed, the deep vertical structure of forests provides insulation that buffers the forest against changes in environmental conditions (Foley et al. 2003b; Juang et al. 2007; Janssen et al. 2008). Forests can even increase local precipitation by generating turbulence that increases convection of transpired water to altitudes at which condensation occurs (see section 5.3). Forests provide the widest range of services, including food, medical and industrial compounds, wood products, fuel, freshwater, carbon sequestration, and climate moderation, but they also are threatened globally by deforestation and conversion to agriculture. Even reforestation requires decades to replace services (such as decay-resistant heartwood for building materials) that are provided only by older, primary forests.

Aquatic ecosystems differ from most terrestrial ecosystems in lacking conspicuous plant biomass (Figure 5.4). Aquatic ecosystems exposed to the sun support algae that are characterized by very high rates of photosynthesis and NPP. Although algal biomass is low, high NPP supports highly productive zooplankton. Headwater streams that lack solar exposure depend on influx

FIGURE 5.4 (SEE COLOR INSERT.)
Representative aquatic ecosystems. (a) High elevation stream with steep gradient, high flow rate, alternating riffles and pools, and relatively limited exposure to sunlight. (b) Beaver pond created by impediment to streamflow with relatively low gradient, low flow rate, and high exposure to sunlight. (From Schowalter, T. D., *Insect Ecology: An Ecosystem Approach*, Elsevier, San Diego, 2011. With permission.)

of terrestrial litter to provide resources that support invertebrate detritivores, including a variety of immature flies, mayflies, and stoneflies. These invertebrates, in turn, support the large biomass of fish, a primary food source for many human populations. The ability of tiny, but highly productive, organisms to support a large biomass of consumers is particularly evident in the unimaginable numbers of krill eaten by whale populations. Clearly, the capacity of aquatic ecosystems to support fish production depends on maintenance of high productivity by algae and invertebrates, including insects. Freshwater fisheries also depend on influx of terrestrial insects that fall into streams or lakes. For example, salmonids may derive half of their diet from terrestrial insects (Wipfli 1997; Nakano et al. 1999; Wipfli and Musslewhite 2004; Baxter et al. 2005; Lake et al. 2007; Menninger et al. 2008; Pray et al. 2009). Unfortunately, aquatic ecosystems also receive all of the fertilizers, insecticides, antibiotics, hormones, and industrial wastes that drain from terrestrial ecosystems, leading to trophic collapse and destruction of fisheries. Changes in aquatic ecosystems that affect detrital resources for aquatic insects (Eggert and Wallace 2003) also affect fish production. Disruption of detrital processing in headwater streams results in accumulation of unprocessed detritus (Cuffney et al. 1990; Wallace et al. 1991, 1995; Eggert and Wallace 2003) and affect water quality.

5.2 Interactions among Species

Ecosystem processes are largely controlled by the network of interactions among species that result in the exchange of energy and matter. The sequence of energy and material transfer from plant to herbivore to predator to decomposer is called a food chain. The integration of all food chains involving all species in the ecosystem constitutes a food web (Figure 5.5). The members of food webs mutually regulate each others' populations through positive (mutualism) and negative (competition, predation, parasitism) feedbacks that contribute to sustainable primary production and the ecosystem services it supports (Schowalter 2011). Feeding relationships between pairs of species (e.g., herbivory, predation, parasitism, and mutualism) represents direct interaction involving contact between organisms. Such interactions are obvious and amenable to experimental manipulation to assess their effects on population and community dynamics. Insects are critical components of food webs, consuming resources at trophic levels below them and providing food for a variety of insectivorous invertebrates and vertebrates at higher tropic levels.

Indirect interactions do not involve contact between organisms and, therefore, are less obvious but can have effects at least as important as direct effects. For example, the near presence of a predator can discourage foraging activity by multiple potential prey, thereby reducing consumption of their resources to a greater extent than would the killing and consumption of a single individual. Obviously, indirect interactions are more difficult to identify and measure than are direct interactions. Examples of direct and indirect interactions are described in the following two subsections.

5.2.1 Direct Interactions

Direct interactions involve physical contact between individuals. Examples include contest competition and various feeding relationships, including herbivory, predation, parasitism, and mutualism.

Contest competition involves direct encounters between individuals fighting for a portion of a shared resource. Such encounters can occur among members of the same species or different species. Examples include territorial grasshoppers battling over access to available females (Schowalter and Whitford 1979) and multiple species of bark beetles competing for available food resources in a dying tree (Flamm et al. 1993). Most competitive interactions are asymmetric, meaning that one species is a superior competitor and is capable of preempting a shared resource at the expense of other competitors

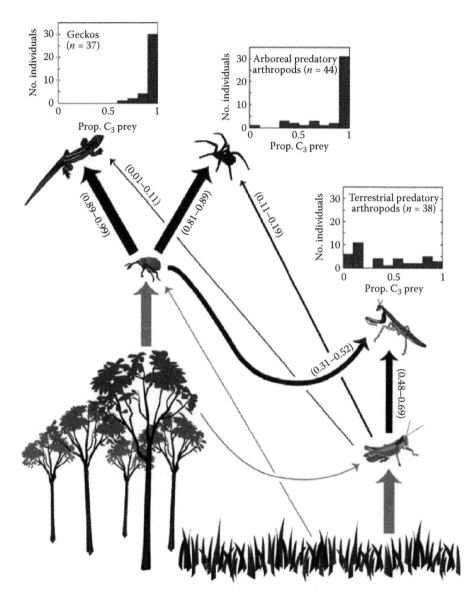

FIGURE 5.5
Diagram of interaction strengths linking canopy and understory food webs in a savanna ecosystem in central Kenya. Values in parentheses represent the range of three mean dietary proportions of C3-feeding prey for each link based on δ ^{13}C trophic discrimination factors (or Δ^{13}C). Links between predators and prey (black lines) are proportional in width to the mean proportion of diet constituted by that link. Histograms illustrate the proportion of C3-feeding prey in the diets of individual predators. The width of the links between prey and plants (gray lines) reflect estimated, not calculated, diet proportions. (From Pringle, R. M. and K. Fox-Dobbs, *Ecology Letters*, 11, 1328–1337, 2008. With permission.)

(Denno et al. 1995). Unless predation, disturbance, or other factors operate to limit population growth by the superior competitor, it may competitively exclude (eliminate) other competitors from its range (Paine 1966, 1969a, b; Denno et al. 1995). Competition also can be indirect, such as when one species occurs earlier and exploits resources in advance of a later-occurring species, thereby depriving the later species of adequate resources (Hunter 1987).

Predation is a major factor regulating prey populations (see Chapter 4). Generalist predators typically capture prey in proportion to prey species abundance but may show preferences for particular size classes or defensive abilities among potential prey. Examples include insectivorous birds, spiders, and predaceous wasps. Specialist predators tend to focus on particular prey types and may show specialized anatomy or behavior to improve their ability to capture selected prey. For example, the carabid beetle, *Promecognathus* sp., a specialist predator on *Harpaphe* spp. and other polydesmid millipedes, avoids the cyanide defense of its prey by quickly biting through the ventral nerve cord at the neck and inducing paralysis (Parsons et al. 1991).

5.2.2 Indirect Interactions

Indirect interactions do not involve direct contact among species. These interactions are less obvious and typically are mediated through chemical communication or through other species that mediate the interaction. Examples in which insects are primary components are described as follows.

Herbivores feeding aboveground frequently deplete resources that would be translocated to roots, thereby negatively affecting root-feeding herbivores or mycorrhizae (Masters et al. 1993; Rodgers et al. 1995; Salt et al. 1996). Early season herbivores can deplete plant resources available for late-season herbivores (Hunter 1987). Furthermore, plant defenses induced by early season herbivory (see Chapter 3, Box 3.1) affect the quality of plant resources available to late-season herbivores (Hunter 1987). Injured plants also produce volatile chemical elicitors, particularly jasmonic acid and ethylene, that induce production of proteinase inhibitors and other defenses among neighboring plants, including unrelated species, that receive these signals (Farmer and Ryan 1990; McCloud and Baldwin 1997; Sticher et al. 1997; Thaler 1999a; Dolch and Tscharntke 2000; Karban et al. 2000; Karban 2001; Thaler et al. 2001; Tscharntke et al. 2001; Karban and Maron 2002; Schmelz et al. 2002; Stout et al. 2006).

Predators often use the volatile plant chemicals induced by herbivores as cues that indicate prey availability (Turlings et al. 1990, 1993, 1995; Chamberlain et al. 2001; Kessler and Baldwin 2001). Thaler (1999b) demonstrated that tomato, *Lycopersicon esculentum*, defenses induced by jasmonate treatment doubled the rate of parasitism of armyworm, *Spodoptera exigua*, by the wasp, *Hyposoter exiguae*. On the other hand, plant defenses sequestered by herbivores can protect them from predators and pathogens (Brower et al. 1968; Traugott and Stamp 1996; Stamp et al. 1997; Tallamy et al. 1998). Predation on pollinators interferes with pollinator-plant interactions (Louda 1982; Knight et al. 2005).

Ants attracted to plant domatia (hollow stems or other features that provide nest sites for ants), to floral or extrafloral nectories, or to aphid honeydew commonly prey on herbivores or predators encountered during foraging for honeydew, thereby indirectly affecting herbivore-plant interactions (Tilman 1978; Fritz 1983; Cushman and Addicott 1991; Oliveira and Brandâo 1991; Jolivet 1996). The strength of this interaction varies inversely with distance from ant nests. Tilman (1978) reported that ant visits to extrafloral nectaries declined with the distance between cherry trees and ant nests. The associated predation on tent caterpillars by nectar-foraging ants also declined with distance from the ant nest.

Many predators prey indiscriminately on competing prey species, consuming prey as they are encountered, thereby indirectly mediating competitive interactions and preventing the most abundant prey species from competitively suppressing others. Paine (1966, 1969a, b) introduced the term "keystone species" to describe such predators that maintain balanced populations of competing prey species. Although the common perception of herbivores may be challenged by considering herbivory as predation (or parasitism) on plants, herbivorous insects that selectively feed on the most abundant plant species function in the same way as keystone predators to maintain higher plant diversity. Because diversity ensures maximum use of available resources and maintenance of ecosystem processes that provide services, trophic regulation of diversity is a key to sustainability of ecosystem services (see Chapters 7 and 8).

Herbivore behavior can be affected by the presence of predators to a greater extent than the actual rate of predation (Johnson et al. 2006; Hawlena et al. 2012). Batzer et al. (2000) reported that indirect effects of predaceous fish on invertebrate predators and competitors of midge prey had a greater effect on midge abundance than did direct predation on midges. Competition or predation among predators can benefit prey indirectly by reducing the number of predators. Predators can be distracted by alternate prey that are less suitable. Meisner et al. (2007) evaluated the effect of the spotted alfalfa aphid, *Therioaphis maculata*, on two parasitoids, the native *Praon pequodorum* and introduced *Aphidius ervi*, of the pea aphid, *Acyrthosiphum pisum*. The spotted alfalfa aphid had a greater distraction effect on the more common *A. ervi*, thereby contributing to persistence of *P. pequodorum* in this system. Hawlena et al. (2012) reported that predation risk caused grasshopper prey to alter nutrient assimilation, reducing their nitrogen content sufficiently to alter soil food webs in ways that slowed the decomposition of plant litter. Because decomposition provides nutrients for plant growth (Christensen et al. 2002; Frost and Hunter 2007; Wood et al. 2009), this indirect effect of predation could also slow plant growth.

A major indirect effect with consequences for insect management involves trophic cascades. Increased abundance at one trophic level increases resources available to the next higher trophic level, which increases abundance at that level but consumes more resources from the lower trophic level, reducing abundance at that level. As this effect cascades through the food chain, reduced abundance at the lower trophic level reduces its control over the second lower trophic level, which increases in abundance and reduces abundance at the third lower trophic level. In short, predators that reduce herbivore abundances sufficiently can increase plant biomass (Mooney et al. 2010).

For example, Mooney (2007) excluded birds and/or ants, primarily *Formica podzolica*, from experimental mature ponderosa pine, *Pinus ponderosa*, trees. She found that birds and ants reduced abundances of folivorous and predaceous arthropods in an additive manner, with the effect of ants stronger than that of birds. Abundance of ants doubled and abundance of tended aphids, *Cinara* spp., tripled when birds were excluded. However, ants only increased the abundance of tended aphids in the absence of birds, whereas birds only reduced aphid abundance in the presence of ants, apparently because of bird disruption of the aphid-ant interaction. Predation by birds, but not ants, resulted in increased wood and foliage growth.

Dyer and Letourneau (1999a, b) and Letourneau and Dyer (1998) described a complex trophic cascade in a neotropical rainforest community. Clerid beetle, *Tarsobaenus letourneauae*, predation on ants, especially *Pheidole bicornis*, reduced ant abundance and increased herbivore abundance and herbivory on *Piper cenocladum* ant plants. Where this beetle was absent and spiders were a less effective top predator, ant abundance was higher and reduced herbivore abundance. Manipulation of top-down and bottom-up effects indicated that increased resources (light and nutrients) directly increased plant biomass but had no indirect effect on predators or top predators, but ant exclusion indirectly affected plant biomass by increasing herbivory (Dyer and Letourneau 1999a).

5.3 Climate Control

Among the most important, and generally unappreciated, services provided by ecosystems is global climate control. The rise of photosynthetic plants on the planet was responsible for the regulation of atmospheric oxygen and carbon dioxide. Carbon dioxide and other greenhouse gases are responsible for the moderate temperatures that characterize earth and permit life as we know it. However, whereas changes in atmospheric CO_2 and temperature in the past have occurred over thousands of years, comparable changes currently are occurring on the scale of decades. Anthropogenic reduction in vegetation cover (e.g., through deforestation and desertification) has substantially interfered with climate regulation, exacerbating the effects of global warming due to fossil fuel combustion, and potentially threatening other ecosystem services, including future agricultural and forestry production (Bale et al. 2002; Millenium Ecosystem Assessment 2005; Williams and

Jackson 2007; Jepsen et al. 2008; Raffa et al. 2008; Council for Agricultural Science and Technology 2011; Hatfield et al. 2011; Izaurralde et al. 2011).

Vegetation modifies local climate conditions in several ways (Figure 5.6). Even the thin (3 mm) biological crusts—composed of cyanobacteria, green algae, lichens, and mosses—on the surface of soils in arid and semiarid regions are capable of modifying surface conditions and reducing erosion (Belnap and Gillette 1998). During the day, vegetation shades the surface of the ground, reducing surface temperature (Lewis 1998). Vegetation also absorbs solar radiation to drive photosynthesis and evapotranspiration (Parker 1995), further cooling the near-surface boundary zone (see next paragraph). At night, vegetation absorbs reradiated infrared energy from the ground, maintaining warmer nocturnal temperatures, compared with nonvegetated areas. As a result, vegetation reduces variation in diurnal and annual temperature ranges. Vegetation also intercepts precipitation and reduces the impact of raindrops on the soil surface, although this effect depends on precipitation volume and droplet size (Calder 2001), and impedes the downslope movement of water, thereby reducing erosion and loss of soil from the ecosystem. Soil organic matter retains water, increasing soil moisture-holding capacity, and reducing temperature change. Resistance to airflow by vegetation reduces wind speeds and increases turbulence, contributing to deposition of airborne particles and aerosols and generating convection that increases local precipitation. Exposure of individual organisms to damaging or lethal wind speeds is reduced as a result of buffering by surrounding individuals.

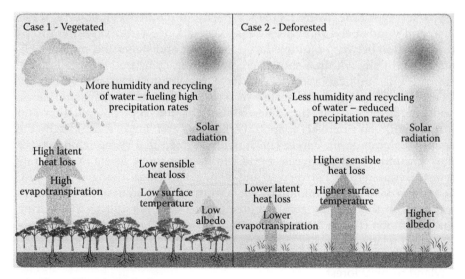

FIGURE 5.6 (SEE COLOR INSERT.)
Diagrammatic representation of the effects of vegetation on climate and atmospheric variables. The capacity of vegetation to modify climate depends on vegetation density and vertical height and complexity. (From Foley, J. A., M. H. Costa, C. Delire, N. Ramankutty, and P. Snyder, *Frontiers in Ecology and the Environment*, 1, 38–44, 2003. With permission.)

When vegetation development or moisture is limited, as in deserts, the soil surface is exposed fully to sunlight and contains insufficient water to restrict temperature change (Lewis 1998). The reflectivity of the soil surface (albedo) determines absorption of solar energy and heat. Soils with high organic content have lower albedo (0.10) than does desert sand (0.30) (Monteith 1973). Albedo also declines with increasing soil water content. In the absence of vegetation cover, surface temperatures can reach 60–70°C during the day (Seastedt and Crossley 1981) but fall rapidly at night as a result of long wavelength (infrared) radiation from the surface. Exposure to high wind speeds dries soil and moves soil particles into the atmosphere. Soil desiccation reduces infiltration of precipitation, leading to greater runoff and erosion. These altered soil characteristics increase surface heating and advective flux of moist air, leading to increased surface warming and drying (Foley et al. 2003b).

The degree of climate modification depends on vegetation height, density, and "roughness" (the degree of unevenness of canopy topography). Albedo is inversely related to vegetation height and roughness, declining from 0.25 for vegetation less than 1.0 m tall to 0.10 for vegetation greater than 30 m tall, and generally reaches lowest values in vegetation with an uneven canopy surface (e.g., tropical forest) and highest values in vegetation with a smooth canopy surface (e.g., agricultural crops) (Monteith 1973). Canopy surface roughness creates turbulence in air flow, contributing to surface cooling by wind (sensible heat loss) and evapotranspiration (latent heat loss) (Foley et al. 2003b; Juang et al. 2007).

Sparse vegetation clearly has less capacity to modify temperature, water flux, and wind speed than does dense vegetation. Shorter vegetation traps less radiation between multiple layers of leaves and stems and modifies climatic conditions within a shorter column of air, compared with taller vegetation. Tall, multicanopied forests have the greatest capacity to modify local and regional climate because the stratified layers of foliage and dense understory successively trap filtered sunlight, intercept precipitation and throughfall, contribute to evapotranspiration, and impede airflow in the deepest column of air. Parker (1995) demonstrated that rising temperatures during midday had the greatest effect in upper canopy levels in temperate forests. Temperatures between 40 and 50 m canopy height ranged from 16°C at night to 38°C during midafternoon (a diurnal fluctuation of 22°C); relative humidity in this canopy zone declined from >95% at night to 50% during mid-afternoon (Parker 1995). Below 10 m, temperature fluctuation was only 10°C, and relative humidity was constant at greater than 95%. Windsor (1990) and Madigosky (2004) reported similar gradients in canopy environment in a lowland tropical forest.

Vegetation can control local and regional precipitation patterns to a significant extent through evapotranspiration. Surface cooling by vegetation lowers the altitude at which moisture condenses, while vegetation-generated evapotranspiration, turbulence, and latent heat flux combine to elevate moist air to the height of condensation, increasing local precipitation (Trenberth 1999; Juang et al. 2007; Janssen et al. 2008). Higher rates of local recycling (>20%) occur where rates of evapotranspiration and convective flux are high and advective moisture flux is low (Trenberth 1999). As much as 30% of precipitation in tropical rainforests in the Amazon basin is generated locally by evapotranspiration (Salati 1987; Trenberth 1999).

The effects of large-scale vegetation changes resulting from insect outbreaks on regional climatic conditions have not been evaluated, although Classen et al. (2005) reported that increased soil temperature and moisture caused by manipulated levels of herbivory were of sufficient magnitude to drive changes in ecosystem processes. Clark et al. (2010) and Kurz et al. (2008) reported that outbreaks significantly reduced net ecosystem productivity and can transform forests from carbon sinks to carbon sources, potentially contributing to further climate change. However, Brown et al. (2010) reported that defoliation or mortality to insects in more diverse forests that retain substantial surviving vegetation results in less carbon loss than seen in harvested forests.

5.4 Regulation of Ecosystem Conditions

Most people have difficulty understanding or appreciating the self-regulatory capacity of ecosystems. What consciousness determines the equilibrium point, what communication system could convey departure from equilibrium, and what mechanisms provide feedback to return the system to equilibrium? Some biologists also challenge this concept because fitness is thought to reflect selfish competition for reproduction, rather than altruistic cooperation to maintain a stable environment. As a result, we disrupt (either intentionally or unintentionally) the communication and feedback mechanisms that maintain equilibrium, thereby jeopardizing the processes that are critical to our survival. This section outlines the mechanisms underlying ecosystem self-regulation, the roles of insects, and the importance of protecting these mechanisms to ensure sustainability of ecosystem services.

The concept of self-regulation is a cornerstone of ecosystem ecology. This concept explains why nutrients are cycled more efficiently and conserved to a greater extent in ecosystems in which nutrient availability is more limited. Just as the size of individual populations is regulated by a combination of resource limitation and predation (see Chapter 4), so the community of individual populations in an ecosystem appear to mutually regulate each other, thereby maintaining long-term stability of ecosystem processes and conditions that enhance their individual fitnesses. Clearly, manipulating ecosystems without regard to natural regulatory mechanisms, including insects, jeopardizes this long-term stability and often leads to insect outbreaks that can function to reverse anthropogenic changes, as described in Section 5.4.3.

5.4.1 Evidence for Self-Regulation

The concept of self-regulating (cybernetic) ecosystems has appeared to be inconsistent with evolutionary theory (emphasizing selection of "selfish" attributes) (Pianka 1974). In particular, stabilizing regulation at the ecosystem level requires self-sacrifice among individuals or species, whereas evolution would seem to favor cheaters that would maximize their fitness at the expense of others, regardless of consequences. Nevertheless, examples of apparent cooperation or altruism among individuals and species abound. The self-sacrificing stinging behavior of worker bees (which die after stinging) in defense of the colony supported the concept of kin selection, in which an individual's inclusive fitness (passing on its genes to its own offspring plus offspring of close relatives that share most of its genes) is increased by sacrifices that benefit near kin with shared genes (Hamilton 1964; Smith 1964). Mutualistic interaction could be explained as payoff, with resources provided by one species (for example, nectar provided by a flowering plant) paying for service by the other (pollination by nectar-feeding insects, birds, or bats). This hypothesis is supported by examples of reduced nectar production when pollinator abundance is reduced (Chittka et al. 1997). More general expression of altruism among species may be favored by reciprocal cooperation (for example, mutual suppression of overexploitation of resources) when interacting species have a high probability of future interaction (Axelrod and Hamilton 1981). However, this explanation required evidence that cheaters would be "punished."

Recent studies have demonstrated that cheaters may, in fact, be punished for selfish acts that could result in destabilization of interactions. Smith et al. (2009) found that reproductive cheaters in ant, *Aphaenogaster cockerelli*, colonies can be reliably identified by cuticular hydrocarbons, particularly pentacosane, that are associated only with fertile individuals. Experimental application of pentacosane to workers elicited aggression by nest mates in colonies with queens but not in colonies without queens, in which workers had begun to reproduce. Although cheaters might benefit from suppressing the hydrocarbon profile, they are prevented from doing so by the reproductive physiology of hydrocarbon biosynthesis.

Jandér and Herre (2010) experimentally excluded pollen from a number of fig wasp species associated with actively and passively pollinated fig species and found that actively pollinated (but not passively pollinated) fig species reduced fitness of nonpollinating fig wasps through a combination of increased abortion of nonpollinated figs that had wasp eggs and reduced production of wasp offspring in figs that were not aborted. The relative proportion of unpollinated figs that matured and the relative number of wasp offspring that matured in unaborted figs both were significantly lower in actively pollinated figs, compared with passively pollinated figs. All fig species aborted figs that received neither pollen nor wasp eggs. The strength of sanctions against nonpollinating wasps (measured as 1-wasp relative fitness) varied from 0.33 to 1.0 for actively pollinated fig species, compared with 0 for passively pollinated species. Pollen-free wasps occurred only among actively pollinating species, and their prevalence was less than 5% and negatively correlated with sanction strength. These data demonstrated substantial selection against cheaters as a means of stabilizing obligate mutualisms when providing a benefit to a host is costly, in terms of wasp time and energy.

These studies demonstrate that selection at supraorganismal levels should be viewed as contributing to the inclusive fitness of individuals derived from their contributions to stabilizing interactions. Cooperating individuals have demonstrated greater ability in finding or exploiting uncommon or aggregated resources, defending shared resources, and mutual protection (Hamilton 1964). Cooperating predators (e.g., wolves and ants) have higher capture efficiency and can acquire larger prey, compared with solitary predators. The mass attack behavior of bark beetles is critical to successful colonization of living trees.

Inclusive fitness also can accrue via feedback from individual effects on the environment. Accumulating evidence demonstrates genetic variation, subject to selection, in biotic effects on ecosystem properties (Whitham et al. 2003, 2006; Bailey et al. 2004; Schweitzer et al. 2004, 2005; Fischer et al. 2006; Shuster et al. 2006; Classen et al. 2007; Wimp et al. 2007), as well as responses to changing environmental conditions (Bradshaw and Holzapfel 2001; Balanyá et al. 2006; Hsu et al. 2006, 2008; Ralph et al. 2006; Edelaar et al. 2008). For example, Fischer et al. (2006) found that high tannin production in foliage requires increased root growth to compensate for tannin inhibition of soil nitrogen cycling (Figure 5.7). Thus, reduced nitrogen availability at the ecosystem level provides feedback on plant fitness, along with other factors, including herbivory, that favor increased tannin production (Whitham et al. 2006).

The concept of self-regulation does not require efficient feedback by all ecosystems or ecosystem components. Just as some organisms (recognized as cybernetic systems) have greater homeostatic ability than do others (e.g., homeotherms versus heterotherms), some ecosystems demonstrate greater homeostatic ability than do others (Webster et al. 1975). Frequently disturbed ecosystems may be reestablished by relatively random assemblages of opportunistic colonists, providing little opportunity for repeated interaction that could lead to stabilizing cooperation (Axelrod and Hamilton 1981). Some species that are favored by post-disturbance conditions appear to increase the likelihood of disturbance (e.g., brittle or flammable plant species such as the easily toppled *Cecropia* and flammable *Eucalyptus*). Insect outbreaks increase variation in some ecosystem parameters (Romme et al. 1986), often in ways that promote regeneration of resources (e.g., Schowalter et al. 1981a). On the other hand, relatively stable environments, such as tropical rainforests, also might not select for stabilizing interactions. However, stable environmental conditions should favor consistent species interactions and the evolution of reciprocal cooperation, such as demonstrated by a greater

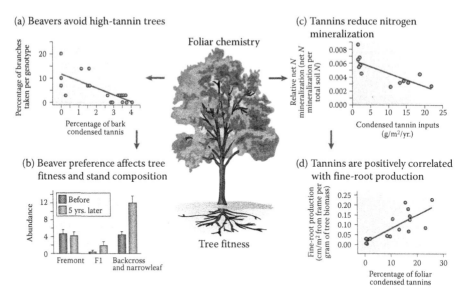

(a) Beavers avoid high-tannin trees

(b) Beaver preference affects tree fitness and stand composition

Foliar chemistry

Tree fitness

(c) Tannins reduce nitrogen mineralization

(d) Tannins are positively correlated with fine-root production

FIGURE 5.7 (SEE COLOR INSERT.)
Selection for species phenotype affects interactions with other species that in turn feedback to affect fitness of the individuals that produced that phenotype. In this example, condensed tannin phenotype in poplar affects herbivore foraging, nutrient turnover, and nutrient acquisition. (a) and (b) Beavers, *Castor canadensis*, select trees with low condensed tannin concentrations, which leads to increased abundance of trees with higher condensed tannin concentrations and reduced abundance of trees with lower concentrations, affecting other species that depend on this tree for survival. (c) and (d) Increased foliar tannin concentration inhibits nitrogen mineralization from litter and soil, requiring the tree to produce more fine roots to acquire more limited soil nitrogen. (From Whitham, T. G., J. K. Bailey, J. A. Schweitzer, S. M. Shuster, R. K. Bangert, C. J. LeRoy, E. V. Lonsdorf, G. J. Allan, S. P. DiFazio, B. M. Potts, et al., *Nature Reviews Genetics*, 7, 510–523. With permission.) [(a) Data from Bailey, J. K., J. A. Schweitzer, B. J. Rehill, R. L. Lindroth, G. D. Martinsen, and T. G. Whitham, *Ecology*, 85, 603–608, 2004. With permission. (c) Data from Schweitzer, J. A., J. K. Bailey, B. J. Rehill, G. D. Martinsen, S. C. Hart, R. L. Lindroth, P. Keim, and T. G. Whitham, *Ecology Letters*, 7, 127–134, 2004. With permission. (d) Data from Fischer, D. G., S. C. Hart, B. J. Rehill, R. L. Lindroth, P. Keim, and T. G. Whitham, *Oecologia*, 149, 668–675, 2006. With permission.]

diversity of mutualistic interactions in tropical forests. Selection for stabilizing interactions should be greatest in ecosystems characterized by intermediate levels of environmental variation. Interactions that reduce variation in such ecosystems would contribute to individual inclusive fitnesses and be favored by natural selection.

Among the parameters that could be stabilized as a result of species interactions, net primary production and biomass structure (living and dead) may be particularly important because many other parameters—including energy, water, and nutrient fluxes, trophic interactions, species diversity, population sizes, climate, and soil development—are directly or indirectly determined by net primary production or biomass structure (Boulton et al. 1992). In particular, the ability of ecosystems to modify internal microclimate, protect and modify soils, and provide stable resource bases for primary and secondary producers depends on NPP and biomass structure. Therefore, natural selection over long periods of coevolution should favor individuals whose interactions stabilize these ecosystem parameters. NPP may be stabilized over long time periods as a result of compensatory community dynamics and biological interactions, such as those resulting from biodiversity and herbivory, as described in the following two subsections.

5.4.2 Biodiversity

A key mechanism of stability in ecosystems is the diversity of species or functional groups (Frank and McNaughton 1991; Tilman and Downing 1994; Naeem and Li 1997; Naeem 1998; Hooper et al. 2005; Spehn et al. 2005). Although this relationship is still debated among ecologists (Hooper et al. 2005), a number of manipulative experiments have demonstrated that more diverse communities maintain more constant rates of ecological processes during environmental changes than do less diverse communities because diversity increases the probability that species tolerant of various environmental changes are represented. Consequently, abundances of tolerant species increase and maintain primary production, vegetation cover, and soil structure as environmental changes or disturbances reduce abundances of less tolerant species, thereby minimizing departures from normal conditions, as shown by the following examples.

In 1982, Tilman and Downing (1994) established replicated plots in which the number of plant species was altered through different rates of nitrogen addition. These plots subsequently (1987–1988) were subjected to a record drought. During the drought, plots with more than nine species averaged about half of their pre-drought biomass, but plots with fewer than five species averaged only about 12% of their pre-drought biomass. Hence, the more diverse plots were better buffered against this disturbance because they were more likely to include drought-tolerant species, compared with less diverse plots. More diverse plots also recovered biomass more quickly following the drought. When biomass was measured in 1992, plots with more than six species had biomass equivalent to pre-drought levels, but plots with less than five species had significantly lower biomass, with deviations of 8–40% (Figure 5.8). Tilman and Downing (1994) and Tilman et al. (1997) concluded that more diverse ecosystems represented a greater variety of ecological strategies that confer greater resistance and greater resilience to environmental variation. However, the contribution of diversity to ecosystem stability may be related to environmental heterogeneity, that is, diversity does not necessarily increase stability in more homogeneous environments (see earlier).

Spehn et al. (2005) manipulated plant diversity in multiple European grassland ecosystems. They found that more diverse communities were more productive and utilized resources more completely through greater

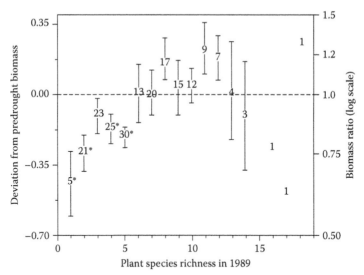

FIGURE 5.8

Relationship between plant species diversity and deviation in 1992 biomass (following drought) from mean (1982–1986) predrought biomass in experimental grassland plots planted with different species diversities. Mean, standard error, and number of plots with given species richness are shown. Negative values indicate 1992 biomass lower than predrought mean. Biomass ratio is 1992 biomass/predrought biomass. Plots with one, two, four, or five species (but not plots with more than five species) differed significantly from predrought means. (From Tilman, D. and J. A. Downing, *Nature*, 367, 363–365, 2004. With permission.)

occupation of available space and uptake and retention of nitrogen than did less diverse communities. Reusch et al. (2005) demonstrated that experimentally increased genetypic diversity of a common sea grass, *Zostera marina*, enhanced biomass production, plant density, and faunal abundances, buffering the coastal community against extreme temperatures. Ewel (1986) and Ewel et al. (1991) experimentally manipulated plant diversity in a tropical forest and demonstrated that soil fertility was significantly higher in more diverse plots.

A number of studies have demonstrated that ecosystem resistance to elevated herbivory or plant disease is positively correlated to vegetation diversity (McNaughton 1985; Schowalter and Turchin 1993; Garrett and Mundt 1999; Knops et al. 1999; Johnson et al. 2006; Jactel and Brockerhoff 2007). As vegetation diversity increases, relative to the host range of any particular herbivore, the ability of herbivores to find and exploit their hosts decreases (Jactel and Brockerhoff 2007), leading to increasing stability of herbivore-plant interactions. Even genetic variation in a dominant plant species can affect herbivory (Hochwender and Fritz 2004; Johnson et al. 2006; Wimp et al. 2007). Similarly, a diverse pollinator community is more likely than a simple community to maintain pollination function as changing environmental conditions favor some species over others (Klein et al. 2003).

A particularly valuable consequence of biodiversity, from a human perspective, is the regulation of reservoir and vector species that transmit human diseases. In particular, Allan et al. (2009) found that the prevalence of West Nile virus (WNV) infection in mosquitoes and humans increased with decreasing bird species diversity and increasing proportion of primary reservoir species within the local bird community. Higher bird diversity distributed mosquito feeding among more bird species, many of which are not reservoir hosts, thereby interrupting the disease cycle. Unfortunately, bird diversity is declining as a result of increasing conversion of natural habitats to agricultural and urban uses. Ironically, those bird species best adapted to persisting in human-dominated ecosystems, such as crows and cardinals, are the most susceptible reservoir hosts for WNV. Similarly, rapid deforestation and conversion to agricultural uses in Peru during the 1980s increased mosquito habitat and the incidence of malaria among humans in an area from which malaria had been eliminated (Vittor et al. 2006). In other words, protection of natural ecosystems and their native biodiversity can reduce our need to employ dangerous insecticides to control disease vectors.

5.4.3 Herbivory

The perspective that herbivores affect vegetation negatively and should destabilize primary production is so entrenched that any contribution of herbivory to ecosystem stability has been difficult to accept. However, accumulating experimental evidence indicates that outbreaks of native insect herbivores are triggered by elevated host plant abundance or stress that indicate departure from carrying capacity. Outbreaks function to reduce primary production to carrying capacity, contributing to long-term stability in this ecosystem process. In fact, low intensity of herbivory, when host density is low or condition good, can stimulate primary production, whereas high levels of herbivory, when host density is high or condition poor, reduce host production, a density-dependent (that is, regulatory) effect. As a result, primary production often peaks at low-to-moderate intensities of pruning and thinning (Figure 5.9), practices used to increase commodity production. These data support the grazing optimization hypothesis (Williamson et al. 1989; Belovsky and Slade 2000). Elevated primary productivity following outbreaks suggests alleviation of stressful conditions that triggered the outbreak and could lead to instability but also reflects greater species turnover as stressed hosts are replaced by species better able to tolerate prevailing conditions (Ritchie et al. 1998; Chase et al. 2000; Belovsky and Slade 2000). In other words, herbivores affect their host plants in the same way that predators affect their prey, by cropping weak individuals and regulating population sizes near ecosystem carrying capacity (Schowalter 2011).

By stabilizing primary production, herbivores also stabilize ecosystem services controlled by primary production. Romme et al. (1986) reported

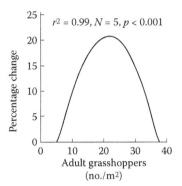

FIGURE 5.9
Relationship between grasshopper, *Melanoplus sanguinipes*, density and primary production in experimental field microcosms at the National Bison Range in Montana. Peak natural densities at the site were 4–36 adults m². (From Belovsky, G. E. and J. B. Slade, *Proceedings of the National Academy of Sciences USA*, 97, 14412–14417, 2000. With permission.)

that mountain pine beetle, *Dendroctonus ponderosae*, outbreaks appeared to increase variation (destabilization) of some ecosystem properties, but some stands recovered pre-outbreak NPP within ten years following the outbreak. Significantly, these outbreaks represented a biotic response to deviation in primary production resulting from human intervention (i.e., increased tree density resulting from fire suppression). Although primary production may be depressed temporarily, the departure is less than would occur if there were no feedback to prevent plant growth in excess of long-term carrying capacity. Herbivore outbreaks also reduce host biomass and increase diversity and abundance of nonhost plant species that are more tolerant of prevailing conditions (Wickman 1980; Alfaro and Shepherd 1991; Ritchie et al. 1998). In other words, outbreaks of native herbivorous insects function to crop excess plant growth, primarily of stressed plants, just as top predators (such as lions and wolves) are necessary regulators of prey population size and condition, via cropping of excess, primarily weakened individuals. In fact, herbivores and predators use odors indicative of suitable hosts (e.g., age and condition) to focus foraging on weak or stressed individuals (Lonnstedt et al. 2012; see also Chapter 3). Stabilization of community structure by herbivorous insects is illustrated by the following example.

Conifer forests dominate much of the montane and high latitude region of western North America. The large, contiguous, lower elevation zone is characterized by relatively arid conditions and frequent droughts that historically maintained a sparse woodland dominated by drought- and fire-tolerant (but shade-intolerant) pine trees and a ground cover of grasses and shrubs, with little understory (Figure 5.10a). Low intensity ground fires occurred frequently, at intervals of 15–25 years, and covered large areas (Agee 1993), minimizing drought-intolerant vegetation and litter accumulation. The relatively isolated higher elevation and riparian zones were more mesic and supported shade-tolerant (but fire- and drought-intolerant) fir and spruce forests. Fire was less frequent (every 150–1,000 years), but more catastrophic, at higher elevations as a result of the greater tree densities and understory development that facilitated fire access to tree canopies (Agee 1993; Veblen et al. 1994).

As a result of fire suppression during the past century, much of the lower elevation zone has undergone succession from pine forest to later successional fir forest (Figure 5.10b), with greatly increased plant biomass, a conspicuous deviation from historic conditions. Outbreaks of a variety of folivore and bark beetle species have become more frequent in these altered forests. During mesic periods and in more mesic locations (e.g., riparian corridors and higher elevations) the mountain pine beetle has advanced succession by facilitating the replacement of competitively stressed pines by more competitive firs. However, during inevitable drought periods, such as occurred during the 1980s, moisture limitation increases the vulnerability of stressed firs to herbivores (Figure 5.11). Insect-induced mortality of the firs favored drought- and fire-tolerant pines. Tree mortality can increase the severity and scale of catastrophic fires, which historically were rare in these forests, unless litter decomposition reduces fuel accumulation before fire occurs.

FIGURE 5.10 (SEE COLOR INSERT.)
(a) The relatively arid interior forest region of North America was characterized by open canopied forests dominated by drought- and fire-tolerant pines and by sparse understories prior to fire suppression beginning in the late 1800s. (b) Fire suppression has transformed forests into dense, multistoried ecosystems stressed by competition for water and nutrients. (From Goyer, R. A., M. R. Wagner, and T. D. Schowalter, *Journal of Forestry*, 96, 29–33, 1998. With permission.)

FIGURE 5.11 (SEE COLOR INSERT.)
Herbivore modification of succession in central Sierran mixed conifer ecosystems during 1998. Understory white fir (*Abies concolor*), the late successional dominant, is increasingly stressed by competition for water in this arid forest type. An outbreak of the Douglas-fir tussock moth, *Orgyia pseudotsugata*, has completely defoliated the white fir (gray trees, bottom), restoring the ecosystem to a more stable condition dominated by earlier successional, drought- and fire-tolerant, sequoias and pines (dark, foliated, trees) (From Schowalter, T. D., *Insect Ecology: An Ecosystem Approach*, Elsevier, San Diego, 2011. With permission.)

Van Langevelde et al. (2003) also suggested a cycle of alternating vegetation states maintained by interaction of fire and herbivores in African savanna. Although the temporary reduction in timber production may be undesirable for maximizing provisioning services, the lower density of site-adapted species that prevailed prior to forest management (Schowalter 2008) is a more sustainable condition and the long-term management recommendation for these forests (North et al. 2007).

Herbivores also affect ecosystem capacity to regulate disease transmission. Jones et al. (1998) reported that human risk of contracting Lyme disease in oak forests reflected complex interactions among tick vectors, deer and rodent reservoirs that feed on acorns, gypsy moth (*Lymantria dispar*) outbreaks that affect acorn availability, and rodent predation on gypsy moth pupae. Manipulation of acorn abundance demonstrated that gypsy moth abundance increased at low acorn abundances, due to reduced rodent abundance and predation, and incidence of Lyme disease increased at high acorn abundances, due to increased reservoir density.

5.4.4 Predation

Predation is a primary regulatory factor that (together with resource limitation) prevents uncontrolled population growth (see Chapter 4). A nearly inconceivable diversity of arthropod predators prey on various insects and other arthropods (Figure 5.12) and even some vertebrates (Reagan et al. 1996; Schowalter 2011). Many vertebrates, including fish, amphibians, reptiles, birds, and mammals, feed primarily or exclusively on insects. The role of insectivous birds in regulating insect populations has been recognized at least as early as the 1870s (Riley 1878). Important freshwater fisheries, including salmonids, depend on adequate availability of insect prey and commonly strike at artificial lures that resemble particular insect species. Baxter et al. (2005), Lake et al. (2007), Wipfli (1997), and Wipfli and Musslewhite (2004) reported that aquatic and terrestrial insects falling into streams from riparian vegetation are equally important for sustaining salmonid populations. Stewart and Woolbright (1996) calculated that tree frog (*Eleutherodactylus coqui*) adults, at densities of about 3300 ha^{-1}, consumed 10,000 insects ha^{-1} per night in a Puerto Rican rainforest; 17,000 pre-adult frogs ha^{-1} ate an additional 100,000 insects ha^{-1} per night. During dry periods, when the availability of insect prey declined, many frogs were emaciated and had empty guts, suggesting vulnerability of insectivores to reduced abundance of insects.

FIGURE 5.12 (SEE COLOR INSERT.)
Insect populations are regulated by a diverse assemblage of invertebrate and vertebrate (and even carnivorous plants) predators and parasites. Examples include (a) a predaceous stinkbug nymph feeding on the forest tent caterpillar, *Malacosoma disstria*; (b) spiders, such as this golden silk orb-weaver, *Nephila clavipes*, feeding on a scarab beetle, *Phyllophaga* sp.; (c) a lizard, *Anolis carolinensis*, feeding on a tipulid; and (d) a fungal pathogen sporulating from its stinkbug host. [Photos (c) and (d) from Schowalter, T. D., *Insect Ecology: An Ecosystem Approach*, Elsevier, San Diego, 2011. With permission.]

FIGURE 5.12 (CONTINUED)

Insect populations are regulated by a diverse assemblage of invertebrate and vertebrate (and even carnivorous plants) predators and parasites. Examples include (a) a predaceous stinkbug nymph feeding on the forest tent caterpillar, *Malacosoma disstria*; (b) spiders, such as this golden silk orb-weaver, *Nephila clavipes*, feeding on a scarab beetle, *Phyllophaga* sp.; (c) a lizard, *Anolis carolinensis*, feeding on a tipulid; and (d) a fungal pathogen sporulating from its stinkbug host. [Photos (c) and (d) from Schowalter, T. D., *Insect Ecology: An Ecosystem Approach*, Elsevier, San Diego, 2011. With permission.]

5.5 Ecosystem Development

Ecosystems typically develop through a process called ecological succession. Primary succession is initiated on new substrates resulting from volcanic deposits, glacial movement, sediment deposition, or erosion, whereas secondary succession is initiated following disturbances that leave legacies of soils, seed banks, and surviving individuals from the previous community. Succession is exemplified by the sequential colonization and replacement of species on abandoned cropland: weedy annual to perennial grass to forb, to shrub, to shade-intolerant tree and, finally, to shade-tolerant tree stages (Odum 1969). Succession on new substrates or following fire or other disturbances shows a similar sequence of stages (Salo et al. 1986). Pioneer species typically are short but widespread; have high light, water, and nutrient requirements; and reproduce rapidly in order to produce offspring capable of colonizing other disturbed sites before larger, more competitive species overtop and suppress them. They play an important role in securing and protecting soil from erosion, as any home gardener who hand pulls "weeds" can attest. Later successional vegetation typically is taller, often more resistant to disturbances, has lower light, water, and nutrient requirements, and is capable of competitively suppressing the earlier vegetation.

Succession can take varying lengths of time to produce a community that is self-perpetuating—up to centuries for some forests (Christensen et al. 2000). Although successional transitions can be explained as the replacement of less competitive or less tolerant individuals by more competitive or more tolerant individuals, insects frequently facilitate or retard transitions by focusing herbivory on stressed hosts and accelerating their decline (Schowalter 1981, 2011; Torres 1992; Davidson 1993). For example, bark beetles are instrumental in controlling successional transitions between pine forests and later successional fir or hardwood forests (Schowalter et al. 1981a; Coleman et al. 2008). After suppressing the target species, the insect population declines (Cairns et al. 2008), often with the aid of predators and parasites that represent a higher level of regulation (see earlier).

Ironically, most plant species valued for crop production are early successional species with high demands for light, water, and nutrients. These plants are easily stressed by insufficient resources and become vulnerable to herbivorous insects that would facilitate their replacement by later successional species in the absence of protection by humans.

Arthropod communities also change during vegetative succession (Shelford 1907; Weygoldt 1969; Brown 1984). Evans (1988) found that grasshopper assemblages showed predictable changes following fire in North American grassland. The relative abundance of grass-feeding species initially increased following fire, reflecting increased grass growth, and subsequently declined as the abundance of forbs increased.

Schowalter (1995), Schowalter and Ganio (2003), and Schowalter et al. (1981b) reported that sap-sucking insects (primarily Hemiptera) and ants dominated early successional temperate and tropical forests, whereas folivores, predators, and detritivores dominated later successional forests. This trend likely reflects the abundance of young, succulent tissues with high translocation rates that favor sapsuckers and tending ants during early regrowth.

Animals affect succession in a variety of ways (MacMahon 1981; Schowalter 1981; Davidson 1993; Willig and McGinley 1999; Quesada et al. 2009), and Blatt et al. (2001) showed that incorporation of herbivory into an old field successional model helped to explain the multiple successional pathways that could be observed. Generally, herbivory and granivory during early seres halt or advance succession (Schowalter 1981; Brown 1984; Torres 1992), whereas herbivory during later seres halts or reverses succession (Davidson 1993). Torres (1992) reported that a sequence of Lepidoptera species appeared and reached outbreak levels on a corresponding sequence of early successional plant species during the first six months following Hurricane Hugo (1989) in Puerto Rico but disappeared after depleting their resources. Schowalter (2011) observed this process repeated following Hurricane Georges (1998). Bishop (2002) reported that insect herbivores limited the persistence and spread of early successional lupines during primary succession on Mount St. Helens following the 1980 eruption. Ants can advance succession through dispersal of plant seeds and manipulation of soil and vegetation around nest sites (Jonkman 1978; Inouye et al. 1980; Guo 1998).

Animals that construct burrows or mounds or that wallow or compact soils can kill all vegetation in small (several m²) patches and/or provide suitable germination habitat and other resources for pioneer plant species (MacMahon 1981; Andersen and MacMahon 1985). Several studies have demonstrated that ant and termite nests create unique habitats, typically with elevated nutrient concentrations, that support distinct vegetation when the colony is active and facilitated succession following colony abandonment (King 1977a, b; Brenner and Silva 1995; Garrettson et al. 1998; Guo 1998; Lesica and Kannowski 1998; Mahaney et al. 1999). Jonkman (1978) reported that the collapse of leaf-cutter ant, *Atta vollenweideri*, nests following colony abandonment provided small pools of water that facilitated plant colonization and accelerated development of woodlands in South American grasslands.

Predators also can affect succession. Hodkinson et al. (2001) observed that spiders often are the earliest colonizers of glacial moraine or other newly exposed habitats. Spiderwebs trap living and dead prey and other organic debris. In systems with low organic matter, nutrient availability, and microbial decomposer activity, spider digestion of prey may accelerate nutrient incorporation into the developing ecosystem. Spiderwebs are composed of structural proteins and may distribute nitrogen over the surface. In addition, webs physically stabilize the surface and increase surface moisture through condensation from the atmosphere. These effects of spiders may facilitate development of cyanobacterial crusts and early successional vegetation.

5.6 Differences among Ecosystems

Ecosystems vary in their regulatory ability. Kratz et al. (1995) compiled data on the variability of climatic, edaphic, plant, and animal variables from 12 Long-Term Ecological Research (LTER) Sites in North America, representing forest, grassland, desert, lake, and stream ecosystems. Unfortunately, comparison was limited because different variables and measurement techniques were represented among these sites. Nevertheless, Kratz et al. offered several important conclusions concerning variability.

First, the level of species combination (e.g., species, family, guild, total plants or animals) had a greater effect on observed variability in community structure than did spatial or temporal extent of data. For plant parameters, species- and guild-level data were more variable than were data for total plants; for animal parameters, species-level data were more variable than were guild-level data, and both were more variable than were total animal data. Detection of long-term trends or spatial patterns depends on data collection for parameters sufficiently sensitive to show significant differences but not so sensitive that their variability hinders detection of differences. Several studies have shown that functional diversity is more important for stability than is species diversity (Naeem 1998). However, species diversity is important for ensuring continuity of ecological functions during environmental changes that favor some species over others (Tilman and Downing 1994; Naeem 1998). Loss of any species could jeopardize this continuity.

Second, spatial variability exceeded temporal variability. This result indicates that individual sites are inadequate to describe the range of variation among ecosystems within a landscape. Variability must be examined over larger spatial scales. Edaphic data were more variable than were climatic data, indicating high spatial variation in substrate properties, whereas common weather across landscapes homogenizes microclimatic conditions. This result also could be explained as the result of greater biotic modification of climatic variables, compared with substrate variables (see following paragraphs).

Third, biotic data were more variable than were climatic or edaphic data. Organisms can exhibit exponential responses to incremental changes in abiotic conditions (see Chapter 4). The ability of animals to move and alter their spatial distribution quickly in response to environmental changes is reflected in greater variation in animal data compared with plant data. However, animals also have greater ability to hide or escape sampling devices.

Finally, two sites, a desert and a lake, provided a sufficiently complete array of biotic and abiotic variables to permit comparison. These two ecosystem types represent contrasting properties. Deserts are exposed to highly variable and harsh abiotic conditions but are interconnected within landscapes, whereas lakes exhibit relatively constant abiotic conditions (buffered from thermal change by mass and latent heat capacity of water, from pH change by bicarbonates, and from biological invasions by their isolation) but are isolated by land barriers. Comparison of variability between these contrasting ecosystems supported the hypothesis that deserts are more variable than lakes among years, but lakes are more variable than deserts among sites.

Ecosystems also vary in range and stability with regard to the particular services they can provide. Forests are particularly valued as sources of fresh water because their accumulated biomass in plants and soil maximize interception and channeling of precipitation through layers of foliage (reducing impact on the ground), litter, and soil (filtering water and minimizing sedimentation in streams), processes critical to providing clear water for downstream uses. Grasslands have rich soils recognized for millennia as particularly suitable for crop and livestock production, as well as herds of game animals. Deserts suffer from high exposure to abiotic conditions but produce adequate food resources for hunter-gatherer societies. However, crop production in deserts or marginal ecosystems requires irrigation that not only provides stagnant water for mosquito production but also tends to increase soil salinity, eventually limiting crop production.

5.7 Predicting Change

Effects of changing environmental conditions, human activities, or biodiversity are difficult to predict. We have sufficient data to anticipate insect responses to environmental changes and anthropogenic or natural disturbances (see Chapter 4). For example, most recent insect outbreaks can be attributed to anthropogenic changes in density of host plants through planting or fire suppression (see Chapter 4). However, we cannot accurately predict which insect species will respond to a particular change or the net effect of their responses for ecosystem services. The difficulty lies in the complexity and feedback loops of even the simplest ecosystems and the increasingly chaotic effects of multiple interacting factors over time (Lorenz 1993).

Consider that each of the thousands of insect species in any ecosystem responds in a unique way, based on its life history attributes, to the particular combination of parameter values at a point in time, including interactions with other species. Some parameter values constitute filters that limit future responses. For example, a drought that eliminates intolerant species prevents these species from sustaining an ecological function during a subsequent wet period (Chase 2007). Models of ecosystems necessarily simplify diversity of species and interactions. A community matrix of interactions among 1,000 species would have one million cells, each varying in strength in response to multiple abiotic and biotic factors (Schowalter 2011). As Lorenz (1993) noted for prediction of climate, small changes in some input parameters can cause large changes in response variables. Therefore, we would need precise information on the current values of millions of parameters, as well as their effects on each other, in order to accurately predict effects of changes in environmental conditions or species abundances for ecosystem services.

Furthermore, we know from archeological evidence that the legacy of human influence on ecosystems can persist for hundreds to thousands of years. Although climate has varied over the past several thousand years, early agricultural practices in the Middle East exacerbated widespread desertification of this once "Fertile Crescent." Maya influence on forest composition surrounding ruins of urban centers remains evident after at least 1,000 years (Ross and Rangel 2011). Overexploitation of forest resources by the Late Classical Maya, after careful forest stewardship for at least 1,200 years, likely contributed to the collapse of the most advanced civilization in the New World (Lentz and Hockaday 2009; Cook et al. 2012; Turner and Sabloff 2012). Agricultural terraces at least several hundred years old are evident in northeastern China and southwestern North America (Schowalter 2011). Selective harvests in forests induce changes in forest floor and stream processes that remain significant after 50–100 years (Harding et al. 1998; Schowalter et al. 2003; Summerville et al. 2009).

Anthropogenic extermination of species can threaten ecosystem services. The disappearance of the North American "megafauna" (the mastodons, giant ground sloths, horses, saber-toothed tigers, etc., that made North America resemble the modern African Serengeti) about 13,000 years ago coincides with the spread of Aboriginal Americans across the continent (Janzen and Martin 1982). This extinction event eliminated the major dispersers of large-seeded tree species, which now are restricted in dispersal ability. More recently, extermination of the dodo from Mauritius in 1680 coincided with the age (300–400 years) of the last naturally regenerated tambalacoque trees, *Sideroxylon sessiliflorum* (= *Calvaria major*). When *S. sessiliflorum* seeds were force-fed to turkeys (approximately the size of the dodo), the seed coats were sufficiently abraded during gut passage to permit germination, demonstrating a potential role of the dodo in dispersal and survival of this once dominant tree (Temple 1977). Although the primacy of the dodo's role in *S. sessiliflorum* survival has been challenged (Witmer and Cheke 1991), it appears that *S. sessiliflorum* and other plant species now require human assistance for their survival. Similarly, disappearance of native ant species as a result of habitat fragmentation or competition from invasive ant species (Suarez et al. 1998) threatens the survival of ant-dispersed plant species. Services provided by these species now depend on human intervention.

Ecosystem ecologists have only recently (during the 1960s) initiated the long-term studies necessary to evaluate effects of experimental manipulation on ecosystem processes and structure (Brown et al. 2001). However, funding and logistical problems prevent manipulation of more than a few factors or large enough plots in either simple or complex ecosystems (Ewel 1986; Ewel et al. 1991; Nepstad et al. 2007; Schowalter et al. 2011). Consequently, models based on state-of-the-art knowledge of complex systems are necessarily used as surrogates for experiments (Thompson et al. 2012). Although models have proven useful for predicting general trends (such as global warming or herbivore effects on primary production (Parton et al. 1993; Throop et al. 2004), the cumulative effects of small differences between modeled and real parameter values can lead to significant differences in predicted temperature and rainfall and their consequences for ecosystem processes and services for particular regions (Lorenz 1993).

5.8 Summary

Ecosystem services do not just appear for our benefit but rather are the result of complex interactions among thousands of species and the abiotic conditions constituting an ecosystem, as well as the interactions among ecosystems that import and export biotic and abiotic resources from other ecosystems. Disruption of key processes, such as net primary production that controls all ecosystem services, and herbivory, predation, and decomposition that regulate primary production and the cycling of limiting water and nutrients, can undermine ecosystem ability to provide services on which we depend. Insects play critical roles in all these processes, and indiscriminant efforts to control insects can disrupt key processes.

Traditionally, ecosystems have been viewed as the community of organisms and the abiotic conditions of a particular site. Such ecosystems are characterized by a particular structure—represented by the distribution of physical features and biomass—and function—represented by the net effects of food web interactions on the degree of retention and internal cycling of various chemicals within the ecosystem. Structure can vary from thin lichen coverings over rocks in early stages of primary succession to the tall, dense, multilayered canopies of tropical forests. Some processes (such as consumption and decomposition) are common to all ecosystems, whereas others (such as primary production) are limited to ecosystems with sufficient light to support photosynthesis.

Our perspective of ecosystems has broadened as we recognize that ecosystems interact across landscapes. Species from early successional ecosystems provide colonists for newly disturbed ecosystems. Some ecosystems depend largely or completely on resources exported from others (e.g., stream ecosystems are heavily subsidized by terrestrial detritus washed in during precipitation events and much of the energy and nutrients in rivers is washed down from upstream).

Various species interact directly through feeding relationships (herbivore-plant and predator-prey interactions) and indirectly through complex feedback loops. Plants wounded by herbivores release volatile chemicals that attract predators that prey on the herbivores; predation reduces herbivory and thereby indirectly increases primary production. Energy and materials transferred during consumption drive processes that support ecosystem services. Transfer from a plant to an herbivore to a predator constitutes a food chain. The network of food chains involving all species in the community is a food web. Disruption of food webs, such as through species loss or invasion, has consequences that are difficult to predict because of the large number of direct and indirect effects, each with cascading impacts on other species and the processes they influence.

Ecosystems vary in their capacity to regulate primary production, soil development, energy and nutrient fluxes, and climate. More stable ecosystems provide more consistent supply of services. Effects of changing environmental conditions on ecosystem processes and services are difficult to predict because of the large number of interactions, each with more or less subtle responses that can cascade through the food web to cause large effects. However, we can measure the effects of such changes on ecosystem services after they occur as a way to anticipate the effects of future changes, as described in the next few chapters.

References

Agee, J. K. 1993. *Fire Ecology of Pacific Northwest Forests.* Washington, DC: Island Press.

Alfaro, R. I. and R. F. Shepherd. 1991. Tree-ring growth of interior Douglas-fir after one year's defoliation by Douglas-fir tussock moth. *Forest Science* 37: 959–964.

Allan, B. F., R. B. Langerhans, W. A. Ryberg, W. J. Landesman, N. W. Griffin, R. S. Katz, B. J. Oberle, M. R. Schutzenhofer, K. N. Smyth, A. de St. Maurice, et al. 2009. Ecological correlates of risk and incidence of West Nile virus in the United States. *Oecologia* 158: 699–708.

Andersen, D. C. and J. A. MacMahon. 1985. Plant succession following the Mount St. Helens volcanic eruption: Facilitation by a burrowing rodent, *Thomomys talpoides. American Midland Naturalist* 114: 63–69.

Arnfield, A. J. 2003. Two decades of urban climate research: A review of turbulence, exchanges of energy and water, and the urban heat island. *International Journal of Climatology* 23: 1–26.

Axelrod, R. and W. D. Hamilton. 1981. The evolution of cooperation. *Science* 211: 1390–1396.

Bailey, J. K., J. A. Schweitzer, B. J. Rehill, R. L. Lindroth, G. D. Martinsen, and T. G. Whitham. 2004. Beavers as molecular geneticists: A genetic basis to the foraging of an ecosystem engineer. *Ecology* 85: 603–608.

Balanyá, J., J. M. Oller, R. B. Huey, G. W. Gilchrist, and L. Serra. 2006. Global genetic change tracks global climate warming in *Drosophila subobscura. Science* 313: 1773–1775.

Bale, J. S., G. J. Masters, I. D. Hodkinson, C. Awmack, T. M. Bezemer, V. K. Brown, J. Butterfield, A. Buse, J. C. Coulson, J. Farrar, et al. 2002. Herbivory in global climate change research: Direct effects of rising temperature on insect herbivores. *Global Change Biology* 8: 1–16.

Batzer, D. P., C. R. Pusateri, and R. Vetter. 2000. Impacts of fish predation on marsh invertebrates: Direct and indirect effects. *Wetlands* 20: 307–312.

Baxter C. V., K. D. Fausch, and W. C. Saunders. 2005. Tangled webs: Reciprocal flows of invertebrate prey link streams and riparian zones. *Freshwater Biology* 50: 201–220.

Belnap, J. and D. A. Gillette. 1998. Vulnerability of desert biological soil crusts to wind erosion: The influences of crust development, soil texture, and disturbance. *Journal of Arid Environments* 39: 133–142.

Belovsky, G. E. and J. B. Slade. 2000. Insect herbivory accelerates nutrient cycling and increases plant production. *Proceedings of the National Academy of Sciences USA* 97: 14412–14417.

Bishop, J. G. 2002. Early primary succession on Mount St. Helens: Impact of insect herbivores on colonizing lupines. *Ecology* 83: 191–202.

Blatt, S. E., J. A. Janmaat, and R. Harmsen. 2001. Modelling succession to include a herbivore effect. *Ecological Modelling* 139: 123–136.

Boulton, A. J., C. G. Peterson, N. B. Grimm, and S. G. Fisher. 1992. Stability of an aquatic macroinvertebrate community in a multiyear hydrologic disturbance regime. *Ecology* 73: 2192–2207.

Bradbury, R. 1952. Sound of thunder. *Colliers Magazine* June 28: 20–21, 60–61.

Bradshaw, W. E. and C. M. Holzapfel. 2001. Genetic shift in photoperiodic response correlated with global warming. *Proceedings of the National Academy of Sciences USA* 98: 14509–14511.

Brenner, A. G. F. and J. F. Silva. 1995. Leaf-cutting ants and forest groves in a tropical parkland savanna of Venezuela: Facilitated succession? *Journal of Tropical Ecology* 11: 651–669.

Brower, L. P., W. N. Ryerson, L. L. Coppinger, and S. C. Glazier. 1968. Ecological chemistry and the palatability spectrum. *Science* 161: 1349–1351.

Brown, J. H., T. G. Whitham, S. K. M. Ernest, and C. A. Gehring. 2001. Complex species interactions and the dynamics of ecological systems: Long-term experiments. *Science* 293: 643–650.

Brown, M., T. A. Black, Z. Nesic, V. N. Foord, D. L. Spittlehouse, A. L. Fredeen, N. J. Grant, P. J. Burton, and J. A. Trofymow. 2010. Impact of mountain pine beetle on the net ecosystem production of lodgepole pine stands in British Columbia. *Agricultural and Forest Meteorology* 150: 254–264.

Brown, V. K. 1984. Secondary succession: Insect-plant relationships. *BioScience* 34: 710–716.

Cairns, D. M., C. L. Lafon, J. D. Waldron, M. Tchakerian, R. N. Coulson, K. D. Klepzig, A. G. Birt, and W. Xi. 2008. Simulating the reciprocal interaction of forest landscape structure and southern pine beetle herbivory using LANDIS. *Landscape Ecology* 23: 403–415.

Calder, I. R. 2001. Canopy processes: Implications for transpiration, interception, and splash induced erosion, ultimately for forst management and water resources. *Plant Ecology* 153: 203–214.

Chamberlain, K., E. Guerrieri, F. Pennacchio, J. Pettersson, J. A. Pickett, G. M. Poppy, W. Powell, L. J. Wadhams, and C. M. Woodcock. 2001. Can aphid-induced plant signals be transmitted aerially and through the rhizosphere? *Biochemical Systematics and Ecology* 29: 1063–1074.

Chase, J. M. 2007. Drought mediates the importance of stochiastic community assembly. *Proceedings of the National Academy of Sciences USA* 104: 17430–17434.

FIGURE 1.1

Food production is a primary ecosystem service necessary to support human populations. (a) Sorghum is a major crop around the world. (b) Manual planting of sugarcane in South Africa.

FIGURE 1.2
Fresh water is a primary ecosystem service necessary to support human populations. (a) Iguaçu Falls, South America. (b) Fishing and laundry from the banks of the Songhua River in China.

FIGURE 1.3
Pollination by insects is a critical ecosystem service that supports reproduction of a large proportion of plant species, especially in the tropics. (From Schowalter, T.D., *Insect Ecology: An Ecosystem Approach*, Elsevier, San Diego, 2011. With permission.)

FIGURE 1.4
Patchwork of old-growth and harvested sites in western Oregon. Each patch can be treated as an ecosystem.

FIGURE 1.5
Interconnected ecosystems. A swamp represents two integrated ecosystems, with terrestrial material providing input to the aquatic ecosystem and flooding providing input to the terrestrial ecosystem. Coastal swamps and marshes also represent important filters that slow water flow and capture sediments and nutrients that otherwise would be exported to estuaries and oceans.

FIGURE 1.6
Examples of disturbances to illustrate variation in type, magnitude, extent, and contrast between disturbed and surrounding landscape patches. (a) Fire in oak savanna (note sites of high and low flame height and intensity); (b) hurricane effect on coastal deciduous forest; (c) landslide resulting from heavy rainfall; (d) volcanic eruption (note fresh lava flow on left and zones of burning and exposure to fumes in vegetation fragment). (From Schowalter, T. D., *Annual Review of Entomology*, 57, 1–20, 2012. With permission.)

FIGURE 1.7
Cockroach in Dominican amber estimated to be approximately 25 million years old.

FIGURE 2.2
Pollination services by honey bees and other insects are necessary for 35% of global crop production.

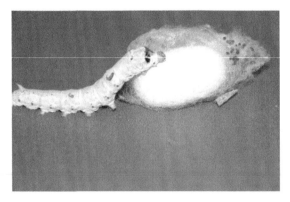

FIGURE 2.3
Silkworm caterpillar and cocoon. Silkworms remain the primary source of commercial silk production.

FIGURE 2.4
Cochineal scale insects on a prickly pear cactus. The scales remain an important source of carmine dye used in cosmetics and food coloring.

FIGURE 2.5
Giant silk moth cocoon rattles worn by Zulu dancer in South Africa. The caterpillars provide valued food, and the pebble-filled cocoons are an important cultural instrument. (From Schowalter, T. D., *Insect Ecology: An Ecosystem Approach*, Elsevier/Academic, San Diego, 2011. With permission.)

FIGURE 2.6
Chinese cricket box to hold fighting crickets. Note the ornate inlay, hidden sliding doors to each of three cricket compartments, and plug for holes used to insert rice grains as food into each compartment. The box is the size of a deck of playing cards for ease of transport in a shirt pocket.

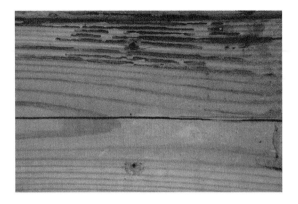

FIGURE 2.7
Termite damage to wood beams.

FIGURE 2.8
Trail sign in Fushan Botanical Garden, Taiwan, warning hikers to be alert for venomous giant centipedes, hornets, and snakes.

FIGURE 2.9
The saddleback caterpillar, *Sibene stimulea,* is perhaps the most potent stinging caterpillar in North America.

FIGURE 3.1
Complete metamorphosis is exemplified by the monarch butterfly, *Danaus plexippus*, with nearly completed development of adult features visible through the pupal exoskeleton (chrysalis).

FIGURE 3.2
Phoretic mites attached to legs of a scarab beetle. The flightless mites disperse to new dung pads by hitching a ride on dung beetles capable of dispersing to new dung pads over long distances. (Photo courtesy of A. Tishechkin. From Schowalter, T. D., *Insect Ecology: An Ecosystem Approach*, Elsevier/Academic, San Diego, 2011. With permission.)

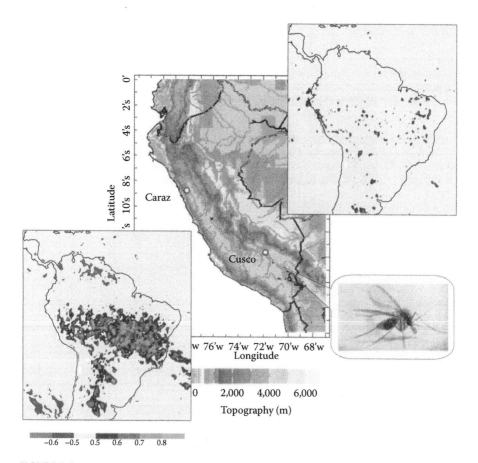

FIGURE 3.4

Topography of Peru (center), comparison of TRMM TMI rainfall at Cuzco (lower left) and Caraz (upper right) relative to their surroundings and the sand fly, *Lutzomyia verrucarum*, vector of bartonellosis that shows increased spread associated with higher rainfall during El Niño events. (From Zhou, J., W. K.-M. Lau, P. M. Masuoka, R. G. Andre, J. Chamberlin, P. Lawyer, and L. W. Laughlin, *EOS, Transactions, American Geophysical Union*, 83, 157, 160–161, 2002. With permission.)

FIGURE 3.6
Representative insect defenses against predators. (a) Cryptic coloration (camouflage) by a grasshopper, *Syrbula admirabilis*, from western North America. (b) Deception by a Taiwanese moth, *Phalera angustipennis*, that resembles a piece of broken wood, and (c) combination of camouflage, warning stripes, and venomous spines in the Io moth caterpillar, *Automeris io*, from eastern North America. (d) Brightly colored insects, such as the red and black cicada, *Huechys sanguinea*, from Southeast Asia attract our attention and appreciation, but such coloration either advertises toxic or distasteful properties or mimics a distasteful model to would-be predators. [Photo (d) from Schowalter, T. D., *Insect Ecology: An Ecosystem Approach*, Elsevier, San Diego, 2011. With permission.]

FIGURE 3.6 (CONTINUED)
Representative insect defenses against predators. (a) Cryptic coloration (camouflage) by a grasshopper, *Syrbula admirabilis*, from western North America. (b) Deception by a Taiwanese moth, *Phalera angustipennis*, that resembles a piece of broken wood, and (c) combination of camouflage, warning stripes, and venomous spines in the Io moth caterpillar, *Automeris io*, from eastern North America. (d) Brightly colored insects, such as the red and black cicada, *Huechys sanguinea*, from Southeast Asia attract our attention and appreciation, but such coloration either advertises toxic or distasteful properties or mimics a distasteful model to would-be predators. [Photo (d) from Schowalter, T. D., *Insect Ecology: An Ecosystem Approach*, Elsevier, San Diego, 2011. With permission.]

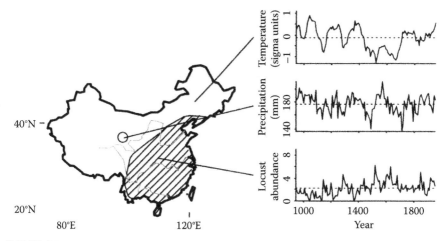

FIGURE 4.1
One thousand years of locust abundance and climate data from China. The present distribution of *Locusta migratoria manilensis* is shown by hatched lines on the map. Large rivers (blue lines) from north to south are the Yellow, Yangtze, and Pearl. Temperature data are a composite index for all of China. Precipitation data are from northeastern Qinghai (circle). Locust data are for all of China. (From Stige, L. C., K.-S. Chan, Z. Zhang, D. Frank, and N. C. Stenseth, *Proceedings of the National Academy of Sciences USA*, 104, 16188–16193, 2007. With permission.)

FIGURE 4.4
Predation tends to concentrate on the most abundant prey, including conspecifics under crowded conditions, as in the case of this dragonfly preying on another dragonfly.

FIGURE 5.1

Ecosystem structure varies widely among terrestrial ecosystem types. (a) Deserts have harsh environmental conditions with relatively small biomass and structure, but dominant plants, such as organ pipe (*Stenocereus therberi*) and saguaro (*Carnegiea gigantea*) cacti, often are hundreds of years old. (b) Grasslands and savannas have moderate environmental conditions, but frequent fire prevents most tree development. Intermediate plant biomass is able to reproduce aboveground plant parts quickly following loss, a characteristic necessary to support the large biomass of herbivores, such as the white rhinoceros (*Ceratotherium simum*) and nearly equal biomass of less conspicuous invertebrates. Rich soils are particularly valued for agriculture. (c) Forests have the largest biomass and most complex structure. Note the heavily fire-scarred base of the giant sequoia, *Sequoiadendron giganteum*, which depends on infrequent fire to maintain optimal soil conditions and minimal competition for water and nutrients.

FIGURE 5.4
Representative aquatic ecosystems. (a) High elevation stream with steep gradient, high flow rate, alternating riffles and pools, and relatively limited exposure to sunlight. (b) Beaver pond created by impediment to streamflow with relatively low gradient, low flow rate, and high exposure to sunlight. (From Schowalter, T. D., *Insect Ecology: An Ecosystem Approach*, Elsevier, San Diego, 2011. With permission.)

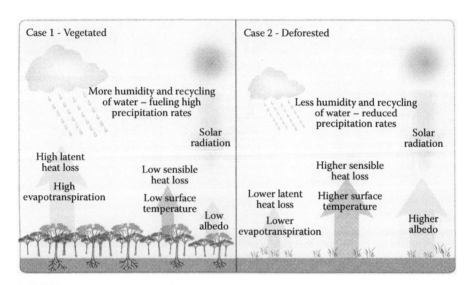

FIGURE 5.6
Diagrammatic representation of the effects of vegetation on climate and atmospheric variables. The capacity of vegetation to modify climate depends on vegetation density and vertical height and complexity. (From Foley, J. A., M. H. Costa, C. Delire, N. Ramankutty, and P. Snyder, *Frontiers in Ecology and the Environment*, 1, 38–44, 2003. With permission.)

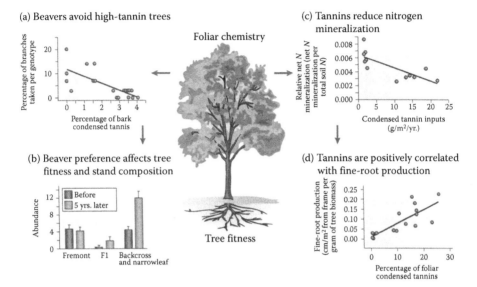

(a) Beavers avoid high-tannin trees

Foliar chemistry

(c) Tannins reduce nitrogen mineralization

(b) Beaver preference affects tree fitness and stand composition

Tree fitness

(d) Tannins are positively correlated with fine-root production

FIGURE 5.7

Selection for species phenotype affects interactions with other species that in turn feedback to affect fitness of the individuals that produced that phenotype. In this example, condensed tannin phenotype in poplar affects herbivore foraging, nutrient turnover, and nutrient acquisition. (a) and (b) Beavers, *Castor canadensis*, select trees with low condensed tannin concentrations, which leads to increased abundance of trees with higher condensed tannin concentrations and reduced abundance of trees with lower concentrations, affecting other species that depend on this tree for survival. (c) and (d) Increased foliar tannin concentration inhibits nitrogen mineralization from litter and soil, requiring the tree to produce more fine roots to acquire more limited soil nitrogen. (From Whitham, T. G., J. K. Bailey, J. A. Schweitzer, S. M. Shuster, R. K. Bangert, C. J. LeRoy, E. V. Lonsdorf, G. J. Allan, S. P. DiFazio, B. M. Potts, et al., *Nature Reviews Genetics*, 7, 510–523. With permission.) [(a) Data from Bailey, J. K., J. A. Schweitzer, B. J. Rehill, R. L. Lindroth, G. D. Martinsen, and T. G. Whitham, *Ecology*, 85, 603–608, 2004. With permission. (c) Data from Schweitzer, J. A., J. K. Bailey, B. J. Rehill, G. D. Martinsen, S. C. Hart, R. L. Lindroth, P. Keim, and T. G. Whitham, *Ecology Letters*, 7, 127–134, 2004. With permission. (d) Data from Fischer, D. G., S. C. Hart, B. J. Rehill, R. L. Lindroth, P. Keim, and T. G. Whitham, *Oecologia*, 149, 668–675, 2006. With permission.]

FIGURE 5.10
(a) The relatively arid interior forest region of North America was characterized by open cano-pied forests dominated by drought- and fire-tolerant pines and by sparse understories prior to fire suppression beginning in the late 1800s. (b) Fire suppression has transformed forests into dense, multistoried ecosystems stressed by competition for water and nutrients. (From Goyer, R. A., M. R. Wagner, and T. D. Schowalter, *Journal of Forestry*, 96, 29–33, 1998. With permission.)

FIGURE 5.11
Herbivore modification of succession in central Sierran mixed conifer ecosystems during 1998. Understory white fir (*Abies concolor*), the late successional dominant, is increasingly stressed by competition for water in this arid forest type. An outbreak of the Douglas-fir tussock moth, *Orgyia pseudotsugata*, has completely defoliated the white fir (gray trees, bottom), restoring the ecosystem to a more stable condition dominated by earlier successional, drought- and fire-tolerant, sequoias and pines (dark, foliated, trees). (From Schowalter, T. D., *Insect Ecology: An Ecosystem Approach*, Elsevier, San Diego, 2011. With permission.)

FIGURE 5.12
Insect populations are regulated by a diverse assemblage of invertebrate and vertebrate (and even carnivorous plants) predators and parasites. Examples include (a) a predaceous stinkbug nymph feeding on the forest tent caterpillar, *Malacosoma disstria*; spiders, such as this golden silk orb-weaver, *Nephila clavipes*, feeding on a scarab beetle, *Phyllophaga* sp.; (c) a lizard, *Anolis carolinensis*, feeding on a tipulid; and (d) a fungal pathogen sporulating from its stinkbug host. [Photos (c) and (d) from Schowalter, T. D., *Insect Ecology: An Ecosystem Approach*, Elsevier, San Diego, 2011. With permission.]

FIGURE 5.12 (CONTINUED)

Insect populations are regulated by a diverse assemblage of invertebrate and vertebrate (and even carnivorous plants) predators and parasites. Examples include (a) a predaceous stinkbug nymph feeding on the forest tent caterpillar, *Malacosoma disstria*; spiders, such as this golden silk orb-weaver, *Nephila clavipes*, feeding on a scarab beetle, *Phyllophaga* sp.; (c) a lizard, *Anolis carolinensis*, feeding on a tipulid; and (d) a fungal pathogen sporulating from its stinkbug host. [Photos (c) and (d) from Schowalter, T. D., *Insect Ecology: An Ecosystem Approach*, Elsevier, San Diego, 2011. With permission.]

FIGURE 6.1
Contrast between post-disturbance conditions: (a) clear-cut harvest, (b) hurricane, and (c) fire. Note the relatively intact soil conditions and scattered surviving vegetation following natural disturbances, compared with the highly disturbed and fully exposed surface of the clear-cut.

FIGURE 6.2
Contrast between edges in anthropogenic versus natural landscapes. (a) Straight edges among patches in anthropogenic landscape. (b) Gradual transitions among patches in natural landscape.

FIGURE 6.3
Deforestation in Panama. Removal of tropical rainforest cover has exposed soil to solar heating and severe erosion, leading to continued ecosystem deterioration and to altered regional temperature and precipitation patterns. (From Schowalter, T. D., *Insect Ecology: An Ecosystem Approach*, San Diego, Elsevier, 2011.)

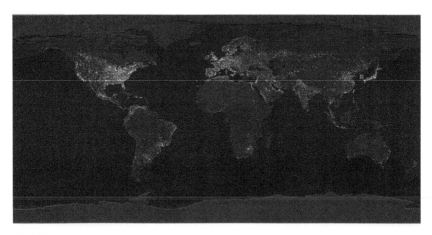

FIGURE 6.5
The global distribution of light. The brightest areas are the most urbanized but not necessarily the most populated; for example, compare western Europe with China. Cities are concentrated along coastlines and transportation networks. The U.S. interstate highway system appears as a lattice connecting the brighter dots of city centers. The Russian Trans-Siberian Railroad is a thin line stretching from Moscow through the center of Asia to Vladivostok. The Nile River, from the Aswan Dam to the Mediterranean Sea, is another bright thread through an otherwise dark region. Created with data from the Defense Meteorological Satellite Program (DMSP) Operational Linescan System (OLS). (Data courtesy of Marc Imhoff of NASA GSFC and Christopher Elvidge of NOAA NGDC; image courtesy of Craig Mayhew and Robert Simmon, NASA GSFC.)

FIGURE 6.7

Apple orchard established in desert. Proximity to river water permits the required irrigation, but this conversion represents a dramatic alteration of the desert ecosystem and cannot be sustained without adequate water supply.

FIGURE 6.8

Former temperate coniferous rainforest converted to urban ecosystem with virtually no remaining vegetation. Replacing vegetation cover and soil with paving greatly reduces albedo due to the preponderance of dark surfaces, thereby increasing diurnal temperature and temperature range, and eliminates water storage capacity, thereby increasing runoff and downstream flooding.

FIGURE 6.9
Urban effects on surrounding ecosystems. (a) Continued fragmentation of natural ecosystems at the urban fringe reduces remaining area for ecosystem services and increases the area affected by higher urban temperatures. (b) Connecting roads increase avenues for introduction and spread of invasive pests.

FIGURE 7.1
Extensive defoliation of soybeans by the soybean looper, *Chrysodeixis includens*, can reduce soybean production substantially.

FIGURE 7.2
Detail of jade carving with two orthopterans on cabbage, Qing Dynasty, 19th century.

FIGURE 7.3
Series of photos with manipulated variation in extent of pine crown mortality shown randomly to survey participants to evaluate visual preference for forests affected by insects or diseases. Note the progressively greater crown defoliation of the central tree from (a) through (c). (Courtesy of F. Baker. From P. Barbosa, D. K. Letourneau, and A. A. Agrawal, eds., *Insect Outbreaks Revisited*, Wiley/Blackwell, Hoboken, NJ, 2012. With permission.)

FIGURE 7.7
Insect feces collected on a 1 m^2 sheet during an outbreak of cypress leaf roller, *Archips goyerana*, and forest tent caterpillar, *Malacosoma disstria*, in cypress–tupelo swamp in southern Louisiana.

FIGURE 7.8
Termite castle formed by transportation of belowground material to build the aboveground structure. (From Schowalter, T. D., *Insect Ecology: An Ecosystem Approach*, Elsevier, San Diego, 2011. With permission.)

FIGURE 8.2
Brightly colored morpho butterflies, *Morpho* spp., delight visitors to a commercial butterfly house. Butterfly houses have become popular tourist destinations in many cities.

FIGURE 8.5
Fouling of pasture grasses by cattle dung in the absence of effective processing by dung beetles. In addition to covering forage grasses, dung also may adhere to grass blades, thereby reducing palatability (inset).

Chase, J. M., M. A. Leibold, A. L. Downing, and J. B. Shurin. 2000. The effects of productivity, herbivory, and plant species turnover in grassland food webs. *Ecology* 81: 2485–2497.

Chittka, L., A. Gumbert, and J. Kunze. 1997. Foraging dynamics of bumble bees: Correlates of movements within and between plant species. *Behavioral Ecology* 8: 239–249.

Christensen, N. L., Jr., S. V. Gregory, P. R. Hagenstein, T. A. Heberlein, J. C. Hendee, J. T. Olson, J. M. Peek, D. A. Perry, T. D. Schowalter, K. Sullivan, et al. 2000. *Environmental Issues in Pacific Northwest Forest Management.* Washington, DC: National Academy Press.

Christenson, L. M., G. M. Lovett, M. J. Mitchell, and P. M. Groffman. 2002. The fate of nitrogen in gypsy moth frass deposited to an oak forest floor. *Oecologia* 131: 444–452.

Clark, K. L., N. Skowronski, and J. Hom. 2010. Invasive insects impact forest carbon dynamics. *Global Change Biology* 16: 88–101.

Classen, A. T., S. K. Chapman, T. G. Whitham, S. C. Hart, and G. W. Koch. 2007. Genetic-based plant resistance and susceptibility traits to herbivory influence needle and root litter nutrient dynamics. *Journal of Ecology* 95: 1181–1194.

Classen, A. T., S. C. Hart, T. G. Whitham, N. S. Cobb, and G. W. Koch. 2005. Insect infestations linked to changes in microclimate: Important climate change implications. *Soil Science Society of America Journal* 69: 2049–2057.

Coleman, D. C., D. A. Crossley, Jr., and P. F. Hendrix. 2004. *Fundamentals of Soil Ecology,* 2nd ed. Amsterdam: Elsevier.

Coleman, T. W., S. R. Clarke, J. R. Meeker, and L. K. Rieske. 2008. Forest composition following overstory mortality from southern pine beetle and associated treatments. *Canadian Journal of Forest Research* 38: 1406–1418.

Cook, B. I., K. J. Anchukaitis, J. O. Kaplan, M. J. Puma, M. Kelley, and D. Gueyffier. 2012. Pre-Columbian deforestation as an amplifier of drought in Mesoamerica. *Geophysical Research Letters* 39: L16706.

Council for Agricultural Science and Technology. 2011. *Carbon Sequestration and Greenhouse Gas Fluxes in Agriculture: Challenges and Opportunities.* Ames, IA: Council for Agricultural Science and Technology.

Coupland, R. T. 1993. Review. In *Ecosystems of the World: Natural Grasslands,* R. T. Coupland, ed. 471–482. Amsterdam: Elsevier.

Coupland, R. T. and G. M. Van Dyne. 1979. Systems synthesis. In *Grassland Ecosystems of the World: Analysis of Grasslands and Their Uses,* R. T. Coupland, ed. 97–106. Cambridge, UK: Cambridge University Press.

Cuffney, T. F., J. B. Wallace, and G. J. Lugthart. 1990. Experimental evidence quantifying the role of benthic invertebrates in organic matter dynamics in headwater streams. *Freshwater Biology* 23: 281–299.

Cushman, J. H. and J. F. Addicott. 1991. Conditional interactions in ant-plant-herbivore mutualisms. In *Ant-Plant Interactions,* C. R. Huxley and D. F. Cutler, eds. 92–103. Oxford, UK: Oxford University Press.

Davidson, D. W. 1993. The effects of herbivory and granivory on terrestrial plant succession. *Oikos* 68: 23–35.

Denno, R. F., M. S. McClure, and J. R. Ott. 1995. Interspecific interactions in phytophagous insects: Competition reexamined and resurrected. *Annual Review of Entomology* 40: 297–331.

DeWalt, S. J. and J. Chave. 2004. Structure and biomass of four lowland neotropical forests. *Biotropica* 36: 7–19.

Dolch, R. and T. Tscharntke. 2000. Defoliation of alders (*Alnus glutinosa*) affects herbivory by leaf beetles on undamaged neighbors. *Oecologia* 125: 504–511.

Dyer, L. A. and D. K. Letourneau. 1999a. Relative strengths of top-down and bottom-up forces in a tropical forest community. *Oecologia* 119: 265–274.

Dyer, L. A. and D. K. Letourneau. 1999b. Trophic cascades in a complex terrestrial community. *Proceedings of the National Academy of Sciences USA* 96: 5072–5076.

Edelaar, P., A. M. Siepielski, and J. Clobert. 2008. Matching habitat choice causes directed gene flow: A neglected dimension in evolution and ecology. *Evolution* 62: 2462–2472.

Eggert, S. L. and J. B. Wallace. 2003. Reduced detrital resources limit *Pycnopsyche gentilis* (Trichoptera: Limnephilidae) production and growth. *Journal of the North American Benthological Society* 22: 388–400.

Endo, W., C. A. Peres, E. Salas, S. Mori, J.-L. Sanchez-Vega, G. H. Shepard, V. Pacheco, and D. W. Yu. 2010. Game vertebrate densities in hunted and nonhunted forest sites in Manu National Park, Peru. *Biotropica* 42: 251–261.

Evans, E. W. 1988. Community dynamics of prairie grasshoppers subjected to periodic fire: Predictable trajectories or random walks in time? *Oikos* 52: 283–292.

Ewel, J. J. 1986. Designing agricultural ecosystems for the humid tropics. *Annual Review of Ecology and Systematics* 17: 245–271.

Ewel, J. J., M. J. Mazzarino, and C. W. Berish. 1991. Tropical soil fertility changes under monocultures and successional communities of different structure. *Ecological Applications* 1: 289–302.

Farmer, E. E. and C. A. Ryan. 1990. Interplant communication: Airborne methyl jasmonate induces synthesis of proteinase inhibitors in plant leaves. *Proceedings of the National Academy of Sciences USA* 87: 7713–7716.

Fischer, D. G., S. C. Hart, B. J. Rehill, R. L. Lindroth, P. Keim, and T. G. Whitham. 2006. Do high-tannin leaves require more roots? *Oecologia* 149: 668–675.

Fisher, S. G. 1986. Structure and dynamics of desert streams. In *Pattern and Process in Desert Ecosystems*, W. G. Whitford, ed. 119–139. Albuquerque: University of New Mexico Press.

Fitzgerald, T. D. 1995. *The Tent Caterpillars*. Ithaca, NY: Cornell University Press.

Flamm, R. O., P. E. Pulley, and R. N. Coulson. 1993. Colonization of disturbed trees by the southern pine beetle guild (Coleoptera: Scolytidae). *Environmental Entomology* 22: 62–70.

Foley, J. A., M. T. Coe, M. Scheffer, and G. Wang. 2003a. Regime shifts in the Sahara and Sahel: Interactions between ecological and climatic systems in northern Africa. *Ecosystems* 6: 524–539.

Foley, J. A., M. H. Costa, C. Delire, N. Ramankutty, and P. Snyder. 2003b. Green surprise? How terrestrial ecosystems could affect earth's climate. *Frontiers in Ecology and the Environment* 1: 38–44.

Frank, D. A. and S. J. McNaughton. 1991. Stability increases with diversity in plant communities: Empirical evidence from the 1988 Yellowstone drought. *Oikos* 62: 360–362.

Fritz, R. S. 1983. Ant protection of a host plant's defoliator: Consequences of an ant-membracid mutualism. *Ecology* 64: 789–797.

Frost, C. J. and M. D. Hunter. 2007. Recycling of nitrogen in herbivore feces: Plant recovery, herbivore assimilation, soil retention, and leaching losses. *Oecologia* 151: 42–53.

Gardiner, T., J. Hill, and D. Chesmore. 2005. Review of the methods frequently used to estimate the abundance of Orthoptera in grassland ecosystems. *Journal of Insect Conservation* 9: 151–173.

Garrett, K. A. and C. C. Mundt. 1999. Epidemiology in mixed host populations. *Phytopathology* 89: 984–990.

Garrettson, M., J. F. Stetzel, B. S. Halpern, D. J. Hearn, B. T. Lucey, and M. J. McKone. 1998. Diveristy and abundance of understorey plants on active and abandoned nests of leaf-cutting ants (*Atta cephalotes*) in a Costa Rican rain forest. *Journal of Tropical Ecology* 14: 17–26.

Golley, F. B. 1968. Secondary productivity in terrestrial communities. *American Zoologist* 8: 53–59.

Gosz, J. R., R. T. Holmes, G. E. Likens, and F. H. Bormann. 1978. The flow of energy in a forest ecosystem. *Scientific American* 238(3): 92–102.

Goyer, R. A., M. R. Wagner, and T. D. Schowalter. 1998. Current and proposed technologies for bark beetle management. *Journal of Forestry* 96(12): 29–33.

Guo, Q. 1998. Microhabitat differentiation in Chihuahuan Desert plant communities. *Plant Ecology* 139: 71–80.

Hamilton, W. D. 1964. The genetic evolution of social behavior. I. and II. *Journal of Theoretical Biology* 7: 1–52.

Harding, J. S., E. F. Benfield, P. V. Bolstad, G. S. Helfman, and E. B. D. Jones III. 1998. Stream biodiversity: The ghost of land use past. *Proceedings of the National Academy of Sciences USA* 95: 14843–14847.

Hatfield, J. L., K. J. Boote, B. A. Kimball, L. H. Ziska, R. C. Izaurralde, D. Ort, A. M. Thomson, and D. Wolfe. 2011. Climate impacts on agriculture: Implications for crop production. *Agronomy Journal* 103: 351–370.

Hawlena, D., M. S. Strickland, M. A. Bradford, and O. J. Schmitz. 2012. Fear of predation slows plant-litter decomposition. *Science* 336: 1434–1438.

Hochwender, C. G. and R. S. Fritz. 2004. Plant genetic differences influence herbivore community structure: Evidence from a hybrid willow system. *Oecologia* 138: 547–557.

Hodkinson, I. D., S. J. Coulson, J. Harrison, J. Moores, and N. R. Webb. 2001. What a wonderful web they weave: Spiders, nutrient capture, and early ecosystem development in the high Arctic—some counter-intuitive ideas on community assembly. *Oikos* 95: 349–352.

Hooper, D. U., F. S. Chapin, III, J. J. Ewel, A. Hector, P. Inchausti, S. Lavorel, J. H. Lawton, D. M. Lodge, M. Loreau, S. Naeem, B. et al. 2005. Effects of biodiversity on ecosystem functioning: A concensus of current knowledge. *Ecological Monographs* 75: 3–35.

Hsu, J.-C., D. S. Haymer, W.-J. Wu, and H.-T. Feng. 2006. Mutations in the acetylcholinesterase gene of *Bactrocera dorsalis* associated with resistance organophosphorus insecticides. *Insect Biochemistry and Molecular Biology* 38: 396–402.

Hsu, J.-C., W.-J. Wu, D. S. Haymer, H.-Y. Liao, and H.-T. Feng. 2008. Alterations of the acetylcholinesterase enzyme in the oriental fruit fly *Bactrocera dorsalis* are correlated with resistance to the organophosphate insecticide fenitrothion. *Insect Biochemistry and Molecular Biology* 38: 146–154.

Hunter, M. D. 1987. Opposing effects of spring defoliation on late season oak caterpillars. *Ecological Entomology* 12: 373–382.

Inouye, R. S., G. S. Byers, and J. H. Brown. 1980. Effects of predation and competition on survivorship, fecundity, and community structure of desert annuals. *Ecology* 61: 1344–1351.

Izaurralde, R. C., A. M. Thomson, J. A. Morgan, P. A. Fay, H. W. Polley, and J. L. Hatfield. 2011. Climate impacts on agriculture: Implicatons for forage and rangeland production. *Agronomy Journal* 103: 371–381.

Jactel, H. and E. G. Brockerhoff. 2007. Tree diversity reduces herbivory by forest insects. *Ecology Letters* 10: 835–848.

Jandér, K. C. and E. A. Herre. 2010. Host sanctions and pollinator cheating in the fig tree-fig wasp mutualism. *Proceedings of the Royal Society B* 277: 1481–1488.

Janssen, R. H. H., M. B. J. Meinders, E. H. van Nes, and M. Scheffer. 2008. Microscale vegetation-soil feedback boosts hysteresis in a regional vegetation-climate system. *Global Change Biology* 14: 1104–1112.

Janzen, D. H. and P. S. Martin. 1982. Neotropical anachronisms: The fruits the gomphotheres ate. *Science* 215: 19–27.

Jepsen, J. U., S. B. Hagen, R. A. Ims, and N. G. Yaccoz. 2008. Climate change and outbreaks of the geometrids *Operophthera brumata* and *Epirrita autumnata* in subarctic birch forest: Evidence of a recent outbreak range expansion. *Journal of Animal Ecology* 77: 257–264.

Johnson, M. L., S. Armitage, B. C. G. Scholz, D. J. Merritt, B. W. Cribb, and M. P. Zalucki. 2006. Predator presence moves *Helicoverpa armigera* larvae to distraction. *Journal of Insect Behavior* 20: 1–18.

Johnson, M. T. J., M. J. Lajeunesse, and A. A. Agrawal. 2006. Additive and interactive effects of plant genotypic diversity on arthropod communities and plant fitness. *Ecology Letters* 9: 24–34.

Jolivet, P. 1996. *Ants and Plants: An Example of Coevolution*. Leiden, The Netherlands: Backhuys Publishers.

Jones, C. G., R. S. Ostfeld, M. P. Richard, E. M. Schauber, and J. O. Wolff. 1998. Chain reactions linking acorns to gypsy moth outbreaks and Lyme disease risk. *Science* 279: 1023–1026.

Jonkman, J. C. M. 1978. Nests of the leaf-cutting ant *Atta vollenweideri* as accelerators of succession in pastures. *Zeitschrift für angewandte Entomologie* 86: 25–34.

Juang, J.-Y., G. G. Katul, A. Porporato, P. C. Stoy, M. S. Sequeira, M. Detto, H.-S. Kim, and R. Oren. 2007. Eco-hydrological controls on summertime convective rainfall triggers. *Global Change Biology* 13: 887–896.

Karban, R. 2001. Communication between sagebrush and wild tobacco in the field. *Biochemical Systematics and Ecology* 29: 995–1005.

Karban, R., I. T. Baldwin, K. J. Baxter, G. Laue, and G. W. Felton. 2000. Communication between plants: Induced resistance in wild tobacco plants following clipping of neighboring sagebrush. *Oecologia* 125: 66–71.

Karban, R. and J. Maron. 2002. The fitness consequences of interspecific eavesdropping between plants. *Ecology* 83: 1209–1213.

Kessler, A. and I. T. Baldwin. 2001. Defensive function of herbivore-induced plant volatile emissions in nature. *Science* 291: 2141–2144.

King, T. J. 1977a. The plant ecology of ant-hills in calcareous grasslands. I. Patterns of species in relation to ant-hills in southern England. *Journal of Ecology* 65: 235–256.

King, T. J. 1977b. The plant ecology of ant-hills in calcareous grasslands. II. Succession on the mounds. *Journal of Ecology* 65: 257–278.

Klein, A.-M., I. Steffan-Dewenter, and T. Tscharntke. 2003. Fruit set of highland coffee increases with the diversity of pollinating bees. *Proceedings of the Royal Society of London B* 270: 955–961.

Knight, T. M., M. W. McCoy, J. M. Chase, K. A. McCoy, and R. D. Holt. 2005. Trophic cascades across ecosystems. *Nature* 437: 880–883.

Knops, J. M. H., D. Tilman, N. M. Haddad, S. Naeem, C. E. Mitchell, J. Haarstad, M. E. Ritchie, K. M. Howe, P. B. Reich, E. Siemann, and J. Groth. 1999. Effects of plant species richness on invasion dynamics, disease outbreaks, insect abundances, and diversity. *Ecology Letters* 2: 286–293.

Kratz, T. K., J. J. Magnuson, P. Bayley, B. J. Benson, C. W. Berish, C. S. Bledsoe, E. R. Blood, C. J. Bowser, S. R. Carpenter, G. L. Cunningham, et al. 1995. Temporal and spatial variability as neglected ecosystem properties: Lessons learned from 12 North American ecosytems. In *Evaluating and Monitoring the Health of Large-Scale Ecosystems*, D. J. Rapport, C. L. Gaudet, and P. Calow, eds. NATO ASI Series, Vol. 128. Berlin: Springer-Verlag.

Kurz, W. A., C. C. Dymond, G. Stinson, G. J. Rampley, E. T. Neilson, A. L. Carroll, T. Ebata, and L. Safranyik. 2008. Mountain pine beetle and forest carbon feedback to climate change. *Nature* 452: 987–990.

Lake, P. S., N. Bond, and P. Reich. 2007. Linking ecological theory with stream restoration. *Freshwater Biology* 52: 597–615.

Lamotte, M. and F. Bourliére. 1983. Energy flow and nutrient cycling in tropical savannas. In *Ecosystems of the World: Tropical Savannas*, F. Bourliére, ed. 583–603. Amsterdam: Elsevier.

Lavelle, P. and B. Pashanasi. 1989. Soil macrofauna and land management in Peruvian Amazonia. *Pedobiologia* 33: 283–291.

Lentz, D. L. and B. Hockaday. 2009. Tikal timbers and temples: Ancient Maya forestry and the end of time. *Journal of Archaeology* 36: 1342–1353.

Lesica, P. and P. B. Kannowski. 1998. Ants create hummocks and alter structure and vegetation of a Montana fen. *American Midland Naturalist* 139: 58–68.

Letourneau, D. K. and L. A. Dyer. 1998. Density patterns of *Piper* ant-plants and associated arthropods: Top-predator trophic cascades in a terrestrial system? *Biotropica* 30: 162–169.

Lewis, T. 1998. The effect of deforestation on ground surface temperatures. *Global and Planetary Change* 18: 1–13.

Lonnstedt, O. M., M. I. McCormick, and D. P. Chivers. 2012. Well-informed foraging: Damage-released chemical cues of injured prey signal quality and size to predators. *Oecologia* 168: 651–658.

Lorenz, E. N. 1993. *The Essence of Chaos*. Seattle: University of Washington Press.

Louda, S. M. 1982. Inflorescence spiders: A cost/benefit analysis for the host plant, *Haplopappus venetus* Blake (Asteraceae). *Oecologia* 55: 185–191.

MacMahon, J. A. 1981. Successional processes: Comparisons among biomes with special reference to probable roles of and influences on animals. In *Forest Succession: Concepts and Application*, D. C. West, H. H. Shugart, and D. B. Botkin, eds. 277–304. New York: Springer-Verlag.

Madigosky, S. R. 2004. Tropical microclimatic considerations. In *Forest Canopies*, 2nd ed., M. D. Lowman and H. B. Rinker, eds. 24–48. San Diego: Elsevier/Academic.

Mahaney, W. C., J. Zippin, M. W. Milner, K. Sanmugadas, R. G. V. Hancock, S. Aufreiter, S. Campbell, M. A. Huffman, M. Wink, D. Malloch, et al. 1999. Chemistry, mineralogy, and microbiology of termite mound soil eaten by the chimpanzees of the Mahale Mountains, western Tanzania. *Journal of Tropical Ecology* 15: 565–588.

Maser, C. 2009. *Earth in Our Care: Ecology, Economy, and Sustainability*. New Brunswick: Rutgers University Press.

Masters, G. J., V. K. Brown, and A. C. Gange. 1993. Plant mediated interactions between above- and below-ground insect herbivores. *Oikos* 66: 148–151.

McCloud, E. S. and I. T. Baldwin. 1997. Herbivory and caterpillar regurgitants amplify the wound-induced increases in jasmonic acid but not nicotine in *Nicotiana sylvestris*. *Planta* 203: 430–435.

McNaughton, S. J. 1985. Ecology of a grazing system: The Serengeti. *Ecological Monographs* 55: 259–294.

Meisner, M., J. P. Harmon, and A. R. Ives. 2007. Presence of an unsuitable host diminishes the competitive superiority of an insect parasitoid: A distraction effect. *Population Ecology* 49: 347–355.

Menninger, H. L., M. A. Palmer, L. S. Craig, and D. C. Richardson. 2008. Periodical cicada detritus impacts stream ecosystem metabolism. *Ecosystems* 11: 1306–1317.

Millenium Ecosystem Assessment. 2005. *Ecosystems and Human Well-Being: Biodiversity Synthesis*. Washington, DC: World Resources Institute.

Monk, C. D. and F. P. Day, Jr. 1988. Biomass, primary production, and selected nutrient budgets for an undisturbed watershed. In *Forest Hydrology and Ecology at Coweeta*, W. T. Swank and D. A. Crossley, Jr., eds. 151–159. New York: Elsevier.

Monteith, J. L. 1973. *Principles of Environmental Physics*. New York: American Elsevier.

Mooney, K. A. 2007. Tritrophic effects of birds and ants on a canopy food web, tree growth, and phytochemistry. *Ecology* 88: 2005–2014.

Mooney, K. A., D. S. Gruner, N. A. Barber, S. A. Van Bael, S. M. Philpott, and R. Greenberg. 2010. Interactions among predators and the cascading effects of vertebrate insectivores on arthropod communities and plants. *Proceedings of the National Academy of Sciences USA* 107: 7335–7340.

Naeem, S. 1998. Species redundancy and ecosystem reliability. *Conservation Biology* 12: 39–45.

Naeem, S. and S. Li. 1997. Biodiversity enhances ecosystem reliability. *Nature* 390: 507–509.

Nakano, S., H. Miyasaka, and N. Kuhara. 1999. Terrestrial-aquatic linkages: Riparian arthropod inputs alter trophic cascades in a stream food web. *Ecology* 80: 2435–2441.

Nepstad, D. C., I. M. Tohver, D. Ray, P. Moutinho, and G. Cardinot. 2007. Mortality of large trees and lianas following experimental drought in an Amazon forest. *Ecology* 88: 2259–2269.

North, M., J. Innes, and H. Zald. 2007. Comparison of thinning and prescribed fire restoration treatments to Sierran mixed-conifer historic conditions. *Canadian Journal of Forest Research* 37: 331–342.

Odum, E. P. 1969. The strategy of ecosystem development. *Science* 164: 262–270.

Odum, H. T. 1957. Trophic structure and productivity of Silver Springs, Florida. *Ecological Monographs* 27: 55–112.

Odum, H. T. 1970. Summary: An emerging view of the ecological system at El Verde. In *A Tropical Rain Forest*, H. T. Odum and R. F. Pigeon, eds. I191–I289. Washington, DC: U.S. Atomic Energy Commission.

Oliveira, P. S.and C. R. F. Brandâo. 1991. The ant community associated with extrafloral nectaries in the Brazilian cerrados. In *Ant-Plant Interactions*, C. R. Huxley and D. F. Cutler, eds. 198–212. Oxford, UK: Oxford University Press.

Paine, R. T. 1966. Food web complexity and species diversity. *American Naturalist* 100: 65–75.

Paine, R. T. 1969a. The *Pisaster-Tegula* interaction: Prey patches, predator food preference, and intertidal community structure. *Ecology* 50: 950–961.

Paine, R. T. 1969b. A note on trophic complexity and community stability. *American Naturalist* 103: 91–93.

Parker, G. G. 1995. Structure and microclimate of forest canopies. In *Forest Canopies*, M. D. Lowman and N. M. Nadkarni, eds. 73–106. San Diego: Academic.

Parsons, G. L., G. Cassis, A. R. Moldenke, J. D. Lattin, N. H. Anderson, J. C. Miller, P. Hammond, and T. D. Schowalter. 1991. *Invertebrates of the H. J. Andrews Experimental Forest, Western Cascade Range, Oregon. V: An Annotated List of Insects and Other Arthropods.* General Technical Report PNW-GTR-290. Portland, OR: USDA Forest Service, Pacific Northwest Research Station.

Parton, W. J., J. M. O. Scurlock, D. S. Ojima, T. G. Gilmanov, R. J. Scholes, D. S. Schimel, T. Kirchner, J-C. Menaut, T. Seastedt, E. G. Moya, et al. 1993. Observations and modeling of biomass and soil organic matter dynamics for the grassland biome worldwide. *Global Biogeochemical Cycles* 7: 785–809.

Perry, D. A., R. Oren, and S. C. Hart 2008. *Forest Ecosystems*, 2nd ed. Baltimore, MD: Johns Hopkins University Press.

Phillipson, J. 1981. Bioenergetic options and phylogeny. In *Physiological Ecology: An Evolutionary Approach to Resource Use*, C. R. Townsend and P. Calow, eds. 20–45. Oxford, UK: Blackwell Scientific.

Pianka, E. R. 1974. *Evolutionary Ecology*. New York: Harper & Row.

Pray, C. L, W. H. Nowlin, and M. J. Vanni. 2009. Deposition and decomposition of periodical cicadas (Homoptera: Cicadidae: Magicicada) in woodland aquatic ecosystems. *Journal of the North American Benthological Society* 28: 181–195.

Pringle, R. M. and K. Fox-Dobbs. 2008. Coupling of canopy and understory food webs by ground-dwelling predators. *Ecology Letters* 11: 1328–1337.

Progar, R. A., T. D. Schowalter, C. M. Freitag, and J. J. Morrell. 2000. Respiration from coarse woody debris as affected by moisture and saprotroph functional diversity in western Oregon. *Oecologia* 124: 426–431.

Quesada, M., G. A. Sanchez-Azofeifa, M. Alvarez-Añorve, K. E. Stoner, L. Avila-Cabadilla, J. Calvo-Alvarado, A. Castillo, M. M. Espírito-Santo, M. Fagundes, G. W. Fernandes, et al. 2009. Succession and management of tropical dry forests in the Americas: Review and new perspectives. *Forest Ecology and Management* 258: 1014–1024.

Raffa, K. F., B. H. Aukema, B. J. Bentz, A. L. Carroll, J. A. Hicke, M. G. Turner, and W. H. Romme. 2008. Cross-scale drivers of natural disturbances prone to anthropogenic amplification: The dynamics of bark beetle eruptions. *BioScience* 58: 501–517.

Ralph, S. G., H. Yueh, M. Friedmann, D. Aeschliman, J. A. Zeznik, C. C. Nelson, Y. S. N. Butterfield, R. Kirkpatrick, J. Liu, S. J. M. Jones, et al. 2006. Conifer defence against insects: Microarray gene expression profiling of Sitka spruce (*Picea sitchensis*) induced by mechanical wounding or feeding by spruce budworms (*Choristoneura occidentalis*) or white pine weevils (*Pissodes strobi*) reveals large-scale changes of the host transcriptome. *Plant, Cell and Environment* 29: 1545–1570.

Reagan, D. P., G. R. Camilo, and R. B. Waide. 1996. The community food web: Major properties and patterns of organization. In *The Food Web of a Tropical Rain Forest*, D. P. Reagan and R. B. Waide, eds. 462–488. Chicago, IL: University of Chicago Press.

Reusch, T. B. H., A. Ehlers, A. Hämmerli, and B. Worm. 2005. Ecosystem recovery after climatic extremes enhanced by genotypic diversity. *Proceedings of the National Academy of Sciences USA* 102: 2826–2831.

Riley, C. V. 1878. *First Annual Report of the United States Entomological Commission for the Year 1877 Relating to the Rocky Mountain Locust and the Best Methods of Preventing Its Injuries and of Guarding Against Its Invasions, in Pursuance of an Appropriation Made by Congress for This Purpose.* Washington, DC: U.S. Department of Agriculture.

Ritchie, M. E., D. Tilman, and J. M. H. Knops. 1998. Herbivore effects on plant and nitrogen dynamics in oak savanna. *Ecology* 79: 165–177.

Rodgers, H. L., M. P. Brakke, and J. J. Ewel. 1995. Shoot damage effects on starch reserves of *Cedrela odorata*. *Biotropica* 27: 71–77.

Romme, W. H., D. H. Knight, and J. B. Yavitt. 1986. Mountain pine beetle outbreaks in the Rocky Mountains: Regulators of primary productivity? *American Naturalist* 127: 484–494.

Ross, N. J. and T. F. Rangel. 2011. Ancient Maya agroforestry echoing through spatial relationships in the extant forest of NW Belize. *Biotropica* 43: 141–148.

Ryszkowski, L. 1979. Consumers. In *Grassland Ecosystems of the World: Analysis of Grasslands and Their Uses*, R. T. Coupland, ed. 309–318. Cambridge, UK: Cambridge University Press.

Salati, E. 1987. The forest and the hydrologic cycle. In *The Geophysiology of Amazonia: Vegetation and Climate Interactions*, R. E. Dickinson, ed. 273–296. New York: John Wiley & Sons.

Salo, J., R. Kallioloa, I. Häkkinen, Y. Mäkinen, P. Niemelä, M. Puhakka, and P. D. Coley. 1986. River dyanamics and the diversity of Amazon lowland forest. *Nature* 322: 254–258.

Salt, D. T., P. Fenwick, and J. B. Whittaker. 1996. Interspecific herbivore interactions in a high CO_2 environment: Root and shoot aphids feeding on *Cardamine*. *Oikos* 77: 326–330.

Schlesinger, W. H., J. F. Reynolds, G. L. Cunningham, L. F. Huenneke, W. M. Jarrell, R. A. Virginia, and W. G. Whitford. 1990. Biological feedbacks in global desertification. *Science* 247: 1043–1048.

Schmelz, E. A., R. J. Brebeno, T. E. Ohnmeiss, and W. S. Bowers. 2002. Interactions between *Spinacia oleracea* and *Bradysia impatiens*: A role for phytoecdysteroids. Archives of *Insect Biochemistry and Physiology* 51: 204–221.

Schowalter, T. D. 1981. Insect herbivore relationship to the state of the host plant: Biotic regulation of ecosystem nutrient cycling through ecological succession. *Oikos* 37: 126–130.

Schowalter, T. D. 1995. Canopy arthropod communities in relation to forest age and alternative harvest practices in western Oregon. *Forest Ecology and Management* 78: 115–125.

Schowalter, T. D. 2008. Insect herbivore responses to management practices in conifer forests in North America. *Journal of Sustainable Forestry* 26: 204–222.

Schowalter, T. D. 2011. *Insect Ecology: An Ecosystem Approach*, 3rd ed. San Diego: Elsevier/Academic.

Schowalter, T. D., R. N. Coulson, and D. A. Crossley, Jr. 1981a. Role of southern pine beetle and fire in maintenance of structure and function of the southeastern coniferous forest. *Environmental Entomology* 10: 821–825.

Schowalter, T. D. and L. M. Ganio. 2003. Diel, seasonal, and disturbance-induced variation in invertebrate assemblages. In *Arthropods of Tropical Forests*, Y. Basset, V. Novotny, S. E. Miller, and R. L. Kitching, eds. 315–328. Cambridge, UK: Cambridge University Press.

Schowalter, T. D. and P. Turchin. 1993. Southern pine beetle infestation development: Interaction between pine and hardwood basal areas. *Forest Science* 39: 201–210.

Schowalter, T. D., J. W. Webb, and D. A. Crossley, Jr. 1981b. Community structure and nutrient content of canopy arthropods in clearcut and uncut forest ecosystems. *Ecology* 62: 1010–1019.

Schowalter, T. D. and W. G. Whitford. 1979. Territorial behavior of *Bootettix argentatus* Bruner (Orthoptera: Acrididae). *American Midland Naturalist* 102: 182–184.

Schowalter, T. D., Y. L. Zhang, and J. J. Rykken. 2003. Litter invertebrate responses to variable density thinning in western Washington forest. *Ecological Applications* 13: 1204–1211.

Schweitzer, J. A., J. K. Bailey, S. C. Hart, G. M. Wimp, S. K. Chapman, and T. G. Whitham. 2005. The interaction of plant genotype and herbivory decelerate leaf litter decomposition and alter nutrient dynamics. *Oikos* 110: 133–145.

Schweitzer, J. A., J. K. Bailey, B. J. Rehill, G. D. Martinsen, S. C. Hart, R. L. Lindroth, P. Keim, and T. G. Whitham. 2004. Genetically based trait in a dominant tree affects ecosystem processes. *Ecology Letters* 7: 127–134.

Seastedt, T. R. and D. A. Crossley, Jr. 1981. Microarthropod response following cable logging and clear-cutting in the southern Appalachians. *Ecology* 62: 126–135.

Shelford, V. E. 1907. Preliminary note on the distribution of the tiger beetles (*Cicindela*) and its relation to plant succession. *Biological Bulletin* 14: 9–14.

Shuster, S. M., E. V. Lonsdorf, G. M. Wimp, J. K. Bailey, and T. G. Whitham. 2006. Community heritability measures the evolutionary consequences of indirect genetic effects on community structure. *Evolution* 60: 991–1003.

Smith, A. A., B. Hölldobler, and J. Liebig. 2009. Cuticular hydrocarbons reliably identify cheaters and allow enforcement of altruism in social insects. *Current Biology* 19: 78–81.

Smith, J. M. 1964. Group selection and kin selection. *Nature* 201: 1145–1147.

Spehn, E. M., A. Hector, J. Joshi, M. Scherer-Lorenzen, B. Schmid, E. Bazeley-White, C. Beierkuhnlein, M. C. Caldeira, M. Diemer, P. G. Dimitrakopoulos, et al. 2005. Ecosystem effects of biodiversity manipulations in European grasslands. *Ecological Monographs* 75: 37–63.

Stamp, N. E., Y. Yang, and T. L. Osier. 1997. Response of an insect predator to prey fed multiple allelochemicals under representative thermal regimes. *Ecology* 78: 203–214.

Stewart, M. M. and L. L. Woolbright. 1996. Amphibians. In *The Food Web of a Tropical Rain Forest*, Reagan, D. P. and R. B. Waide, eds., 273–320. University of Chicago Press, Chicago, IL.

Sticher, L., B. Mauch-Mani, and M. P. Métraux. 1997. Systematic acquired resistance. *Annual Review of Phytopathology* 35: 235–270.

Stout, M. J., J. S. Thaler, and B. P. H. J. Thomma. 2006. Plant-mediated interactions between pathogenic microorganisms and herbivorous insects. *Annual Review of Entomology* 51: 663–689.

Suarez, A. V., D. T. Bolger, and T. J. Case. 1998. Effects of fragmentation and invasion on native ant communities in coastal southern California. *Ecology* 79: 2041–2056.

Summerville K. S., D. Courard-Houri, and M. M. Dupont. 2009. The legacy of timber harvest: Do patterns of species dominance suggest recovery of lepidopteran communities in managed hardwood stands? *Forest Ecology and Management* 259: 8–13.

Tallamy, D. W., D. P. Whittington, F. Defurio, D. A. Fontaine, P. M. Gorski, and P. W. Gothro. 1998. Sequestered cucurbitacins and pathogenicity of *Metarhizium anisopliae* (Moniliales: Moniliaceae) on spotted cucumber beetle eggs and larvae (Coleoptera: Chrysomelidae). *Environmental Entomology* 27: 366–372.

Temple, S. A. 1977. Plant-animal mutualism: Coevolution with dodo leads to near extinction of plant. *Science* 197: 885–886.

Thaler, J. S. 1999a. Jasmonic acid mediated interactions between plants, herbivores, parasitoids, and pathogens: A review of field experiments with tomato. In *Induced Plant Defenses Against Pathogens and Herbivores: Biochemistry, Ecology, and Agriculture*, A. A. Agrawal, S. Tuzun, and E. Bent, eds. 319–334. St. Paul, MN: American Phytopathological Society.

Thaler, J. S. 1999b. Jasmonate-inducible plant defenses cause increased parasitism of herbivores. *Nature* 399: 686–687.

Thaler, J. S., M. J. Stout, R. Karban, and S. S. Duffey. 2001. Jasmonate-mediated induced plant resistance affects a community of herbivores. *Ecological Entomology* 26: 312–324.

Thompson, J. R., A. Wiek, F. J. Swanson, S. R. Carpenter, N. Fresco, T. Hollingsworth, T. Spies, and D. R. Foster. 2012. Scenario studies as a synthetic and integrative research activity for long-term ecological research. *BioScience* 62: 367–376.

Throop, H. L., E. A. Holland, W. J. Parton, D. S. Ojima, and C. A. Keough. 2004. Effects of nitrogen deposition and insect herbivory on patterns of ecosystem-level carbon and nitrogen dynamics: Results from the CENTURY model. *Global Change Biology* 10: 1092–1105.

Tilman, D. 1978. Cherries, ants, and tent caterpillars: Timing of nectar production in relation to susceptibility of caterpillars to ant predation. *Ecology* 59: 686–692.

Tilman, D. and J. A. Downing. 1994. Biodiversity and stability in grasslands. *Nature* 367: 363–365.

Tilman, D., J. Knops, D. Wedin, P. Reich, M. Ritchie, and E. Siemann. 1997. The influence of functional diversity and composition on ecosystem processes. *Science* 277: 1300–1302.

Torres, J. A. 1992. Lepidoptera outbreaks in response to successional changes after the passage of Hurricane Hugo in Puerto Rico. *Journal of Tropical Ecology* 8: 285–298.

Traugott, M. S. and N. E. Stamp. 1996. Effects of chlorogenic acid- and tomatine-fed prey on behavior of an insect predator. *Journal of Insect Behavior* 9: 461–476.

Trenberth, K. E. 1999. Atmospheric moisture recycling: role of advection and local evaporation. *Journal of Climate* 12: 1368–1381.

Tscharntke, T., S. Thiessen, R. Dolch, and W. Boland. 2001. Herbivory, induced resistance, and interplant signal transfer in *Alnus glutinosa*. *Biochemical Systematics and Ecology* 29: 1025–1047.

Turcek, F. J. 1971. On vertebrate secondary production of forests. In *Productivity of Forest Ecosystems*, P. Duvigneaud, ed. 379–385. Paris: UNESCO.

Turlings, T. C. J., J. H. Loughrin, P. J. McCall, U. S. R. Röse, W. J. Lewis, and J. H. Tumlinson. 1995. How caterpillar-damaged plants protect themselves by attracting parasiting wasps. *Proceedings of the National Academy of Sciences USA* 92: 4169–4174.

Turlings, T. C. J., P. J. McCall, H. T. Alborn, and J. H. Tumlinson. 1993. An elicitor in caterpillar oral secretions that induces corn seedlings to emit chemical signals attractive to parasitic wasps. *Journal of Chemical Ecology* 19: 411–425.

Turlings, T. C. J., J. H. Tumlinson, and W. J. Lewis. 1990. Exploitation of herbivore-induced plant odors by host-seeking parasitic wasps. *Science* 250: 1251–1253.

Turner, B. L., III and J. A. Sabloff. 2012. Classic Period collapse of the Central Maya Lowlands: Insights about human-environment relationships for sustainability. *Proceedings of the National Academy of Sciences USA* 109: 13908–13914.

Van Langevelde, R., C. A. D. M. van de Vijver, L. Kumar, J. van de Koppel, N. de Ridder, J. van Andel, A. K. Skidmore, J. W. Hearne, L. Stroosnijder, W. J. Bond, et al. 2003. Effects of fire and herbivory on the stability of savanna ecosystems. *Ecology* 84: 337–350.

Veblen, T. T., K. S. Hadley, E. M. Nel, T. Kitzberger, M. Reid, and R. Villalba. 1994. Disturbance regime and disturbance interactions in a Rocky Mountain subalpine forest. *Journal of Ecology* 82: 125–135.

Vittor, A. Y., R. H. Gilman, J. Tielsch, G. Glass, T. Shields, W. S. Lozano, V. Pinedo-Cancino, and J. A. Patz. 2006. The effect of deforestation on the human-biting rate of *Anopheles darlingi*, the primary vector of falciparum malaria in the Peruvian Amazon. *American Journal of Tropical Medicine and Hygiene* 74: 3–11.

Wallace, J. B., T. F. Cuffney, J. R. Webster, G. J. Lugthart, K. Chung, and G. S. Goldwitz. 1991. Export of fine organic particles from headwater streams: Effects of season, extreme discharges, and invertebrate manipulation. *Limnology and Oceanography* 36: 670–682.

Wallace, J. B., M. R. Whiles, S. Eggert, T. F. Cuffney, G. J. Lugthart, and K. Chung. 1995. Long-term dynamics of coarse particulate organic matter in three Appalachian Mountain streams. *Journal of the North American Benthological Society* 14: 217–232.

Waring, R. H. and S. W. Running. 2007. *Forest Ecosystems: Analysis at Multiple Scales*, 3rd ed. San Diego: Academic.

Watts, J. G., E. W. Huddleston, and J. C. Owens. 1982. Rangeland entomology. *Annual Review of Entomology* 27: 283–311.

Webster, J. R., J. B. Waide, and B. C. Patten. 1975. Nutrient recycling and the stability of ecosystems. In *Mineral Cycling in Southeastern Ecosystems*, CONF-740513, F. G. Howell, J. B. Gentry, and M. H. Smith, eds. 1–27. Washington, DC: USDOE Energy Research and Development Administration.

Weygoldt, P. 1969. *The Biology of Pseudoscorpions*. Cambridge, MA: Harvard University Press.

Whitham, T. G., J. K. Bailey, J. A. Schweitzer, S. M. Shuster, R. K. Bangert, C. J. LeRoy, E. V. Lonsdorf, G. J. Allan, S. P. DiFazio, B. M. Potts, et al. 2006. A framework for community and ecosystem genetics: From genes to ecosystems. *Nature Reviews Genetics* 7: 510–523.

Whitham, T. G., W. P. Young, G. D. Martinsen, C. A. Gehring, J. A. Schweitzer, S. M. Shuster, G. M. Wimp, D. G. Fischer, J. K. Bailey, R. L. Lindroth, et al. 2003. Community and ecosystem genetics: A consequence of the extended phenotype. *Ecology* 84: 559–573.

Whittaker, R. H. 1970. *Communities and Ecosystems*. London: Macmillan.

Wickman, B. E. 1980. Increased growth of white fir after a Douglas-fir tussock moth outbreak. *Journal of Forestry* 78: 31–33.

Wiegert, R. G. and C. E. Petersen. 1983. Energy transfer in insects. *Annual Review of Entomology* 28: 455–486.

Williams, J. W. and S. T. Jackson. 2007. Novel climates, no-analog communities, and ecological surprises. *Frontiers in Ecology and the Environment* 5: 475–482.

Williamson, S. C., J. K. Detling, J. L. Dodd, and M. I. Dyer. 1989. Experimental evaluation of the grazing optimization hypothesis. *Journal of Range Management* 42: 149–152.

Willig, M. R. and M. A. McGinley. 1999. Animal responses to natural disturbance and roles as patch generating phenomena. In *Ecosystems of the World: Ecosystems of Disturbed Ground*, L. R. Walker, ed. 667–689. Amsterdam, The Netherlands: Elsevier Science.

Wimp, G. M., S. Wooley, R. K. Bangert, W. P. Young, G. D. Martinsen, P. Keim, B. Rehill, R. L. Lindroth, and T. G. Whitham. 2007. Plant genetics predicts intra-annual variation in phytochemistry and arthropod community structure. *Molecular Ecology* 16: 5057–5069.

Windsor, D. M. 1990. *Climate and Moisture Variability in a Tropical Forest: Long-Term Records from Barro Colorado Island, Panamá*. Washington, DC: Smithsonian Institution Press.

Wipfli, M. S. 1997. Terrestrial invertebrates as salmonid prey and nitrogen sources in streams: Contrasting old-growth and young-growth riparian forests in southeastern Alaska, U.S.A. *Canadian Journal of Fisheries and Aquatic Science* 54: 1259–1269.

Wipfli, M. S. and J. Musslewhite. 2004. Density of red alder (*Alnus rubra*) in headwaters influences invertebrate and detritus subsidies to downstream fish habitats in Alaska. *Hydrobiologia* 520: 153–163.

Witmer, M. C. and A. S. Cheke. 1991. The dodo and the tambalocoque tree: An obligate mutualism reconsidered. *Oikos* 61: 133–137.

Wood, T. E., D. Lawrence, D. A. Clark, and R. L. Chazdon. 2009. Rain forest nutrient cycling and productivity in response to large-scale litter manipulation. *Ecology* 90: 109–121.

Xue, Y., K. N. Liou, and A. Kashahara. 1990. Investigation of biophysical feedback on the African climate using a two-dimensional model. *Journal of Climate* 3: 337–352.

Zhang, D. D., P. Brecke, H. F. Lee, Y. Q. He, and J. Zhang. 2007. Global climate change, war, and population decline in recent human history. *Proceedings of the National Academy of Sciences USA* 104: 19214–19219.

Zheng, X. and E. A. B. Eltahir. 1998. The role of vegetation in the dynamics of West African monsoons. *Journal of Climate* 11: 2078–2096.

6

Effects of Anthropogenic Changes and Management

"It grieves me to witness the extravagance that pervades this country" said the Judge, "where the settlers trifle with the blessings they might enjoy . . . I earnestly beg you to remember, that they [the trees] are the growth of centuries, and when gone, none living will see their loss remedied."

"Why, I don't know, Judge . . . It seems to me, if there's plenty of any thing in this mountaynious country, it's the trees."

James Fenimore Cooper (1823)

Humans have engaged in a number of activities over the millennia that affect the sustainability of ecosystem services, either directly or indirectly via effects on insects. Even prehistoric societies had considerable influence on ecosystem development and capacity to provide services through their exploitation of resources and manipulation of fire. Paleohuman colonization of North America about 15,000 years ago may have been responsible for the concomitant disappearance of most large mammals that included mastodons, giant ground sloths, and herds of horses, antelope, and oxen that made North American grasslands the equivalent of the modern African Serengeti (Janzen and Martin 1982). A similar disappearance of large animals occurred in Australia and New Zealand shortly after colonization by humans (Diamond 1999). In addition, aboriginal Americans and Australians intentionally burned grasslands and savannahs to stimulate plant growth and attract game animals (Agee 1993; Hill et al. 1999; Murphy and Bowman 2007). Additional unintentional burning likely reflects escape of campfires (Morrison and Swanson 1990; Agee 1993).

The earliest agricultural civilizations developed in fertile valleys of the Nile (Egypt), Tigris/Euphrates (Mesopotamia), Indus (Harappan), and Yellow (China) rivers. Even with relatively crude technologies, those civilizations were capable of sufficient overexploitation of vegetation and soil resources to cause desertification (ecosystem deterioration through overuse that leads to desert-like conditions) of their regions and collapse of their civilizations (Diamond 1999; Foley et al. 2003a; Juang et al. 2007; Janssen et al. 2008; Perry et al. 2008).

Modern chainsaws and earthmoving machinery have accelerated our ability to remove vegetation and impound rivers over large areas, greatly affecting albedo, evapotranspiration, and turbulence of airflow that determine regional and global temperature and precipitation patterns. The global warming and drying effected by these changes are exacerbated by the conversion of fossil fuels into atmospheric carbon dioxide and other greenhouse gases (Keeling et al. 1995). Between 1850 and 1980, an estimated 90 to 120 billion tons of carbon entered the atmosphere as a result of tropical deforestation, compared with 165 billion tons as a result of fossil fuel combustion (Perry et al. 2008). Transportation of species across geographic barriers, either intentionally—as in the case of crop plants, livestock, and honey bees—or unintentionally—as in the case of many crop pests and disease vectors that travel unnoticed with humans or commercial products—has further altered natural communities. As a result, no part of the globe remains unaffected by human activity. Anthropogenic changes in ecosystem conditions and efforts to control insects are among the most pervasive factors affecting the services on which we depend.

6.1 How Humans Affect Ecosystems

6.1.1 Disturbances

The most immediate effects of humans on ecosystem services are through anthropogenic disturbances and introduction of invasive species. Anthropogenic disturbances, including harvest and replanting, soil disruption, altered fire regimes, road construction, river impoundment, and release of toxic materials (such as oil spills and industrial effluents) have become a pervasive environmental factor (Figure 6.1). The impact of such disturbances on insects reflects the degree to which direct and indirect effects resemble those of natural disturbances. For example, forest harvest may affect insects in ways similar to those of other canopy opening disturbances, depending on the degree of exposure (Paquin and Coderre 1997; Gandhi et al. 2004; Buddle et al. 2006); vegetation conversion to crop production elicits insect responses to changes in density and apparency of early successional hosts; river impoundment may elicit responses similar to landslides that also alter drainage pattern. Anthropogenic disturbances also introduce novel conditions such as large numbers of cut stumps with exposed fresh surfaces and inground root systems that provide unique habitats for root-feeding insects (Witcosky et al. 1986), sharp boundaries that alter the steepness of environmental gradients between disturbed and undisturbed patches and restrict dispersal of species intolerant of exposed conditions (Chen et al. 1995; Haynes and Cronin 2003), and toxic chemicals to which species are not adapted (Figure 6.2). Oil spills and urban sewage in streams affect not only the aquatic fauna but also terrestrial fauna in seasonally flooded habitats

FIGURE 6.1 (SEE COLOR INSERT.)
Contrast between post-disturbance conditions: (a) clear-cut harvest, (b) hurricane, and (c) fire. Note the relatively intact soil conditions and scattered surviving vegetation following natural disturbances, compared with the highly disturbed and fully exposed surface of the clear-cut.

FIGURE 6.2 (SEE COLOR INSERT.)
Contrast between edges in anthropogenic versus natural landscapes. (a) Straight edges among patches in anthropogenic landscape. (b) Gradual transitions among patches in natural landscape.

(Couceiro et al. 2007). Catastrophic wildfires in western North America that have resulted from fuel accumulation during a century of fire suppression create conditions quite different from those resulting from low intensity ground fires to which species have adapted (Agee 1993; Christensen et al. 2000). Paving previously vegetated surfaces in urban settings has created the most extreme changes in habitat conditions for organisms sensitive to high temperature and desiccation.

Anthropogenic disturbances differ from natural disturbances in their frequency, duration, and scale. Whereas the return frequency of stand-replacing fire in coniferous forests of the Pacific Northwest was about 500 years, harvest and replanting practices now restart forest recovery every 70 to 100 years (Christensen et al. 2000). In northern Australia, natural ignition would come from lightning during storm events at the onset of monsoon rains, whereas prescribed fires often are set during drier periods to maximize fuel reduction (Braithwaite and Estbergs 1985). Consequently, prescribed fires burn hotter, are more homogeneous in their severity, and cover larger areas than do lower intensity, more patchy fires burning during cooler, moister periods.

Anthropogenic changes also may exacerbate the effect of natural disturbances. Clearing land and impounding reservoirs have occurred simultaneously over large areas of the globe. These practices alter surface albedo in ways that increase regional warming and storm intensity (Foley et al. 2003b; Hossain et al. 2009) and increase runoff and stream discharge that lead to downstream flooding (Ehrlich and Mooney 1983). Consequently, the average return time for floods of 100-year severity has shrunk to 30 years (Kundzewicz et al. 1998). Vegetation removal and smoke from fires that accompany forest conversion to agricultural or urban land use reduce cloud cover (from 38% in clean air to 0% in heavy smoke) and increase the altitude at which water condenses, leading to more violent thunderstorms and hail, rather than warm rain (Ackerman et al. 2000; Andreae et al. 2004; Koren et al. 2004). Global warming is increasing the prevalence of extreme weather events (Gleason et al. 2008; Bender et al. 2010; Gutschick and BassiriRad 2010; Lubchenco and Karl 2012). The increased frequency of extreme disturbances will affect insects and other organisms, as well as the ecosystem services they influence, in ways that are difficult to predict (Hossain et al. 2009; Bender et al. 2010; Gutschick and BassiriRad 2010; Kishtawal et al. 2010), but which likely will affect species survival more than will changes in average conditions (Reusch et al. 2005; Jentsch et al. 2007; Gutschick and BassiriRad 2010; Kaushal et al. 2010).

The effects of such changes on insect populations may persist for long periods because local mechanisms are not in place to reverse the effects of extreme alteration of vegetation, substrate, or water conditions. For example, Benstead et al. (2007) reported that effects on insects from low levels of nutrient enrichment in Arctic freshwater streams depended on the duration of enrichment as mediated by bryophyte colonization and subsequent physical disturbances that removed bryophytes. Harding et al. (1998) reported that responses of aquatic insects to restoration treatments reflected differences in community structure among stream segments with different histories of anthropogenic disturbances. Similarly, Schowalter et al. (2003) found that litter arthropod (including insect) responses to variable density thinning of conifer forests for restoration purposes reflected different initial community structures resulting from previous thinning as much as 30 years earlier.

6.1.2 Global Change

Global surface and lower atmospheric temperatures increased progressively during the last quarter of the twentieth century, and the first decade of the twenty-first century has been the warmest on record (Arndt et al. 2010). Detection of biotic responses to global warming is complicated because effects of nonclimatic (especially land use) factors tend to dominate local, short-term responses (Parmesan and Yohe 2003; Raffa et al. 2008). Parmesan and Yohe (2003) conducted a global meta-analysis for 1,700 species. Results showed a 6.1 km decade^{-1} range shift toward the poles (or 6 m decade^{-1} increase in elevation) and a 2.3 day decade^{-1} advance in spring events. This diagnostic pattern was found for 279 species (including plants, insects, and vertebrates), indicating that climate change already is affecting ecosystems. Balanyá et al. (2006) reported that 22 populations of a cosmopolitan fly, *Drosophila subobscura*, on three continents had experienced the equivalent of a one-degree change in latitude toward the Equator and showed a corresponding shift in genotypic composition equivalent to a one degree lower latitude genotype. Bradshaw and Holzapfel (2001) and Mathias et al. (2007) found a detectable shift in genetically controlled response of pitcher plant mosquitoes, *Wyeomyia smithii*, to changing photoperiod as a result of warming temperatures.

A number of studies have documented insect responses to elevated temperature, increased atmospheric or aqueous concentrations of CO_2 or various pollutants (including pesticides), and habitat disturbance and fragmentation (Alstad et al. 1982; Heliövaara and Väisänen 1986, 1993; Lincoln et al. 1993; Arnone et al. 1995; Marks and Lincoln 1996; Kinney et al. 1997; Bezemer and Jones 1998; Valkama et al. 2007; Currano et al. 2008; Zavala et al. 2008; Yu et al. 2009). Although insect herbivores respond to a variety of factors—including photoperiod, relative humidity, and host condition—that interact with effects of temperature (Bale et al. 2002), a number of studies suggest increased likelihood of herbivore outbreaks under future warming and/or drying scenarios (Mattson and Haack 1987; Logan et al. 2003; Breshears et al. 2005; Yu et al. 2009). Currano et al. (2008) found that herbivory was higher during a warmer period 56 million years ago than during the cooler periods before or after (see Figure 4.2). Zavala et al. (2008) demonstrated that elevated atmospheric CO_2 compromised plant defense against herbivores by down-regulating gene expression for defensive compounds (see Chapter 4).

Human use of ecosystem resources also changes conditions for insects, including disease vectors. Diversion of streams for irrigation purposes alters water level, flow rate, and temperature, changing habitat conditions for aquatic insects and altering community structure (Miller et al. 2007). Excavation of shallow pits during brick making in sub-Saharan Africa increased predator-free habitat for, and larval abundance of, *Anopheles gambiae*, the primary vector of malaria in the region (Carlson et al. 2009). Predator diversity in these pools increased with time because disturbance was a major factor in ultimately reducing *A. gambiae* larval populations. Vittor et al. (2006) demonstrated that deforestation and human encroachment into tropical forest in Peru increased contact between humans and mosquito vectors of malaria, increasing the incidence of malaria several orders of magnitude in an area from which malaria had been eradicated.

However, humans are changing environmental conditions in many ways simultaneously, through fossil fuel combustion, industrial effluents, water impoundment and diversion, and land use practices. Large areas have been planted to genetically modified crops or occupied by invasive exotic species. Global atmospheric concentrations of CO_2 and other greenhouse gases are clearly increasing, and global climate has shown a distinct warming trend (Keeling et al. 1995; Beedlow et al. 2004). Acidic precipitation has greatly reduced the pH of many aquatic ecosystems in northern temperate countries. Nitrogen subsidies resulting from increased atmospheric nitrous oxide may provide a short-term fertilization effect in N-limited ecosystems, until pH buffering capacity of the soil is depleted. Dams restrict upstream movement of fish and other aquatic organisms affecting their reproduction and predation on insect populations. Deforestation, desertification, and other changes in regional landscapes are fragmenting habitats and altering habitat suitability for organisms around the globe (Figure 6.3). Mining activities and industrial effluents add highly toxic minerals to terrestrial and aquatic ecosystems.

The effects of these combined anthropogenic changes on ecosystem services are difficult to predict because of complex feedbacks among species and other ecosystem components, many of which we have begun to recognize only recently. Our most sophisticated models for natural ecosystems, such as CENTURY and BIOME-BGC (Parton et al. 1993; Waring and Running 1998), and agroecosystems (Rudd et al. 1984) are subject to large effects of small changes in parameter values (i.e., the "butterfly effect"). Many, perhaps most, people assume that plant hardiness zones will simply move toward the poles, making it possible to grow crops further north. In fact, yields of our most important crops decline precipitously as temperature increases above 30°C (86°F), as already seen in recent years (Schlenker and Roberts 2009; Federoff et al. 2010). In addition, the regional distribution of precipitation

FIGURE 6.3 (SEE COLOR INSERT.)
Deforestation in Panama. Removal of tropical rainforest cover has exposed soil to solar heating and severe erosion, leading to continued ecosystem deterioration and to altered regional temperature and precipitation patterns. (From Schowalter, T. D., *Insect Ecology: An Ecosystem Approach*, San Diego, Elsevier, 2011.)

will change independently of temperature, altering evapotranspiration rates and water balances for many plants, alleviating or exacerbating moisture stress for crops and other plants (Hatfield et al. 2011; Izurralde et al. 2011), thereby altering their vulnerability to herbivores (Mattson and Haack 1987). Various barriers, including roads and intensive agricultural landscapes, will restrict range shifts for many native species (Wilson et al. 2007). Many plant and animal species are likely to disappear, and altered interactions among plants, herbivores, and predators (Van der Putten et al. 2010; Hatfield et al. 2011; see Chapter 4) will affect their associated contributions to ecosystem services (Parmesan 2006; Zvereva and Kozlov 2006; Traill et al. 2010).

Shifts in cropping systems to adjust to changing climate patterns may be difficult. Farmers will need to learn new agricultural practices and strategies for dealing with climate changes and new pests spreading into previously inhospitable habitats. For example, soybeans are one of the most important food crops globally (Hartman et al. 2011). Soybean growers in the northern United States currently are concerned primarily with the invasive soybean aphid, *Aphis glycines* (Ragsdale et al. 2011) but have no experience with the diversity of defoliators, weevils, stink bugs, and aphids, each with unique biology and susceptibility to management options, that southern growers experience (Smith et al. 2009).

Insects certainly will be among the factors affecting plant responses to future climatic conditions. Whereas some insects will disappear or become less widespread, other (especially) invasive species with wider tolerance ranges to environmental conditions will follow shifts in host plant ranges (Figure 6.4) (Williams and Liebhold 1995, 2002). Both native and invasive species will exploit any climate-induced host stress and exacerbate anthropogenic changes and effects on ecosystem services (Schowalter 2011). For example, herbivore outbreaks likely will accelerate the flux of carbon from vegetation to the atmosphere, increasing atmospheric CO_2 and its contribution to global warming (Kurz et al. 2008; Clark et al. 2010). Furthermore, warming of aquatic ecosystems likely will generate a shift in community composition from dominance by insect detritivores to dominance by microbial decomposers that also will accelerate CO_2 flux to the atmosphere (Boyero et al. 2011).

6.1.3 Pollution and Invasive Species

Pollution and invasive species are the most immediate threats to many species and ecosystems. Pollutants threaten the survival of species at all trophic levels (Alstad et al. 1982; Trumble and Jensen 2004), thereby threatening food web interactions and ecosystem processes (Butler and Trumble 2008; Butler et al. 2009). For example, Trumble and Jensen (2004) reported that concentrations of 500–1,000 µg g^{-1} of Cr^{+6} (within the range of environmental contamination) in diet fed to a terrestrial dipteran detritivore, *Megaselia scaleris*, increased development time by 65–100% and reduced survival by 50–94%.

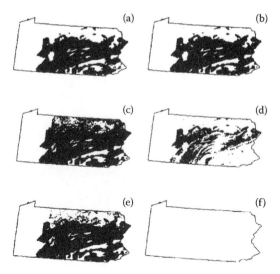

FIGURE 6.4
Potential outbreak areas of gypsy moths in Pennsylvania under climate change scenarios. (a) Current temperature and precipitation, (b) a 2°C increase, (c) a 2°C increase and 0.5 mm d^{-1} precipitation increase, (d) a 2°C increase and 0.5 mm d^{-1} precipitation decrease, (e) a GISS model, and (f) a GFDL model. (From Williams, D. W. and A. M. Liebhold, *Environmental Entomology*, 24, 1–9, 2006. With permission.)

Because females did not discriminate among substrates varying in Cr levels, populations are not likely to persist in contaminated areas, thereby reducing rates of decomposition and nutrient flux. Toxic aerosols and oil spills can spread over wide areas, reducing abundances of species that play key roles in critical ecological processes (Alstad et al. 1982; Heliövaara and Väisänen 1986, 1993; Couceiro et al. 2007).

Light also is becoming a global pollutant (Figure 6.5). Many nocturnal insects and migrating birds orient by moonlight and become confused by the bright lights of urban and industrial sites. Normal dispersal or foraging behavior can be disrupted by such artificial lights. Aquatic insects often are deceived by horizontally polarized light from dark-colored reflective surfaces such as automobiles, asphalt roads, and oil spills, mistaking these for surfaces of aquatic habitats (Kriska et al. 2006; Horváth et al. 2009).

Invasive species can be considered to be biotic pollution because their introduction has dramatic effects on native species and ecosystems. As described in Chapters 4 and 5, populations in their native ecosystems are regulated at sizes near carrying capacity by a combination of predation, parasitism, and availability of suitable food resources. However, species transplanted to new regions typically lack such population regulation. Although many introduced species are unable to find suitable resources or are unable to increase because of difficulty in finding mates, and eventually disappear (Yamanaka and Liebhold 2009), too many have exploited release from natural regulatory

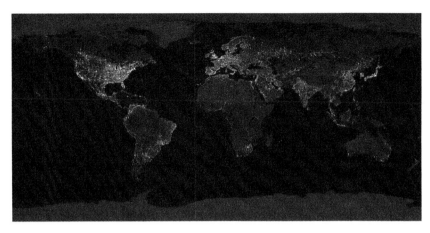

FIGURE 6.5 (SEE COLOR INSERT.)
The global distribution of light. The brightest areas are the most urbanized but not necessarily the most populated; for example, compare western Europe with China. Cities are concentrated along coastlines and transportation networks. The U.S. interstate highway system appears as a lattice connecting the brighter dots of city centers. The Russian Trans-Siberian Railroad is a thin line stretching from Moscow through the center of Asia to Vladivostok. The Nile River, from the Aswan Dam to the Mediterranean Sea, is another bright thread through an otherwise dark region. Created with data from the Defense Meteorological Satellite Program (DMSP) Operational Linescan System (OLS). (Data courtesy of Marc Imhoff of NASA GSFC and Christopher Elvidge of NOAA NGDC; image courtesy of Craig Mayhew and Robert Simmon, NASA GSFC.)

factors and have become invasive, disrupting processes and services in their new ecosystems.

In some cases, species have become invasive after human alteration of their ecosystems provided a stepping-stone to suitable habitats outside their natural range. The Colorado potato beetle, *Leptinotarsa decemlineata*, subsisted on wild solanaceous hosts in western North America until westward movement of settlers brought it into contact with the cultivated potato in the Midwest during the late 1800s (Riley 1883; Stern et al. 1959; Hitchner et al. 2008), allowing it to spread eastward and, eventually, to Europe. Similarly, the cotton boll weevil, *Anthonomus grandis*, co-evolved with scattered wild cotton, *Gossypium* spp., including *G. hirsutum*, in tropical Mesoamerica until citrus cultivation in the 1890s provided overwintering food resources that allowed the insect to spread into subtropical cotton-growing regions of south Texas and northern Argentina (Showler 2009). Subsequently, rapid reproduction in the spring by overwintering adults permitted spread throughout the U.S. Cotton Belt (Showler 2009).

In other cases, introduction by humans has been more direct. Examples of species introduced unintentionally in infested plants or packaging materials include the European wood wasp, *Sirex noctilio*, which has devastated forest resources in Australia and Brazil and is now in North America (Yemshanov et al. 2009); the cactus moth, *Cactoblastis cactorum*, that threatens native cacti in the southwestern United States and Mexico (Pemberton and Cordo 2001); and the red imported fire ant, *Solenopsis invicta*, originally from Argentina, which has caused considerable disruption to ecosystems (and annoyance to humans) in the southern United States, and more recently in Australia, China, and Taiwan (Stiles and Jones 1998; Yang et al. 2009).

Some species have been introduced intentionally for various purposes and have subsequently become invasive. One of the most notorious, albeit valued, invasive species in the Americas is the European honey bee, *Apis mellifera*, and its Africanized hybrids. This insect was introduced by early colonists for honey and wax but has more recently provided pollination service, with thousands of colonies transported from south to north as the growing season progresses. Despite the commercial value of this insect, feral colonies have outcompeted and eliminated many native pollinators that are more efficient pollinators of specialty crops such as squash and melons, as well as native plants of conservation concern (Aizen and Feinsinger 1994; Kremen et al. 2002).

The gypsy moth, *Lymantria dispar*, also was introduced intentionally from Europe to the northeastern United States in the early 1870s as part of early efforts to establish a silk industry in North America (Trouvelot 1867). This species subsequently spread south and west with considerable economic costs in lost forest resources and intensive control efforts (Forbush and Fernald 1896; Johnson et al. 2006).

Many invasive species are favored by habitat disturbance, which promote species with rapid reproduction and dispersal capabilities (see Chapters 3 and 4). Disturbed borders along roads and power lines create particularly suitable corridors for disturbance-adapted species to spread widely, even into "natural" ecosystems (Figure 6.6) (Spencer and Port 1988; Spencer et al. 1988; DeMers 1993; Stiles and Jones 1998; Vasconcelos et al. 2006). Once established, they can alter abundances and interactions among other species

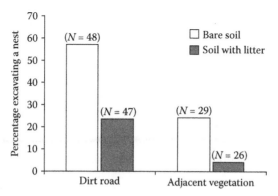

FIGURE 6.6
Percentage of *Atta laevigata* queens excavating nests in bare soil or soil with litter in roadsides or adjacent cerrado vegetation in Brazil. (From Vasconcelos, H. L., E. H. M. Vieira Neto, and F. M. Mundim, *Biotropica*, 38, 661–665, 2006. With permission.)

in the same way as abiotic pollutants. For example, invasive red imported fire ants, *S. invicta*, are most abundant in disturbed habitats (Stiles and Jones 1998; Zettler et al. 2004), where they displace native ants and negatively affect many ground nesting birds, small mammals, and herpetofauna through aggressive foraging behavior, high reproductive rates, and lack of effective predators or parasites (Porter and Savignano 1990; Allen et al. 2004; Zettler et al. 2004). Suarez et al. (1998) found that habitat fragmentation also favored the invasive Argentine ant, *Linepithema humile*, at the expense of native ant species.

On the other hand, the fire ant is an effective predator of the invasive sugar-cane borer, *Diatraea saccharalis*, the most serious sugarcane pest in Louisiana. Where present, this ant significantly reduces the need for insecticides; when these ants are suppressed by myrmecides or hurricane storm surge, borer populations increase substantially and require more intensive insecticide application (Hensley et al. 1961; Beuzelin et al. 2009).

The effects of pollution and invasive species can spread quickly far beyond the point of initial introduction. Agricultural and industrial runoff, including insecticides, entering aquatic ecosystems threatens many aquaculture services (Baldwin et al. 2009; Barbee and Stout 2009). Pollutants flowing down the Mississippi River have created a hypoxic zone in the Gulf of Mexico, threatening the sustainability of important seafood harvests in this area (Howarth et al. 2011). At least 50,000 nonnative insects in the United States now cause $120 billion annually in resource losses and control costs (Pimentel et al. 2005).

Eastern deciduous forests of North America have been dramatically altered by invasive species. These forests were characterized by beech and maple in moister areas, chestnut, oaks, and hickories in drier areas, with eastern hemlock a key component of many sites, especially moist coves and stream banks where it provided denser shade, more acidic soil conditions, and deeper litter that maintain moisture and cooler conditions that affect water runoff and water temperature (Ford and Vose 2007). During the past century, chestnut has been eliminated as a dominant tree by introduced chestnut blight, *Cryphonectria parasitica*, beech has declined, especially in northern forests, as a result of introduced beech bark disease fungus, *Neonectria* spp., vectored by the introduced beech scale insect, *Cryptococcus fagisuga*, and hemlock has declined, especially in southern forests, as a result of introduced hemlock woolly adelgid, *Adelges tsugae*. Oaks and other trees that replace the disappearing species are themselves vulnerable to mortality caused by gypsy moths, sudden oak death fungus, *Phytophthora ramorum*, and spread of the Asian longhorned beetle, *Anoplophora glabripennis*, and emerald ash borer, *Agrilus planipennis*.

These invasive species have already significantly altered forest structure and ability to sustain ecosystem services. Changes in forest composition and rates and pathways of water and nutrients flux (Orwig 2002) have affected food web interactions (Jones et al. 1998) and stream flow and water temperature, thereby affecting the supply of wood products, water yield, and fish production (Ford and Vose 2007; Spaulding and Rieske 2010). For example, chestnut trees were a reliable source of nut crops for Native Americans and later residents, as well as wildlife, in eastern forests and were heavily harvested before this tree species was eliminated as a nut-producer by chestnut blight (Day 1953).

6.1.4 Ecosystem Fragmentation and Conversion

Fragmentation results from harvest activities that introduce gaps into a previously continuous ecosystem. Conversion results from creation of entirely new land uses, such as agriculture, urban, or reservoir. Anthropogenic fragmentation creates more uniform and distinct patches with sharp edges between harvested and natural ecosystems, whereas natural disturbances typically leave remnant islands of the previous ecosystem and broad ecotones of blended communities along the boundaries (see Figures 6.1 and 6.2). Ecosystem fragmentation and conversion raise local and regional temperature and cause soil desiccation (Chen et al. 1995; Briant et al. 2010), thereby increasing the frequency and severity of disturbances such as droughts and floods that have dramatic effects on survival or movement of various insects (Franklin et al. 1992; Rubenstein 1992; Roland 1993; Raffa et al. 2008; Gutschick and BassiriRad 2010; Lubchenco and Karl 2012). Kimberling et al. (2001) reported that physical disturbance related to construction or waste disposal had relatively less effect on invertebrate communities than did conversion of shrub-steppe to agricultural use.

Braschler et al. (2009) found that many orthopterans avoided the mown matrix in a fragmented grassland, likely because of the lack of shelter. Consequently, small populations became increasingly isolated in remnant patches of grassland, increasing their vulnerability to local extinction if large areas are mown simultaneously. Nessimian et al. (2008) found that aquatic insect assemblages in central Amazonia differed significantly between streams passing through forest versus pasture, but not among streams passing through primary forest, forest fragments, or secondary forests. Altered drainage patterns affect temperature and chemical conditions of aquatic ecosystems and opportunities for organisms to disperse upstream or downstream (Pringle et al. 2000).

Perhaps more important for management of insect pests, fragmentation disproportionately restricts predators and parasites, compared with pest species, that often are adapted to colonize disturbed habitats (see section 6.1.5; see also Chapter 4). Roland and Kaupp (1995) found that transmission of nuclear polyhedrosis virus was reduced along forest edges, prolonging outbreaks of the forest tent caterpillar, *Malacosoma disstria*, that were able to disperse across edges. Similar results were found for understory insectivorous birds in tropical forests, suggesting that outbreaks of some insects would be more likely in fragments from which predators have disappeared (Şekercioğlu et al. 2002).

Coastal swamps and marshes provide important ecosystem service as filters for sediments and nutrients that otherwise create imbalances in highly productive estuarine and coastal marine ecosystems (see Figures 1.5 and 5.2). Draining and filling swamps and marshes on a global scale for conversion to urban or agricultural uses and/or control of disease vectors has had minimal effect on disease epidemiology but has increased export of terrestrial materials to estuary and marine ecosystems. Increasing incidence and extent of coastal hypoxia and red tides now threaten important fisheries (Howarth et al. 2011).

Deforestation, overgrazing, and other activities that reduce vegetation cover increase the intensity of flooding and initiate changes in regional precipitation patterns that exacerbate regional warming and drying, leading to further ecosystem degradation and eventual loss of ecosystem services (desertification) (Schlesinger et al. 1990; Whitford 2002; Foley et al. 2003a; Juang et al. 2007; Bonan 2008; Janssen et al. 2008). Even past civilizations with primitive technologies were capable of sufficient deforestation to threaten their survival (Xue et al. 1990; Zheng and Eltahir 1998; Janssen et al. 2008; Lentz and Hockaday 2009; Cook et al. 2012). The collapse of the Mesopotamian and Maya civilizations, in what were originally fertile regions, may have resulted from climate changes induced by deforestation and overgrazing (Whitford 2002; Perry et al. 2008; Lentz and Hockaday 2009; Cook et al. 2012). Although droughts are related to changes in global circulation patterns, anthropogenic deforestation may exacerbate drought severity (Briant et al. 2010). Furthermore, anthropogenic conversion of ecosystems can persist for long periods (Harding et al. 1998; Schowalter et al. 2003; Ross and Rangel 2011). For example, Ross and Rangel (2011) found that the legacy of Maya agroforestry (significant differences in tree species composition between forests surrounding Maya ruins and forests at greater distance) has persisted for at least 1,000 years around Maya ruins along the Belize-Guatemala border. Cook et al. (2012) used climate models to demonstrate that deforestation during the Maya period may have increased the severity of drought effects by 250%.

Conversion of natural ecosystems to agricultural, urban, industrial, or reservoir developments represents the most extreme changes to ecosystems (Kimberling et al. 2001) and their ability to produce services. Although the benefits of intensive agriculture for production of food and other commodities are well known, they also have required enormous inputs of fossil fuels, irrigation water, fertilizers, and pesticides for maintenance of their artificial state and have led to introduction of many associated invasive species (Figure 6.7). Furthermore, such conversion has compromised the ability of the formerly continuous forest or grassland landscapes to provide pollinators, native biocontrol agents, fresh water, fisheries, and other ecosystem services (Klein et al. 2007; Landis et al. 2008).

The crops valued for food and fiber production typically are early successional species that pioneer disturbed sites, grow and reproduce rapidly (the attributes we value), and are replaced over time by later successional

FIGURE 6.7 (SEE COLOR INSERT.)
Apple orchard established in desert. Proximity to river water permits the required irrigation, but this conversion represents a dramatic alteration of the desert ecosystem and cannot be sustained without adequate water supply.

species that are better adapted to competition as the community develops (see Chapter 5). Crop systems typically are monocultures of a particular species, often a particular variety, planted over large areas. Tillage practices leave much bare soil between rows that invite a variety of early successional "weeds" that exploit the available space, as well as plant species that represent advanced stages of ecological succession. The consistent production of particular crop species, and often the same varieties, over large areas (e.g., wheat belt, corn belt, cotton belt) provides an abundant and reliable food source that promotes population growth by associated herbivorous insects and pathogens (Schowalter 2011). In the absence of control efforts, these insects and pathogens would accelerate succession. Abandoned agroecosystems inevitably undergo "old-field" succession toward the ecosystem type (biome) characteristic of the region (Odum 1969).

Conversion of natural ecosystems to urban ecosystems or paved surfaces (e.g., roads) is an even more extreme change in ecosystem conditions. Extreme three-dimensional structure and low albedo surfaces increase local temperature, air turbulence, and vertical sheer (Figure 6.8) (Arnfield 2003; Akbari et al. 2009). Urban heat islands can be as much as 10°C hotter than surrounding ecosystems (Arnfield 2003). If your bare feet burn as you hop across sunlit pavement, consider the plight of small, flightless animals, such as many insects, that have no other way to disperse than over this forbidding surface. Given that about 60% of urban surface is pavement and roofs (Akbari et al. 2009), replacement of vegetation (albedo 0.1–0.3) with asphalt (albedo 0.05–0.10) exacerbates urban warming, whereas concrete (albedo 0.5–0.6) could offset urban warming (Akbari et al. 2009). Industrial effluents negatively affect urban dwellers, as well as ecosystems downwind and downstream. Exotic plant species dominate urban vegetation that

FIGURE 6.8 (SEE COLOR INSERT.)
Former temperate coniferous rainforest converted to urban ecosystem with virtually no remaining vegetation. Replacing vegetation cover and soil with paving greatly reduces albedo due to the preponderance of dark surfaces, thereby increasing diurnal temperature and temperature range, and eliminates water storage capacity, thereby increasing runoff and downstream flooding.

often becomes stressed by soil compaction, nutrient imbalance, and water limitation.

Some native species of insects are capable of exploiting urban habitats, particularly unsanitary conditions and detrital resources. Some species may adapt to higher temperatures and greater distances between potential resources in the urban matrix (San Martin y Gomez and Van Dyck 2012). Urban vegetation stressed by elevated temperatures, pollutants, fertilizer application, soil compaction, impervious surfaces, and altered drainage conditions is vulnerable to a variety of herbivorous insects (Frankie and Ehler 1978; Cregg and Dix 2001; Raupp et al. 2010). Frequent pesticide application to control nuisances also reduces the abundance of desirable insects, such as butterflies, dragonflies, and biological control agents. However, depending on the complexity of urban vegetation and landscaping, some ecosystem services can be maintained in urban ecosystems, including sufficient survival of biological control agents to control populations of some pest species and minimize need for insecticides (Shrewsbury and Raupp 2006; Pouyat et al. 2007).

Human-transported species, such as black and brown rats and associated fleas, German cockroaches, *Blatta germanica*, lice, *Pediculus* spp., and bedbugs, *Cimex* spp., are typical urban pests, enjoying the optimal environmental conditions and food supply provided by air-conditioned structures. Various detritivores, such as silverfish, native and Formosan subterranean termites, *Reticulitermes* spp. and *Coptotermes formosanus*, respectively, and Argentine ants, *L. humile*, become household structural pests. Human transported mosquitoes, such as *Aedes aegypti* and *Anopheles gambiae*, as well as native mosquitoes, breed in rain gutters, neglected pots, fountains, and pools (Carlson et al. 2009) and are capable of initiating disease outbreaks in crowded urban populations. Urban ecosystems become centers for unregulated use of toxic pesticides to control unwanted weeds and insects. These pesticides or their degradation products (often equally toxic) subsequently wash into drainage systems and affect aquatic ecosystems downstream.

Continued expansion of human activity at urban fringes reduces the remaining area of fragmented ecosystems capable of providing their natural services. Although remnant soils and ornamental plantings in urban ecosystems continue to provide some services (e.g., shade, water infiltration, garden produce, habitat for some native insects and other species, and aesthetic values), the range and level of these services is greatly reduced (Figure 6.8). As the proportion of landscapes occupied by urban, suburban, and industrial patches and connecting paved roads increases, their influence on remnant ecosystems increases, further compromising ecosystem processes that support services (Figure 6.9).

6.1.5 Induction of Insect Problems

Anthropogenic activity has created favorable conditions for elevated population sizes of native and nonnative insects. For example, crop breeding has improved flavor and production of food products because crop plants are selected to allocate resources to growth instead of distasteful defenses (Michaud and Grant 2009). Less defended cultivars become more vulnerable to a wider array of generalist herbivores, as well as to higher abundances of specialist herbivores (Kessler and Baldwin 2002), thereby requiring increased efforts to control crop losses. Conversion of natural, diverse vegetation to rapidly growing, commercially valuable crop species on a regional scale has resulted in more severe and widespread outbreaks of adapted insects. Outbreaks of the mountain pine beetle, *Dendroctonus ponderosae*, have become more frequent and severe in western North America as a result of fire exclusion and replanting harvested forests with selected commercially valuable species at higher density (see Figure 4.5) (Schowalter and Turchin 1993; Raffa et al. 2008). Transformation of many agricultural landscapes into expansive monocultures with little remnant natural habitat also has simplified landscapes, reducing natural biological control of prey species (Landis et al. 2008) and promoting spread of insects among suitable patches. Such land use change has favored elevated insect pest abundance on a landscape and regional scale and increased the need for insecticides. Meehan et al. (2011) used remotely sensed land cover data, data from a national census of farm management practices, and data from a regional crop pest monitoring network across a range of cropping systems in the midwestern United States and found, as expected, that the proportion of harvested cropland treated with insecticides (a) increased with the

FIGURE 6.9 (SEE COLOR INSERT.)

Urban effects on surrounding ecosystems. (a) Continued fragmentation of natural ecosystems at the urban fringe reduces remaining area for ecosystem services and increases the area affected by higher urban temperatures. (b) Connecting roads increase avenues for introduction and spread of invasive pests.

proportion and patch size of cropland and (b) decreased with the proportion of seminatural habitat in a county. In other words, loss of uncultivated, seminatural habitats in agricultural landscapes has increased pest problems and the need for insecticides.

Overgrazing by livestock reduces plant cover, leading to soil desiccation that results in further loss of plant cover in a positive feedback cycle that can spread over large areas, as in sub-Saharan Africa (Janssen et al. 2008) and the southwestern United States (Schlesinger et al. 1990). Overgrazing promotes locust outbreaks that contribute to further loss of vegetation cover (Cease et al. 2012).

Anthropogenic effects on climate also are likely to increase the frequency and severity of herbivore outbreaks. Grasshopper, bark beetle, and many lepidopteran populations are favored by warm, dry conditions (Ma 1958; Konishi and Itô 1973; Capinera 1987; Mattson and Haack 1987; Breshears et al. 2005; Schowalter 2011) predicted by climate change models to increase in many regions. Williams and Liebhold (2002) projected increased outbreak area and shift northward for the southern pine beetle, *Dendroctonus frontalis*, but reduced outbreak area and shift to higher elevations for the mountain pine beetle in North America as a result of increasing temperature. Interaction among multiple factors changing simultaneously may affect insects differently than predicted from responses to individual factors (Franklin et al. 1992; Marks and Lincoln 1996).

Zimmerman et al. (1982) suggested that tropical deforestation and conversion to pasture and agricultural land could increase the biomass and methane emissions of fungus-growing and soil-feeding termites. More recent calculations by Sanderson (1996) indicate that termites are responsible for about 2% of global carbon dioxide flux and 4–5% of global methane flux, or about 25–50% of annual emissions from fossil fuel combustion (Khalil et al. 1990).

Disease vectors also are likely to increase in abundance and distribution if current trends continue (Merritt et al. 2005; Vittor et al. 2006). Mosquito vectors of malaria are promoted by deforestation and conversion to agricultural uses that increase the availability of stagnant pool habitats and proximity to susceptible human populations (Póvoa et al. 2003; Vittor et al. 2006).

However, the most serious consequence of continued ecosystem conversion or degradation is restriction of ecosystem services. Crop failures and shortages of food or water have been primary factors leading to social unrest, population displacement, and increase in insect-vectored crowd diseases in stressed, densely populated refugee communities (Riley 1878; Bray 1996; Diamond 1999; Smith 2007; Zhang et al. 2007; Bora et al. 2010; Brouqui 2011; Hsiang et al. 2011). Diseases such as malaria and yellow fever are particular threats, given the resistance of both diseases and their vectors to currently used antibiotics or insecticides (see section 6.2).

Management activities that disrupt natural food webs (see Chapter 5) also release potential pest species from regulation by predators. The importance of predation to population regulation cannot be overstated.

Predators eliminate almost inconceivable numbers of insects. Stewart and Woolbright (1996) calculated that tree frog (*Eleutherodactylus coqui*) adults, at densities of about 3,300 ha^{-1}, consumed 10,000 insects ha^{-1} per night in a Puerto Rican rainforest 17,000 pre-adult frogs ha^{-1} ate an additional 100,000 insects ha^{-1} per night. In other words, this single predator species eats about 40 million insects per year. Lizards, insectivous birds, bats, and other insectivores contribute additional regulation of insect numbers. Habitat fragmentation, insecticides, and other management activities that disproportionately reduce predator populations interfere with their regulatory capability.

6.2 Insecticide Effects

BOX 6.1 INSECTICIDE MODES OF ACTION AND RESISTANCE MECHANISMS

Despite decades of research by thousands of chemists to develop new insecticides, relatively few biochemical modes of action are represented among commercially important compounds (Nauen 2006). For example, chlorinated hydrocarbons introduced in the 1930s, organophosphates in the 1940s, carbamates in the 1950s, synthetic pyrethroids in the 1970s, and neonicotinoids in the 1990s represent the most widely used insecticides, but these act on only three different target sites in insects, all affecting a single pathway, cholinergic nerve transmission (Nauen 2006). Nevertheless, additional modes of action are available, and their increased use would help delay insecticide resistance among target insect species (Sparks 1996). Modes of action of the major classes of insecticides, and insect adaptations, are described as follows.

Dichlorodiphenyltrichloroethane (DDT) and its analogues, such as methoxychlor, are directly toxic to nerves of invertebrates and vertebrates. Although pyrethroids are chemically dissimilar, both groups of insecticides interact with the same or related target sites on the nerve membrane (Beeman 1982). This common mechanism of neurotoxicity can be explained in terms of ion fluxes across membranes in isolated, intact axons. The mode of action is primarily depolarization of the nerve impulse, such that conductance is unimpeded but the channels cannot return to their resting condition (Beeman 1982). This may result in continuous "firing" until conduction is blocked by exhaustion of membrane potential.

Peripheral nerves are the primary target of chlorinated hydrocarbons and pyrethroids, but pyrethroids also affect the central nervous system. This explains the rapid "knock-down" effect of pyrethroids (Beeman 1982).

Cyclodiene hydrocarbons, such as chlordane, lindane, and aldrin, have a somewhat different mode of action. The cyclodienes act on both peripheral and central nervous systems but have a greater effect on the central nervous system, primarily through stimulation of presynaptic release of γ-aminobutyric acid (GABA), a stimulatory neurotransmitter in the insect central nervous system (Beeman 1982; Sparks 1996). Like the DDT analogues and pyrethroids noted above these compounds affect both invertebrates and vertebrates (Sparks 1996).

Organophosphates (such as malathion and guthion) and carbamates (such as carbaryl) inhibit acetylcholinesterase, an enzyme required to terminate nerve pulses, through phosphorylation or carbamoylation, respectively (Sparks 1996; Nauen 2006). The binding of these insecticides with acetylcholinesterase prevents breakdown of acetylcholine, causing overstimulation of the postsynaptic nerve axon by the excess acetylcholine, resulting in paralysis and death (Sparks 1996). These compounds are powerful nerve toxins that affect both invertebrates and vertebrates (Sparks 1996).

Neonicotinoids, such as imidichloprid, bind directly to the acetylcholine receptors on the postsynaptic nerve axon, thereby interrupting transmission of nerve impulses (Sparks 1996; Ishaaya et al. 2007). Given that these chemicals act directly on postsynaptic reception, rather than on breakdown of acetylcholine, insects resistant to the chemicals described above typically are not resistant to neonicotinoids, at least initially (Sparks 1996).

Avermectins—including avermectin, abamectin, and ivermectin— are naturally occurring products of fermentation by *Streptomyces avermitilis* (Ishaaya et al. 2007). Avermectins block the transmission of nerve impulses by opening the chloride channels, stimulating presynaptic release of GABA at nerve endings (Sparks 1996). Although this represents a different mode of action from the chemicals described above varying degrees of cross-resistance to avermectins may develop in insects resistant to pyrethroids (Sparks 1996).

Spinosyns, primarily spinosad, are naturally occurring chemicals produced as fermentation products by a rare actinomycete, *Saccharopolyspora spinosa* (Sparks 1996; Ishaaya et al. 2007). Its mode of action is apparently unique, causing rapid excitation of the insect nervous system. The chemical must be ingested to be effective, so it is relatively nontoxic to predators. Currently, there is no evidence of cross-resistance (Sparks 1996).

Another group of microbial toxins with a unique mode of action is the pyrroles, primarily dioxapyrrolomycin from a *Streptomyces* bacterium. These compounds uncouple oxidative phosphorylation from the electron transport process in cells, disrupting the production of adenosine triphosphate (ATP) and ending further metabolic activity (Sparks 1996).

A group of crystalline toxins derived from the bacterium, *Bacillus thuringiensis*, has been used as insecticides and, more recently, incorporated into the genome of transgenic crops (Tabashnik et al. 1996, 1997; Carrière et al. 2001; Heuberger et al. 2008a, b, 2011), substantially reducing the use of more toxic and broad-spectrum insecticides (Carriére et al. 2003; Cattaneo et al. 2006). Subspecies of this bacterium produce toxins that are relatively specific to moths, flies, and beetles. A crystalline endotoxin is activated by high pH and proteolytic enzymes in the insect gut. The toxins bind to specific receptors on the brush border membrane of midgut columnar cells (Sparks 1996; González-Cabrera et al. 2006; Ishaaya et al. 2007). The cells of the gut epithelium swell and separate, increasing permeability of the gut-hemocoel barrier and causing death (Sparks 1996). Different endotoxins (e.g., Cry1Ac, Cry1C, and Cry2Ab) bind to different receptors, potentially requiring different mechanisms of insect resistance (Sparks 1996).

Insect growth regulators (IGRs) include juvenile hormone (JH) analogues (juvenoids) and chitin synthesis inhibitors (CSIs). JH is unique to insects. In concert with molting hormone (20-hydroxyecdysone), JH regulates insect development. Juvenoids mimic the action of JH, thereby disrupting the normal process of metamorphosis, leading to incomplete molting or incomplete transition to the new life stage (Harborne 1994; Sparks 1996). CSIs, such as diflubenzeron and related benzoylphenyl ureas, inhibit ecdysone-dependent N-acetylglucosamine incorporation into chitin, the protein that forms the insect exoskeleton (Beeman 1982), but the exact mechanism remains unclear (Sparks 1996). This prevents normal exoskeleton formation following molting of immature insects.

Because most target herbivores, detritivores, and blood-feeders have adapted to the variety of plant and animal biochemical defenses, genetic mechanisms are in place for adaptation to insecticides (Hsu et al. 2006, 2008). Four major mechanisms are responsible for resistance (Miller 1996; Ottea and Leonard 2006). *Behavioral resistance* involves inheritance of behaviors that allow the insect to avoid a toxic dose. For example, selection of feeding sites on lower portions of plants could provide protection from aerially applied insecticides. *Penetration resistance* reflects cuticular modifications to reduce penetration of a toxin through the insect cuticle. *Altered site of action resistance* occurs when the target site is somehow altered genetically to prevent the toxic effect. This requires a greater dose of the insecticide to accomplish the desired effect. For example, many insects have developed knock-down resistance (kdr) to pyrethroid insecticides. Insects with this trait either do not respond to a dose that kills susceptible insects or the symptoms take longer to appear. *Metabolic resistance* is demonstrated by the ability of exposed insects to recover from the effects of the insecticide.

As described in Chapter 3 (see Box 3.1), insects have demonstrated considerable capacity to adapt to plant defensive chemicals, representing a wide variety of molecular structures. Many of these structures are common among various insecticides, making mixed-function oxidases, glutathione S-transferases, and hydrolases produced by most insects effective in detoxification (Miller 1996; Schowalter 2011). Most of these enzymes are encoded by cytochrome P450 genes (Feyereisen 1999; Daborn et al. 2002; Karban and Agrawal 2002; Ranson et al. 2002; Mao et al. 2006; Mao et al. 2007). Binding site alteration appears to be the primary mechanism of resistance to Bt toxins (Caccia et al. 2010). For many insects, multiple alleles are associated with increased resistance, and over-expression of genes can be induced by exposure to insecticides (Miller 1996; Feyereisen 1999; Daborn et al. 2002; Ranson et al. 2002; Brun-Barale et al. 2010).

Target species can vary considerably in their resistance to various insecticides. For example, Pridgeon et al. (2008) tested three mosquito species against 19 insecticides with different modes of action and concluded that successful mosquito control requires selection of the most effective chemical for the least susceptible species. Clearly, this complexity adds to the difficulty of controlling insects.

Among the most controversial anthropogenic effects on the environment has been the introduction of insecticides to control insects, starting with arsenical compounds in the late 1800s and continuing through chlorinated hydrocarbons (such as DDT), organophosphates, carbamates, and neonicotinoids during the later 1900s. Insecticides, as well as other pesticides, have had conflicting results. On one hand, insecticides have led to reduced abundances of crop-destroying and disease-transmitting species in the short term, potentially saving millions of lives. On the other hand, insecticides themselves have had devastating, unintended consequences for nontarget species and ecosystem processes fundamental to ecosystem services (Scholz et al. 2012).

DDT is perhaps the most familiar example of an insecticide with serious environmental consequences that negatively affected sustainability of ecosystem services (Carson 1962). When DDT was introduced in the 1940s, it was viewed as a much safer replacement for arsenical compounds that had been widely used since the late 1800s and were toxic to a broader range of species. DDT was recognized as an effective insect control tool in the late 1930s and was instrumental in protecting troops from disease vectors during World War II (Bray 1996). Its success in controlling disease vectors led to its widespread use for crop protection, resulting in widespread exposure of many insect species that subsequently developed resistance to this insecticide. Mosquitoes, in particular, were resistant to DDT by the late 1940s (Carson 1962; Ribeiro and Mexia 1970; Collins and Paskewitz 1995; Vezenegho et al. 2009). Worse, malaria strains also have become resistant to common anti-malaria drugs, including chloroquine (Sidhu et al. 2002). Despite worldwide bans against agricultural use of DDT, in order to maintain some effectiveness against mosquitoes, much of the current inventory allocated for mosquito control is diverted to agricultural uses in many countries (Katima and Mng'anya 2009).

Increased application rates necessary to combat resistant mosquitoes has led to human health problems associated with exposure to DDT. Corin and Weaver (2005) conducted a cost-benefit analysis of DDT for reducing infant mortality in South Africa and concluded that costs outweighed benefits where malaria incidence is low. Furthermore, DDT-resistant mosquitoes showed cross-resistance to pyrethroids due to detoxification mechanisms or reduced sensitivity of their nervous system to insecticides with similar modes of action (see Box 6.1) (Rongsriyam and Busvine 1975; Omer et al. 1980). Resistance to insecticides among target species has become such a serious problem that entomological research is increasingly focused on tactics, especially reduced application of insecticides, to delay genetic selection for resistance. This brings into question the wisdom of attempting to control insects with insecticides.

Effective alternatives are available for preventing malaria where mosquitoes are resistant to DDT or insecticides with similar modes of action (Grieco et al. 2007; van den Berg 2009). In particular, screened windows and doors can prevent mosquitoes from gaining access to human abodes. Pyrethroid impregnated bed nets are highly effective in repelling and killing mosquitoes in homes and protecting children under five years of age, the group most vulnerable to malaria (Goodman and Mills 1999). Furthermore, elimination of mosquito breeding sites in construction debris, discarded containers, and other stagnant pools near human habitations can substantially reduce human proximity and exposure to populations of vectors and pathogens (Carlson et al. 2009). Use of these alternatives can substantially reduce the need for spraying with DDT or other insecticides (Over et al. 2004). Ito et al. (2002) generated transgenic *Anopheles stephensi* that produce antiparasitic genes in their midgut epithelium, making them incapable of vectoring disease organisms. If a safe method for spreading such genes throughout wild mosquito populations can be found, this technique would offer a means of interrupting disease transmission with minimal environmental effect.

Targeted control of larvae around human habitations can increase the effectiveness of mosquito suppression efforts (Killeen et al. 2002). Although treatment of water bodies with broad-spectrum toxins, such as DDT and

Paris Green (copper II acetoarsenite), resulted in devastating consequences for aquatic ecosystems and fisheries, more specific treatment, such as application of *Bacillus thuringiensis* var. *israelensis* (Bti) and *B. sphaericus*, can control mosquito larvae effectively under some conditions with little effect on nontarget species (Russell and Kay 2008; Skovmand et al. 2009). Unfortunately, mosquitoes have shown an ability to become resistant to these bacteria (Singh and Prakash 2009). As stated earlier, elimination of mosquito breeding habitats near human abodes can substantially reduce the need for chemical controls.

Other insects also become resistant quickly to regularly used insecticides. For example, Felland et al. (1990) reported that the soybean looper, *Chrysodeixis includens*, was resistant to permethrin (mortality declining from 99% to 60%) within five years of first application in soybeans in Mississippi. The potential for resistance among target species was observed early. Gypsy moths, for example, were tolerant of much higher exposure to arsenic than were humans (Riley 1885; Forbush and Fernald 1895). The earliest documented case of resistance development was published by Melander (1923), who reported that San Jose scale, *Quadraspidiotus perniciosus*, in Washington was becoming progressively more tolerant of insecticides, particularly sulfur compounds. A number of target fly species were resistant to DDT by the mid-1940s (Carson 1962), and the first report of resistance in a major crop pest, the boll weevil, was in 1957 (Roussel and Clower 1957). Much entomological research is now directed toward noninsecticidal control options and tactics to delay resistance to insecticides among target species (Soderlund and Bloomquist 1990; Feyereisen 1999).

Furthermore, pesticides are rapidly becoming a major cause of human illness and allergic reactions. Victor et al. (2010) reported that pesticides, primarily fungicides, were responsible for 9% of reported cases of contact dermatitis. Tanner et al. (2011) found that rotenone and a number of pyrethroids increased the incidence of Parkinson's disease. Harley et al. (2011) found that exposure to organophosphates increased the risk of premature and low-weight births. Adverse effects of Bt toxins from transgenic crops have not been reported, but Aris and LeBlanc (2011) detected Cry1Ab toxin in blood samples of pregnant and nonpregnant women and fetuses.

More serious for sustainability of ecosystem services, insecticides disrupt food webs that control the sustainability of ecosystem services. Predators are less exposed to natural environmental toxins than are herbivores and tend to be disproportionately affected by insecticide application, resulting in disruption of food webs. A number of studies have demonstrated cascading effects of changes in predation rate (Marquis and Whelan 1994; Letourneau and Dyer 1998; Dyer and Letourneau 1999a, b; Knight et al. 2005; Mooney 2007; Letourneau et al. 2009), including outbreaks of nontarget pests (Luck and Dahlsten 1975) and indirect effects on pollinator function (Knight et al. 2005). Many insectivorous birds and predaceous arthropods are susceptible to insecticide toxicity (Bajwa and Aliniazee 2001; Baldwin et al. 2009), undermining natural regulation of prey/pest populations and triggering nontarget pest outbreaks (Sandquist et al. 1993). Honey bee, fish, and aquaculture production also are negatively affected, representing a more direct effect on provisioning services (Smith 1983; Claudianos et al. 2006; Halm et al. 2006; Barbee and Stout 2009). The disruption of indirect, as well as direct, interactions may have serious consequences for ecosystem services (see Chapter 5).

Unfortunately, insecticides often are used even when insects are absent, with the intent of preventing their appearance. This practice is dangerous, especially in homes, because most insecticides must be reapplied frequently to maintain their effectiveness, and this constant use and human exposure can cause serious medical problems and lead more quickly to insecticide resistance in affected insects, outbreaks of nontarget insects, and mortality in nontarget insectivorous fish and birds that provide important ecosystem services (Scholz et al. 2012).

An example is the growing popularity of automatic insecticide (pyrethroid) misting systems in the United States for protection of homes and outdoor entertainment areas against mosquitoes. Such systems are easily sold to homeowners frightened of potential disease vectors and nuisances but are ultimately counterproductive. Although mosquitoes flying directly into the mist when released may be killed, many more are exposed to sublethal levels that promote adaptation for resistance, shortening the time until most mosquitoes will no longer be killed by recommended concentrations of the insecticide.

6.3 Alternative Control Options

A number of options are available to prevent or reduce insect populations, as necessary. Preventative measures that ensure maintenance of natural regulatory mechanisms have the fewest undesirable consequences, but many remedial alternatives also are available, often with other ill effects.

Transgenic crops offer a means of minimizing crop losses to target insects without applying more toxic pesticides. A number of crops—including wheat, corn, maize, rice, and soybeans—have been genetically transformed to include genes from the bacterium, *B. thuringiensis* (Bt), that enable the plant to produce Bt toxins that are effective against a number of caterpillars or beetles that feed on these crops. Since the first planting of transgenic crops on a large scale in 1996, their adoption increased to a global total of more than 20 million hectares in 2005 (Tabashnik and Carriére 2008). The widespread and prolonged exposure to Bt toxins in transgenic crops represents one of the largest and most acute selections for resistance ever experienced by herbivorous insects (Tabashnik et al. 2003). A few target insects have developed resistance to single-Bt toxin-producing crops in the field, but the requirements for non-Bt refuges to minimize breeding among resistant individuals has led to less resistance than would be expected from laboratory experiments (Tabashnik et al. 2003, 2008). Nevertheless, a second generation of transgenic crops that express multiple Bt toxins with different modes of action should impede development of resistance.

Although these transformed crops have substantially reduced the application of chemical insecticides for insect control and have fewer nontarget effects than do applied insecticides (Marvier et al. 2007), they do have unintended, and often undesirable, consequences. For example, honey bees and other pollinators moving pollen among transformed and nontransformed cultivars and wild relatives not only facilitate adaptation of target species to the bacterial toxins but introduce the genes into populations of wild relatives of the crop species, with largely unknown consequences for food webs and ecosystem processes in the surrounding landscape (Sisterson et al. 2007; Heuberger et al. 2008a, b).

Although strains of Bt have been developed to be relatively specific to Lepidoptera, Coleoptera, or Diptera, many nontarget species are susceptible. Among Lepidoptera, a number of nontarget butterflies are among the species whose populations can be reduced by exposure to Bt application (Losey et al. 1999; Hansen Jesse and Obrycki 2000; Sears et al. 2001; Zangerl et al. 2001; O'Callaghan et al. 2005). Black swallowtail caterpillars (*Papilio polyxenes*) that feed on plants in the carrot family, including dill, fennel, and parsley, giant swallowtail caterpillars (*Papilio cresphontes*) that feed on plants in the citrus family, and tiger swallowtail caterpillars (*Papilio glaucus*) that feed on plants in the rose family, including cherry and other fruit trees, are among the colorful butterflies that can be exposed to Bt, leading to a decline in abundance of some of North America's most popular insects. Other nontarget species are the giant silk moths (Saturniidae)—including the cecropia moth, *Hyalophora cecropia*, and luna moth, *Actias luna*—both of which feed on a variety of tree species and are among North America's most popular moths. Many of these species also are pollinators. Reduced abundances would threaten pollination services for some plant species. Laboratory studies have demonstrated that Bt-laden plant detritus can affect soil organisms or processes susceptible to Bt toxins (Tarkalson et al. 2008) and wash into aquatic ecosystems where they can affect other organisms (Rosi-Marshall et al. 2007; Swan et al. 2009), but effects have not been observed in the field (Frouz et al. 2008; Hönnemann et al. 2008; Lehman et al. 2008).

Augmentation of natural plant defenses represents another means of increasing natural regulation of pest populations. Application of jasmonic acid (see Chapter 5) can induce increased production of chemical defenses by crop plants, thereby reducing pest damage but perhaps with some effect on food flavor (Thaler et al. 2001; Mészáros et al. 2012).

Introduction or augmentation of biological control agents can impose greater population regulation by predators. Whereas augmentation of native predators would reduce the time delay before initiation of normal regulatory responses (Landis et al. 2000, 2008; Tscharntke et al. 2005, 2007), introduction of exotic predators raises the same concerns as introduction of their target pests. Although predators generally are not as likely to be invasive as their hosts, and biological control avoids the concerns of using insecticides, not all consequences of introducing new species to control other species have been anticipated, and once introduced, the results may be irreversible (Louda et al. 2003; McCoy and Frank 2010). Despite improved selection and quarantine procedures, undesirable effects can follow biocontrol agent introductions. In some cases, introduced predators have attacked native species related to the target host or switched hosts to native species when the target host was rare (Louda et al. 2003). In other cases, biological control agents can improve the competitive ability of their target host, perhaps because defensive responses to the biocontrol agent negatively affect competitors (Callaway et al. 1999; see Chapter 5).

6.4 Summary

Humans influence ecosystems in a variety of ways that affect insect populations, as well as the sustainability of ecosystem services. Some of these changes have acute and dramatic effects, such as landscape fragmentation and conversion to agricultural, industrial, or urban uses, whereas others will have more subtle or gradual effects, such as changes in atmospheric chemistry.

Landscape fragmentation and conversion alter habitat distribution and the ability of many species to find or reach suitable habitats or resources from refuges in remnant fragments of natural habitat. However, for mobile species capable of exploiting new concentrations of host plants in agricultural landscapes, conditions could not be more favorable, often inducing unprecedented population growth and spread. Introduction of agricultural chemicals, especially fertilizers and pesticides, has led to adaptation for resistance in target insects, with far-reaching effects on nontarget species that control ecosystem processes, even leading to hypoxia in coastal zones. These changes threaten ecosystem services, such as water quality and yield and wildlife and fisheries production. Alternative insect control options with new modes of action may have fewer nontarget effects but may have largely unknown consequences for ecosystem services.

Changes in atmospheric chemistry resulting from industrial effluents and fossil fuel combustion have led to greenhouse warming of the planet with largely unknown effects for ecosystem services. Unfortunately, the combined effects of these various changes are difficult to predict, given that our most sophisticated models are subject to large effects of small changes in parameter values—the "butterfly effect," as described in the previous chapter.

References

Ackerman, A. S., O. B. Toon, D. E. Stevens, A. J. Heymsfield, V. Ramanathan, and E. J. Welton. 2000. Reduction of tropical cloudiness by soot. *Science* 288: 1042–1047.

Agee, J. K. 1993. *Fire Ecology of Pacific Northwest Forests*. Washington, DC: Island Press.

Aizen, M. A. and P. Feinsinger. 1994. Habitat fragmentation, native insect pollinators, and feral honey bees in Argentine "Chaco Serano." *Ecological Applications* 4: 378–392.

Akbari, H., S. Menon, and A. Rosenfeld. 2009. Global cooling: Increasing world-wide urban albedos to offset CO_2. *Climate Change* 94: 275–286.

Allen, C. R., D. M. Epperson, and A. S. Garmestani. 2004. Red imported fire ant impacts on wildlife: A decade of research. *American Midland Naturalist* 152: 88–103.

Alstad, D. N., G. F. Edmunds, Jr., and L. H. Weinstein. 1982. Effects of air pollutants on insect populations. *Annual Review of Entomology* 27: 369–384.

Andreae, M. O., D. Rosenfeld, P. Artaxo, A. A. Costa, G. P. Frank, K. K. Longo, and M. A. F. Silva-Dias. 2004. Smoking rain clouds over the Amazon. *Science* 303: 1337–1342.

Aris, A. and S. LeBlanc. 2011. Maternal and fetal exposure to pesticides associated to genetically modified foods in Eastern Townships of Quebec, Canada. *Reproductive Toxicology* 31: 528–533.

Arndt, D. S., M. O. Baringer, and M. R. Johnson, eds. 2010. State of the climate in 2009. *Bulletin of the American Meteorological Society* 91: S1–S224.

Arnfield, A. J. 2003. Two decades of urban climate research: A review of turbulence, exchanges of energy and water, and the urban heat island. *International Journal of Climatology* 23: 1–26.

Arnone, J. A. III, J. G. Zaller, C. Ziegler, H. Zandt, and C. Körner. 1995. Leaf quality and insect herbivory in model tropical plant communities after long-term exposure to elevated atmospheric CO_2. *Oecologia* 104: 72–78.

Bajwa, W. I. and M. T. Aliniazee. 2001. Spider fauna in apple ecosystem of western Oregon and its field susceptibility to chemical and microbial insecticides. *Journal of Economic Entomology* 94: 68–75.

Balanyá, J., J. M. Oller, R. B. Huey, G. W. Gilchrist, and L. Serra. 2006. Global genetic change tracks global climate warming in *Drosophila subobscura*. *Science* 313: 1773–1775.

Baldwin, D. H., J. A. Spromberg, T. K. Collie, and N. L. Scholz. 2009. A fish of many scales: Extrapolating sublethal pesticide exposures to the productivity of wild salmon populations. *Ecological Applications* 19: 2004–2015.

Bale, J. S., G. J. Masters, I. D. Hodkinson, C. Awmack, T. M. Bezemer, V. K. Brown, J. Butterfield, A. Buse, J. C. Coulson, J. Farrar, et al. 2002. Herbivory in global climate change research: Direct effects of rising temperature on insect herbivores. *Global Change Biology* 8: 1–16.

Barbee, G. C. and M. J. Stout. 2009. Comparative acute toxicity of neonicotinoid and pyrethroid insecticides to non-target crayfish (*Procambarus clarkii*) associated with rice–crayfish crop rotations. *Pest Management Science* 65: 1250–1256.

Beedlow, P. A., D. T. Tingey, D. L Phillips, W. E. Hogset, and D. M. Olszyk. 2004. Rising atmospheric CO_2 and carbon sequestration in forests. *Frontiers in Ecology and the Environment* 2: 315–322.

Beeman, R. W. 1982. Recent advances in mode of action of insecticides. *Annual Review of Entomology* 27: 253–281.

Bender, M. A., T. R. Knutson, R. E. Tuleya, J. J. Sirutis, G. A. Vecchi, S. T. Garner, and I. M. Held. 2010. Modeled impact of anthropogenic warming on the frequency of intense Atlantic hurricanes. *Science* 327: 454–458.

Benstead, J. P., A. C. Green, L. A. Deegan, B. J. Peterson, K. Slavik, W. B. Bowden, and A. E. Hershey. 2007. Recovery of three Arctic stream reaches from experimental nutrient enrichment. *Freshwater Biology* 52: 1077–1089.

Beuzelin, J. M., T. E. Reagan, W. Akbar, H. J. Cormier, J. W. Flanagan, and D. C. Blouin. 2009. Impact of Hurricane Rita storm surge on sugarcane borer (Lepidoptera: Crambidae) management in Louisiana. *Journal of Economic Entomology* 102: 1054–1061.

Bezemer, T. M. and T. H. Jones. 1998. Plant-herbivore interactions in elevated atmospheric CO_2: Quantitative analysis and guild effects. *Oikos* 82: 212–222.

Bonan, G. B. 2008. Forests and climate change: Forcings, feedbacks, and the climate benefits of forests. *Science* 320: 1444–1449.

Bora, S., I. Ceccacci, C. Delgado, and R. Townsend. 2010. *World Development Report 2011: Food Security and Conflict.* Agriculture and Rural Development Department, World Bank.

Boyero, L., R. G. Pearson, M. O. Gessner, L. A. Barmuta, V. Ferreira, M. A. S. Graça, D. Dudgeon, A. J. Boulton, M. Callisto, E. Chauvet, et al. 2011. A global experiment suggests climate warming will not accelerate litter decomposition in streams but might reduce carbon sequestration. *Ecology Letters* 14: 289–294.

Bradshaw, W. E. and C. M. Holzapfel. 2001. Genetic shift in photoperiodic response correlated with global warming. *Proceedings of the National Academy of Sciences USA* 98: 14509–14511.

Braithwaite, R. W. and J. A. Estbergs. 1985. Fire patterns and woody litter vegetation trends in the Alligator Rivers region of northern Australia. In *Ecology and Management of the World's Savannahs*, J. C. Tothill and J. J. Mott, eds. 359–364. Canberra, ACT: Australian Academy of Science.

Braschler, B., L. Marini, G. H. Thommen, and B. Baur. 2009. Effects of small-scale grassland fragmentation and frequent mowing on population density and species diversity of orthopterans: A long-term study. *Ecological Entomology* 34: 321–329.

Bray, R. S. 1996. *Armies of Pestilence: The Impact of Disease on History.* New York: Barnes and Noble.

Breshears, D. D., N. S. Cobb, P. M. Rich, K. P. Price, C. D. Allen, R. G. Balice, W. H. Romme, J. H. Kastens, M. L. Floyd, J. Belnap, et al. 2005. Regional vegetation die-off in response to global-change-type drought. *Proceedings of the National Academy of Sciences USA* 102: 15144–15148.

Briant, G., V. Gond, and S. G. W. Laurance. 2010. Habitat fragmentation and the desiccation of forest canopies: A case study from eastern Amazonia. *Biological Conservation* 143: 2763–2769.

Brouqui, P. 2011. Arthropod-borne diseases associated with political and social disorder. *Annual Review of Entomology* 56: 357–374.

Brun-Barale, A., O. Héma, T. Martin, S. Suraporn, P. Audant, H. Sezutsud, and R. Feyereisen. 2010. Multiple P450 genes overexpressed in deltamethrin-resistant strains of *Helicoverpa armigera*. *Pest Management Science* 66: 900–909.

Buddle, C. M., D. W. Langor, G. R. Pohl, and J. R. Spence. 2006. Arthropod responses to harvesting and wildfire: Implications for emulation of natural disturbance in forest management. *Biological Conservation* 128: 346–357.

Butler, C. D., N. E. Beckage, and J. T. Trumble. 2009. Effects of terrestrial pollutants on insect parasitoids. *Environmental Toxicology and Chemistry* 28: 1111–1119.

Butler, C. D. and J. T. Trumble. 2008. Effects of pollutants on bottom-up and top-down processes in insect-plant interactions. *Environmental Pollution* 156: 1–10.

Caccia, S., C. S. Hernández-Rodríguez, R. J. Mahon, S. Downes, W. James, N. Bautsoens, J. Van Rie, and J. Ferre. 2010. Binding site alteration is responsible for field-isolated resistance to *Bacillus thuringiensis* Cry2A insecticidal proteins in two *Helicoverpa* species. *PLoS ONE* 5(4): e9975.

Callaway, R. M., R. H. DeLuca, and W. M. Belliveau. 1999. Biological-control herbivores may increase competitive ability of the noxious weed *Centaurea maculosa*. *Ecology* 80: 1196–1201.

Capinera, J. L. 1987. Population ecology of rangeland grasshoppers. In *Integrated Pest Management on Rangeland: A Shortgrass Prairie Perspective*, J. L. Capinera, ed. 162–182. Boulder, CO: Westview Press.

Carlson, J. C., L. A. Dyer, F. X. Omlin, and J. C. Beier. 2009. Diversity cascades and malaria vectors. *Journal of Medical Entomology* 46: 460–464.

Carrière, Y., T. J. Dennehy, B. Pedersen, S. Haller, C. Ellers-Kirk, L. Antilla, Y.-B. Liu, E. Willott, and B. E. Tabashnik. 2001. Large-scale management of insect resistance to transgenic cotton in Arizona: Can transgenic insecticidal crops be sustained? *Journal of Economic Entomology* 94: 315–325.

Carrière, Y., C. Ellers-Kirk, M. Sisterson, L. Antilla, M. Whitlow, T. J. Dennehy, and B. E. Tabashnik. 2003. Long-term regional suppression of pink bollworm by *Bacillus thuringiensis cotton*. *Proceedings of the National Academy of Sciences USA* 100: 1519–1523.

Carson, R. 1962. *Silent Spring*. New York: Houghton-Mifflin.

Cease, A. J., J. J. Elser, C. F. Ford, S. Hao, L. Kang, and J. F. Harrison. 2012. Heavy livestock grazing promotes locust outbreaks by lowering plant nitrogen content. *Science* 335: 467–469.

Chen, J., J. F. Franklin, and T. A. Spies. 1995. Growing-season microclimatic gradients from clearcut edges into old-growth Douglas-fir forests. *Ecological Applications* 5: 74–86.

Christensen, N. L., Jr., S. V. Gregory, P. R. Hagenstein, T. A. Heberlein, J. C. Hendee, J. T. Olson, J. M. Peek, D. A. Perry, T. D. Schowalter, K. Sullivan, et al. 2000. *Environmental Issues in Pacific Northwest Forest Management*. Washington, DC: National Academy Press.

Clark, K. L., N. Skowronski, and J. Hom. 2010. Invasive insects impact forest carbon dynamics. *Global Change Biology* 16: 88–101.

Claudianos, C., H. Ranson, R. M. Johnson, S. Biswas, M. A. Schuler, M. R. Berenbaum, R. Feyereisen, and J. G. Oakeshott. 2006. A deficit of detoxification enzymes: Pesticide sensitivity and environmental response in the honey bee. *Insect Molecular Biology* 15: 615–636.

Collins, F. H. and S. M. Paskewitz. 1995. Malaria: Current and future prospects for control. *Annual Review of Entomology* 40: 195–219.

Cook, B. L., K. J. Anchukaitis, J. O. Kaplan, M. J. Puma, M. Kelley, and D. Gueyffier. 2012. Pre-Columbian deforestation as an amplifier of drought in Mesomerica. *Geophysical Research Letters* 39: L16706.

Cooper, J. F. 1823. *The Pioneers*. New York: Charles Wiley.

Corin, S. and S. A. Weaver. 2005. A risk analysis model with an ecological perspective on DDT and malaria control in South Africa. *Journal of Rural and Tropical Public Health* 4: 21–32.

Couceiro, S. R. M., N. Hamada, R. L. M. Ferreira, B. R. Forsberg, and J. O. da Silva. 2007. Domestic sewage and oil spills in streams: Effects on edaphic invertebrates in flooded forest, Manaus, Amazonas, Brazil. *Water, Air and Soil Pollution* 180: 249–259.

Cregg, B. M. and M. E. Dix. 2001. Tree moisture stress and insect damage in urban areas in relation to heat island effects. *Journal of Arboriculture* 27: 8–17.

Currano, E. D., P. Wilf, S. L. Wing, C. C. Labandeira, E. C. Lovelock, and D. L. Royer. 2008. Sharply increased insect herbivory during the Paleocene-Eocene Thermal Maximum. *Proceedings of the National Academy of Sciences USA* 105: 1960–1964.

Daborn, P. J., J. L. Yen, M. R. Bogwitz, G. Le Goff, E. Feil, S. Jeffers, N. Tijet, T. Perry, D. Heckel, P. Batterham, R. Feyereisen, T.G. Wilson, and R. H. ffrench-Constant. 2002. A single P450 allele associated with insecticide resistance in *Drosophila*. *Science* 297: 2253–2256.

Day, G. M. 1953. The Indian as an ecological factor in the northeastern forest. *Ecology* 34: 329–346.

DeMers, M. N. 1993. Roadside ditches as corridors for range expansion of the western harvester ant (*Pogonomyrmex occidentalis* Cresson). *Landscape Ecology* 8: 93–102.

Diamond, J. 1999. *Guns, Germs, and Steel: The Fates of Human Societies*. New York: W. W. Norton.

Dyer, L. A. and D. K. Letourneau. 1999a. Relative strengths of top-down and bottom-up forces in a tropical forest community. *Oecologia* 119: 265–274.

Dyer, L. A. and D. K. Letourneau. 1999b. Trophic cascades in a complex terrestrial community. *Proceedings of the National Academy of Sciences USA* 96: 5072–5076.

Ehrlich, P. R. and H. A. Mooney. 1983. Extinction, substitution, and ecosystem services. *BioScience* 33: 248–254.

Fedoroff, N. V., D. S. Battisti, R. N. Beachy, P. J. M. Cooper, D. A. Fischhoff, C. N. Hodges, V. C. Knauf, D. Lobell, B.J. Mazur, D. Molden, et al. 2010. Radically rethinking agriculture for the 21st century. *Science* 327: 833–834.

Felland, C. M., H. N. Pitre, R. G. Luttrell, and J. L. Hamer. 1990. Resistance to pyrethroid insecticides in soybean looper (Lepidoptera: Noctuidae) in Mississippi. *Journal of Economic Entomology* 83: 35–40.

Feyereisen, R. 1999. Insect P450 enzymes. *Annual Review of Entomology* 44: 507–533.

Foley, J. A., M. T. Coe, M. Scheffer, and G. Wang. 2003a. Regime shifts in the Sahara and Sahel: Interactions between ecological and climatic systems in northern Africa. *Ecosystems* 6: 524–539.

Foley, J. A., M. H. Costa, C. Delire, N. Ramankutty, and P. Snyder. 2003b. Green surprise? How terrestrial ecosystems could affect earth's climate. *Frontiers in Ecology and the Environment* 1: 38–44.

Forbush, E. H. and C. H. Fernald. 1896. *The Gypsy Moth*. Boston; Massachusetts Board of Agriculture.

Ford, C. R. and J. M. Vose. 2007. *Tsuga canadensis* (L.) Carr. mortality will impact hydrologic processes in southern Appalachian forest ecosystems. *Ecological Applications* 17: 1156–1167.

Frankie, G. W. and L. E. Ehler. 1978. Ecology of insects in urban environments. *Annual Review of Entomology* 23: 367–387.

Franklin, J. F., F. J. Swanson, M. E. Harmon, D. A. Perry, T. A. Spies, V. H. Dale, A. McKee, W. K. Ferrell, J. E. Means, S. V. Gregory, et al. 1992. Effects of global climatic change on forests in northwestern North America. In *Global Warming and Biological Diversity*, R. L. Peters and T. E. Lovejoy, eds. 244–257. New Haven, CT: Yale University Press.

Frouz, J., D. Elhottová, M. Helingerová, and F. Kocourek. 2008. The effect of Bt-corn on soil invertebrates, soil microbial community, and decomposition rates of corn post-harvest residues under field and laboratory conditions. *Journal of Sustainable Agriculture* 32: 645–655.

Gandhi, K. J. K., J. R. Spence, D. W. Langor, L. E. Morgantini, and K. J. Cryer. 2004. Harvest retention patches are insufficient as stand analogues of fire residuals for litter-dwelling beetles in northern coniferous forests. *Canadian Journal of Forest Research* 34: 1319–1331.

Gleason, K. L., J. H. Lawrimore, D. H. Levinson, T. R. Karl, and D. J. Karoly. 2008. A revised U.S. climate extremes index. *Journal of Climate* 21: 2124–2137.

González-Cabrera, J., G. P. Farinós, S. Caccia, M. Díaz-Mendoza, P. Castañera, M. G. Leonardi, B. Giordana, and J. Ferre. 2006. Toxicity and mode of action of *Bacillus thuringiensis* Cry proteins in the Mediterranean corn borer, *Sesamia nonagrioides* (Lefebvre). *Applied and Environmental Microbiology* 72: 2594–2600.

Goodman, C. A. and A. J. Mills. 1999. The evidence base on the cost-effectiveness of malaria control measures in Africa. *Health Policy and Planning* 14: 301–312.

Grieco, J. P., N. L. Achee, T. Chareonviriyaphap, W. Suwonkerd, K. Chauhan, M. R. Sardelis, and D. R. Roberts. 2007. A new classification system for the actions of IRS chemicals traditionally used for malaria control. *PLoS One* 2(8): e716.

Gutschick, V. P. and H. BassiriRad. 2010. Biological extreme events: A research framework. *Eos, Transactions of the American Geophysical Union* 91: 85–86.

Halm, M.-P., A. Rortais, G. Arnold, J. N. Taséi, and S. Rault. 2006. New risk assessment approach for systemic insecticides: The case of honey bees and imidacloprid (Gaucho). *Environmental Science and Technology* 40: 2448–2454.

Hansen Jesse, L. C. and J. J. Obrycki. 2000. Field deposition of Bt transgenic corn pollen: Lethal effects on the monarch butterfly. *Oecologia* 125: 241–248.

Harborne, J. B. 1994. *Introduction to Ecological Biochemistry*, 4th ed. London: Academic.

Harding, J.S., E. F. Benfield, P. V. Bolstad, G. S. Helfman, and E. B. D. Jones III. 1998. Stream biodiversity: The ghost of land use past. *Proceedings of the National Academy of Sciences USA* 95: 14843–14847.

Harley, K. G., K. Huen, R. A. Schall, N. T. Holland, A. Bradman, D. B. Barr, and B. Eskenazi. 2011. Association of organophosphate pesticide exposure and paraoxonase with birth outcome in Mexican-American women. *PLoS ONE* 6(8): e23923.

Hartman, G. L., E. D. West, and T. K. Herman. 2011. Crops that feed the world 2. Soybean—worldwide production, use, and constraints caused by pathogens and pests. *Food Security* 3: 5–17.

Hatfield, J. L., K. J. Boote, B. A. Kimball, L. H. Ziska, R. C. Izaurralde, D.Ort, A. M. Thomson, and D. Wolfe. 2011. Climate impacts on agriculture: Implications for crop production. *Agronomy Journal* 103: 351–370.

Haynes, K. J. and J. T. Cronin. 2003. Matrix composition affects the spatial ecology of a prairie planthopper. *Ecology* 84: 2856–2866.

Heliövaara, K. and R. Väisänen. 1986. Industrial air pollution and the pine bark bug, *Aradus cinnamomeus* Panz. (Het., Aradidae). *Zeitschrift für angewandte Entomologie* 101: 469–478.

Heliövaara, K. and R. Väisänen. 1993. *Insects and Pollution*. Boca Raton, FL: CRC Press.

Hensley, S. D., W. H. Long, L. R. Roddy, W. J. McCormick, and E. J. Concienne. 1961. Effects of insecticides on the predaceous arthropod fauna of Louisiana sugarcane fields. *Journal of Economic Entomology* 54: 146–149.

Heuberger, S., C. Ellers-Kirk, C. Yafuso, A.J. Gassmann, B. E. Tabashnik, T. J. Dennehy, and Y. Carriére. 2008a. Effects of refuge contamination by transgenes on Bt resistance in pink bollworm (Lepidoptera: Gelichiidae). *Journal of Economic Entomology* 101: 504–514.

Heuberger, S., C. Yafuso, G. DeGrandi-Hoffman, B. E. Tabashnik, Y. Carriére, and T. J. Dennehy. 2008b. Outcrossed cottonseed and adventitious Bt plants in Arizona refuges. *Environmental Biosafety Research* 7: 87–96.

Heuberger, S. M., D. W. Crowder, T. Brevault, B. E. Tabashnik, and Y. Carriére. 2011. Modeling the effects of plant-to-plant gene flow, larval behavior, and refuge size on pest resistance to Bt cotton. *Environmental Entomology* 40: 484–495.

Hill, R., A. Baird, and D. Buchanan. 1999. Aborigines and fire in the wet tropics of Queensland, Australia: Ecosystem management across cultures. *Society and Natural Resources* 12: 205–223.

Hitchner, E. M., T. P. Kuhar, J. C. Dickens, R. R. Youngman, P. B. Schultz, and D. G. Pfeiffer. 2008. Host plant choice experiments of Colorado potato beetle (Coleoptera: Chrysomelidae) in Virginia. *Journal of Economic Entomology* 101: 859–865.

Hönemann, L., C. Zurbrügg, and W. Nentwig. 2008. Effects of Bt-corn decomposition on the composition of the soil meso- and macrofauna. *Applied Soil Ecology* 40: 203–209.

Horváth, G., G. Kriska, P. Malik, and B. Robertson. 2009. Polarized light pollution: A new kind of ecological photopollution. *Frontiers in Ecology and the Environment* 7: 317–325.

Hossain, F., I. Jeyachandran, and R. Pielke. 2009. Have large dams altered extreme precipitation patterns? *Eos, Transactions of the American Geophysical Union* 90: 453–454.

Howarth, R., F. Chan, D. J. Conley, J. Garnier, S. C. Doney, R. Marino, and G. Billen. 2011. Coupled biogeochemical cycles: Eutrophication and hypoxia in temperate estuaries and coastal marine ecosystems. *Frontiers in Ecology and the Environment* 9: 18–26.

Hsiang, S. M., K. C. Meng, and M. A. Cane. 2011. Civil conflicts are associated with the global climate. *Nature* 476: 438–441.

Hsu, J.-C., D. S. Haymer, W.-J. Wu, and H.-T. Feng. 2006. Mutations in the acetylcholinesterase gene of *Bactrocera dorsalis* associated with resistance organophosphorus insecticides. *Insect Biochemistry and Molecular Biology* 38: 396–402.

Hsu, J.-C., W.-J. Wu, D. S. Haymer, H.-Y. Liao, and H.-T. Feng. 2008. Alterations of the acetylcholinesterase enzyme in the oriental fruit fly *Bactrocera dorsalis* are correlated with resistance to the organophosphate insecticide fenitrothion. *Insect Biochemistry and Molecular Biology* 38: 146–154.

Ishaaya, I., A. Barazani, S. Kontsedalov, and A. R. Horowitz. 2007. Insecticides with novel modes of action: Mechanism, selectivity, and cross-resistance. *Entomological Research* 37: 148–152.

Ito, J., A. Ghosh, L. A. Moreira, E. A. Wimmer, and M. Jacobs-Lorena. 2002. Transgenic anopheline mosquitoes impaired in transmission of a malaria parasite. *Nature* 417: 452–455.

Izaurralde, R. C., A. M. Thomson, J. A. Morgan, P. A. Fay, H. W. Polley, and J. L. Hatfield. 2011. Climate impacts on agriculture: Implications for forage and rangeland production. *Agronomy Journal* 103: 371–381.

Janssen, R. H. H., M. B. J. Meinders, E. H. van Nes, and M. Scheffer. 2008. Microscale vegetation-soil feedback boosts hysteresis in a regional vegetation-climate system. *Global Change Biology* 14: 1104–1112.

Janzen, D. H. and P. S. Martin. 1982. Neotropical anachronisms: The fruits the gomphotheres ate. *Science* 215: 19–27.

Jentsch, A., J. Kreyling, and C. Beierkuhnlein. 2007. A new generation of climate-change experiments: Events, not trends. *Frontiers in Ecology and the Environment* 5: 365–374.

Johnson, D. M., A. M. Liebhold, P. C. Tobin, and O. N. Bjørnstad. 2006. Allee effects and pulsed invasion by the gypsy moth. *Nature* 444: 361–363.

Jones, C. G., R. S. Ostfeld, M. P. Richard, E. M. Schauber, and J. O. Wolff. 1998. Chain reactions linking acorns to gypsy moth outbreaks and Lyme disease risk. *Science* 279: 1023–1026.

Juang, J.-Y., G. G. Katul, A. Porporato, P. C. Stoy, M.S. Sequeira, M. Detto, H.-S. Kim, and R. Oren. 2007. Eco-hydrological controls on summertime convective rainfall triggers. *Global Change Biology* 13: 887–896.

Karban, R. and A. A. Agrawal. 2002. Herbivore offense. *Annual Review of Ecology and Systematics* 33: 641–664.

Katima, J. H. Y. and S. Mng'anya. 2009. African NGOs outline commitment to malaria control without DDT. *Pesticide News* 84: 5.

Kaushal, S. S., M. L. Pace, P. M. Groffman, L. E. Band, K. T. Belt, P. M. Mayer, and C. Welty. 2010. Land use and climate variability amplify contaminant pulses. *Eos, Transactions of the American Geophysical Union* 91: 221–222.

Keeling, C. D., T. P. Whorf, M. Wahlen, and J. van der Pilcht. 1995. Interannual extremes in the rate of rise of atmospheric carbon dioxide since 1980. *Science* 375: 666–670.

Kessler, A. and I. T. Baldwin. 2002. Plant responses to herbivory: The emerging molecular analysis. *Annual Review of Plant Biology* 53: 299–328.

Khalil, M. A. K., R. A. Rasmussen, J. R. J. French, and J. A. Holt. 1990. The influence of termites on atmospheric trace gases: CH_4, CO_2, $CHCl_3$, N_2O, CO, H_2, and light hydrocarbons. *Journal of Geophysical Research* 95: 3619–3634.

Killeen, G. F., U. Fillilnger, I. Kiche, L.C. Gouagna, and B. G. J. Knols. 2002. Eradication of *Anopheles gambiae* from Brazil: Lessons for malaria control in Africa? *Lancet Infectious Diseases* 2: 618–627.

Kimberling, D. N., J. R. Karr, and L. S. Fore. 2001. Measuring human disturbance using terrestrial invertebrates in the shrub-steppe of eastern Washington (USA). *Ecological Indicators* 1: 63–81.

Kinney, K. K., R. L. Lindroth, S. M. Jung, and E. V. Nordheim. 1997. Effects of CO_2 and NO_3^- availability on deciduous trees: Phytochemistry and insect performance. *Ecology* 78: 215–230.

Kishtawal, C. M., D. Niyogi, M. Tewari, R. A. Pielke, Sr., and J. M. Shepherd. 2010. Urbanization signature in the observed heavy rainfall climatology over India. *International Journal of Climatology* 30: 1908–1916.

Klein, A.-M, B. E. Vaissière, J. H. Cane, I. Steffan-Dewenter, S. A. Cunningham, C. Kremen, and T. Tscharntke. 2007. Importance of pollinators in changing landscapes for world crops. *Proceedings of the Royal Society B* 274: 303–313.

Knight, T. M., M. W. McCoy, J. M. Chase, K. A. McCoy, and R. D. Holt. 2005. Trophic cascades across ecosystems. *Nature* 437: 880–883.

Konishi, M. and Y. Itô. 1973. Early entomology in East Asia. In *History of Entomology*, R. F. Smith, T. E. Mittler, and C. N. Smith, eds. 1–20. Palo Alto, CA: Annual Reviews.

Koren, I., Y. J. Kaufman, L. A. Remer, and J. V. Martins. 2004. Measurement of the effect of Amazon smoke on inhibition of cloud formation. *Science* 303: 1342–1345.

Kremen, C., N. M. Williams, and R. W. Thorp. 2002. Crop pollination from native bees as risk from agricultural intensification. *Proceedings of the National Academy of Sciences USA* 99: 16812–16816.

Kriska, G., Z. Csabai, P. Boda, P. Malik, and G. Horváth. 2006. Why do red and dark-coloured cars lure aquatic insects? The attraction of water insects to car paintwork explained by reflection-polarization signals. *Proceedings of the Royal Society B* 273: 1667–1671.

Kundzewicz, Z. W., Y. Hirabayashi, and S. Kanae. 1998. River floods in the changing climate—observations and projections. *Water Resources Management* 24: 2633–2646.

Kurz, W. A., C. C. Dymond, G. Stinson, G. J. Rampley, E. T. Neilson, A. L. Carroll, T. Ebata, and L. Safranyik. 2008. Mountain pine beetle and forest carbon feedback to climate change. *Nature* 452: 987–990.

Landis, D. A., M. M. Gardiner, W. van der Werf, and S. M. Swinton. 2008. Increasing corn for biofuel production reduces biocontrol services in agricultural landscapes. *Proceedings of the National Academy of Sciences USA* 105: 20552–20557.

Landis, D. A., S. D. Wratten, and G. M. Gurr. 2000. Habitat management to conserve natural enemies of arthropod pests in agriculture. *Annual Review of Entomology* 45: 175–201.

Lehman, R. M., S. L. Osborne, and K.A. Rosentrater. 2008. No differences in decomposition rates observed between *Bacillus thuringiensis* and non-*Bacillus thuringiensis* corn residue incubated in the field. *Agronomy Journal* 100: 163–168.

Lentz, D. L. and B. Hockaday. 2009. Tikal timbers and temples: Ancient Maya forestry and the end of time. *Journal of Archaeology* 36: 1342–1353.

Letourneau, D. K. and L. A. Dyer. 1998. Density patterns of Piper ant-plants and associated arthropods: Top-predator trophic cascades in a terrestrial system? *Biotropica* 30: 162–169.

Letourneau, D. K., J. A. Jedlicka, S. G. Bothwell, and C. R. Moreno. 2009. Effects of natural enemy biodiversity on the suppression of arthropod herbivores in terrestrial ecosystems. *Annual Review of Ecology, Evolution and Systematics* 40: 573–592.

Lincoln, D. E., E. D. Fajer, and R. H. Johnson. 1993. Plant-insect herbivore interactions in elevated CO_2 environments. *Trends in Ecology and Evolution* 8: 64–68.

Logan, J. A., J. Régnière, and J. A. Powell. 2003. Assessing the impacts of global warming on forest pest dynamics. *Frontiers in Ecology and the Environment* 1: 130–137.

Losey, J. E., L. S. Rayor, and M. E. Carter. 1999. Transgenic pollen harms monarch larvae. *Nature* 399: 214.

Louda, S. M., R. W. Pemberton, M. T. Johnson, and P. A. Follett. 2003. Non-target effects—the Achilles' heel of biocontrol? Retrospective analyses to assess risk associated with biocontrol introductions. *Annual Review of Entomology* 48: 365–396.

Lubchenco, J. and T. R. Karl. 2012. Predicting and managing extreme weather events. *Physics Today* 65(3): 31–37.

Luck, R.F . and D. Dahlsten. 1975. Natural decline of a pine needle scale (*Chionaspis pinifoliae* [Fitch]), outbreak at South Lake Tahoe, California, following cessation of adult mosquito control with malathion. *Ecology* 56: 893–904.

Ma, S.-C. 1958. The population dynamics of the oriental migratory locust (*Locusta migratoria manilensis* Mayen) in China. *Acta Entomologica Sinica* 8: 1–40.

Mao, W., S. Rupasinghe, A. R. Zangerl, M. A. Schuler, and M. R. Berenbaum. 2006. Remarkable substrate-specificity of CYP6AB3 in *Depressaria pastinacella*, a highly specialized herbivore. *Insect Molecular Biology* 15: 169–179.

Mao, Y.-B., W.-J. Cai, J.-W. Wang, G.-J. Hong, X.-Y. Tao, L.-J. Wang, Y.-P. Huang, and X.-Y. Chen. 2007. Silencing a cotton bollworm P450 monooxygenase gene by plant-mediated RNAi impairs larval tolerance for gossypol. *Nature Biotechnology* 25: 1307–1313.

Marks, S. and D. E. Lincoln. 1996. Antiherbivore defense mutualism under elevated carbon dioxide levels: A fungal endophyte and grass. *Environmental Entomology* 25: 618–623.

Marquis, R. J. and C. J. Whelan. 1994. Insectivorous birds increase growth of white oak through consumption of leaf-chewing insects. *Ecology* 75: 2007–2014.

Marvier, M., C. McCreedy, J. Regetz, and P. Kareiva. 2007. A meta-analysis of effects of Bt cotton and maize on non-target invertebrates. *Science* 316: 1475–1477.

Mathias, D., L. Jacky, W. E. Bradshaw, and C. M. Holzapfel. 2007. Quantitative trait loci associated with photopheriodic response and stage of diapause in the pitcher-plant mosquito, *Wyeomyia smithii*. *Genetics* 176: 391–402.

Mattson, W. J. and R. A. Haack. 1987. The role of drought in outbreaks of plant-eating insects. *BioScience* 37: 110–118.

McCoy, E. D. and J. H. Frank. 2010. How should the risk associated with the introduction of biological control agents be estimated? *Agricultural and Forest Entomology* 12: 1–8.

Meehan, T. D., B. P. Werling, D. A. Landis, and C. Gratton. 2011. Agricultural landscape simplification and insecticide use in the midwestern United States. *Proceedings of the National Academy of Sciences USA* 108: 11500–11505.

Melander, A. L. 1923. *Tolerance of San Jose Scale to Sprays*. State College of Washington Agricultural Experiment Station Bulletin 174. Pullman: State College of Washington.

Merritt, R. W., M. E. Benbow, and P. L. C. Small. 2005. Unraveling an emerging disease associated with disturbed aquatic environments: The case of Buruli ulcer. *Frontiers in Ecology and the Environment* 3: 323–331.

Mészáros, A., J. M. Beuzelin, M .J. Stout, P. L. Bommireddy, M. R. Riggio, and B. R. Leonard. 2012. Jasmonic acid-induced resistance to the fall armyworm, *Spodoptera frugiperda*, in conventional and transgenic cottons expressing *Bacillus thuringiensis* insecticidal proteins. *Entomologia Experimentalis et Applicata* 140: 226–237.

Michaud, J. P. and A. K. Grant. 2009. The nature of resistance to *Dectes texanus* (Col., Cerambycidae) in wild sunflower, *Helianthus annuus*. *Journal of Applied Entomology* 133: 518–523.

Miller, S. W., D. Wooster, and J. Li. 2007. Resistance and resilience of macroinvertebrates to irrigation water withdrawals. *Freshwater Biology* 52: 2494–2510.

Miller, T. A. 1996. Resistance to pesticides: Mechanisms, development and management. In *Cotton Insects and Mites: Characterization and Management*, E. G. King, J. R. Phillips, and R. J. Coleman, eds. 323–378. Memphis, TN: The Cotton Foundation.

Mooney, K. A. 2007. Tritrophic effects of birds and ants on a canopy food web, tree growth, and phytochemistry. *Ecology* 88: 2005–2014.

Morrison, P. H. and F. J. Swanson. 1990. *Fire History and Pattern in a Cascade Range Landscape*. General Technical Report PNW-GTR-254. Portland, OR: U.S. Department of Agriculture, Forest Service, Pacific Northwest Research Station.

Murphy, B. P. and D. M. J. S. Bowman. 2007. The interdependence of fire, grass, kangaroos, and Australian Aborigines: A case study from central Arnhem Land, northern Australia. *Journal of Biogeography* 34: 237–250.

Nauen, R. 2006. Insecticide mode of action: Return of the ryanodine receptor. *Pest Management Science* 62: 690–692.

Nessimian, J. L., E. M. Venticinque, J. Zuanon, P. de Marco, Jr., M. Gordo, L. Fidelis, J. D'arc Batista, and L. Juen. 2008. Land use, habitat integrity, and aquatic insect assemblages in central Amazonian streams. *Hydrobiologia* 614: 117–131.

O'Callaghan, M., T. R. Glare, E. P. J. Burgess, and L. A. Malone. 2005. Effects of plants genetically modified for insect resistance on nontarget organisms. *Annual Review of Entomology* 50: 271–292.

Odum, E. P. 1969. The strategy of ecosystem development. *Science* 164: 262–270.

Omer, S. M., G. P. Georghiou, and S. N. Irving. 1980. DDT/Pyrethroid resistance interrelationships in *Anopheles stephensi*. *Mosquito News* 40: 200–209.

Orwig, D. A. 2002. Ecosystem to regional impacts of introduced pests and pathogens: Historical context, questions, and issues. *Journal of Biogeography* 29: 1471–1474.

Ottea, J. and R. Leonard. 2006. Insecticide/acaricide resistance and management strategies. In *Use and Management of Insecticides, Acaricides, and Transgenic Crops*, J. N. All and M. F. Treacy, eds. 82–92. Lanham, MD: Entomological Society of America.

Over, M., B. Bakote'e, R. Velayudhan, P. Wilikai, and P. M. Graves. 2004. Impregnated nets or DDT residual spraying? Field effectiveness of malaria prevention techniques in Solomon Islands, 1993–1999. *American Journal of Tropical Medicine and Hygiene* 71 (suppl 2): 214–223.

Paquin, P. and D. Coderre. 1997. Deforestation and fire impact on edaphic insect larvae and other macroarthropods. *Environmental Entomology* 26: 21–30.

Parmesan, C. 2006. Ecological and evolutionary responses to recent climate change. *Annual Review of Ecology, Evolution and Systematics* 37: 637–669.

Parmesan, C. and G. Yohe 2003. A globally coherent fingerprint of climate change impacts across natural systems. *Nature* 421: 37–42.

Parton, W. J., J. M. O. Scurlock, D. S. Ojima, T. G. Gilmanov, R. J. Scholes, D. S. Schimel, T. Kirchner, J-C. Menaut, T. Seastedt, E. G. Moya, et al. 1993. Observations and modeling of biomass and soil organic matter dynamics for the grassland biome worldwide. *Global Biogeochemical Cycles* 7: 785–809.

Pemberton, R. W. and H. A. Cordo. 2001. Potential and risks of biological control of *Cactobastis cactorum* (Lepidoptera: Pyralidae) in North America. *Florida Entomologist* 84: 513–526.

Perry, D. A., R. Oren, and S. C. Hart 2008. *Forest Ecosystems*, 2nd ed. Baltimore, MD: Johns Hopkins University Press.

Pimentel D, R. Zuniga, and D. Morrison. 2005. Update on the environmental and economic costs associated with alien-invasive species in the United States. *Ecological Economics* 52: 273–288.

Porter, S. D. and D. A. Savignano. 1990. Invasion of polygyne fire ants decimates native ants and disrupts arthropod community. *Ecology* 71: 2095–2106.

Pouyat, R. V., D. E. Pataki, K. T. Belt, P. M. Groffman, J. Hom, and L. E. Band. 2007. Effects of urban land-use change on biogeochemical cycles. In *Terrestrial Ecosystems in a Changing World*, J. G. Canadell, D. E. Pataki, and L. F. Pitelka, eds. 45–58. Berlin: The IGBP Series, Springer-Verlag.

Póvoa, M. M., J. E. Conn, C. D. Schlichting, J. C. O. F. Amaral, M. N. O. Segura, A. N. M. da Silva, C. C. B. dos Santos, R. N. L. Lacerda, R. T. L. de Souza, D. Galiza, et al. 2003. Malaria vectors, epidemiology, and the re-emergence of *Anopheles darlingi* in Belém, Pará, Brazil. *Journal of Medical Entomology* 40: 379–386.

Pridgeon, J. W., R. M. Pereira, J. J. Becnel, S. A. Allan, G. G. Clark, and K. J. Linthicum. 2008. Susceptibility of *Aedes aegypti*, *Culex quinquefasciatus* Say, and *Anopheles quadrimaculatus* Say to 19 pesticides with different modes of action. *Journal of Medical Entomology* 45: 82–87.

Pringle, C. M., M. C. Freeman, and B. J. Freeman. 2000. Regional effects of hydrologic alternations on riverine macrobiota in the New World: Tropical-temperate comparisons. *BioScience* 50: 807–823.

Raffa, K. F., B. H. Aukema, B. J. Bentz, A. L. Carroll, J. A. Hicke, M. G. Turner, and W. H. Romme. 2008. Cross-scale drivers of natural disturbances prone to anthropogenic amplification: The dynamics of bark beetle eruptions. *BioScience* 58: 501–517.

Ragsdale, D. W., D. A. Landis, J. Brodeur, G. E. Heimpel, and N. Desneux. 2011. Ecology and management of the soybean aphid in North America. *Annual Review of Entomology* 56: 375–399.

Ranson, H., C. Claudianos, F. Ortelli, C. Abgrall, J. Hemingway, M. V. Sharakhova, M. F. Unger, F. H. Collins, and R. Feyereisen. 2002. Evolution of supergene families associated with insecticide resistance. *Science* 298: 179–181.

Raupp, M. J., P. M. Shrewsbury, and D. A. Herms. 2010. Ecology of herbivorous arthropods in urban landscapes. *Annual Review of Entomology* 55: 19–38.

Reusch, T. B. H., A. Ehlers, A. Hämmerli, and B. Worm. 2005. Ecosystem recovery after climatic extremes enhanced by genotypic diversity. *Proceedings of the National Academy of Sciences USA* 102: 2826–2831.

Ribeiro, H. and J. T. Mexia. 1970. Detection of DDT resistance in adult *Anopheles melas* and larval *Anopheles listeria* populations from Lobito, Angola (Portuguese West Africa). *Mosquito News* 30: 611–613.

Riley, C. V. 1878. *First Annual Report of the United States Entomological Commission for the Year 1877 Relating to the Rocky Mountain Locust and the Best Methods of Preventing Its Injuries and of Guarding Against Its Invasions, in Pursuance of an Appropriation Made by Congress for This Purpose.* Washington, DC: U.S. Department of Agriculture.

Riley, C. V. 1883. *Third Report of the United States Entomological Commission, Relating to the Rocky Mountain Locust, the Western Cricket, the Army-worm, Canker Worms, and the Hessian Fly, Together with Descriptions of Larvae of Injurious Forest Insects, Studies on the Embryological Development of the Locust and of Other Insects, and on the Systematic Position of the Orthoptera in Relation to Other Orders of Insects.* Washington, DC: U.S. Department of Agriculture.

Riley, C. V. 1885. *Fourth Report of the United States Entomological Commission, Being a Revised Edition of Bulletin No. 3, and the Final Report on the Cotton Worm, Together with a Chapter on the Boll Worm.* Washington, DC: U.S. Department of Agriculture.

Roland, J. 1993. Large-scale forest fragmentation increases the duration of tent caterpillar outbreak. *Oecologia* 93: 25–30.

Roland, J. and W. J. Kaupp. 1995. Reduced transmission of forest tent caterpillar (Lepidoptera: Lasiocampidae) nuclear polyhedrosis virus at the forest edge. *Environmental Entomology* 24: 1175–1178.

Rongsriyam, Y. and J. R. Busvine. 1975. Cross-resistance in DDT-resistant strains of various mosquitoes (Diptera, Culicidae). *Bulletin of Entomological Research* 65: 459–471.

Rosi-Marshall, E. J., J. L. Tank, T. V. Royer, M. R. Whiles, M. Evans-White, C. Chambers, N. A. Griffiths, J. Pokelsek, and M. L. Stephen. 2007. Toxins in transgenic crop byproducts may affect headwater stream ecosystems. *Proceedings of the National Academy of Sciences USA* 104: 16204–16208.

Ross, N. J. and T. F. Rangel. 2011. Ancient Maya agroforestry echoing through spatial relationships in the extant forest of NW Belize. *Biotropica* 43: 141–148.

Roussel, J. S. and D. F. Clower 1957. Resistance to the chlorinated hydrocarbon insecticides in the boll weevil. *Journal of Economic Entomology* 50: 463–468.

Rubenstein, D. I. 1992. The greenhouse effect and changes in animal behavior: Effects on social structure and life-history strategies. In *Global Warming and Biological Diversity*, R. L. Peters and T. E. Lovejoy, eds. 180–192. New Haven, CT: Yale University Press.

Rudd, W. G., D. C. Herzog, and L. D. Newsom. 1984. Hierarchical models of ecosystems. *Environmental Entomology* 13: 584–587.

Russell, T. L. and B. H. Kay. 2008. Biologically based insecticides for the control of immature Australian mosquitoes: A review. *Australian Journal of Entomology* 47: 232–242.

Sanderson, M. G. 1996. Biomass of termites and their emissions of methane and carbon dioxide: A global database. *Global Biogeochemical Cycles* 10: 543–557.

Sandquist, R. E., D. L. Overhulser, and J. D. Stein. 1993. Aerial applications of esfenvalerate to supress *Contarinia oregonensis* (Diptera: Cecidomyiidae) and *Megastigmus spermotrophus* (Hymenoptera: Torymidae) in Douglas-fir seed orchards. *Journal of Economic Entomology* 86: 470–474.

San Martin y Gomez, G. and H. Van Dyck. 2012. Ecotypic differentiation between urban and rural populations of the grasshopper *Chorthippus brunneus* relative to climate and habitat fragmentation. *Oecologia* 169: 125–133.

Schlenker, W. and M. J. Roberts. 2009. Nonlinear temperature effects indicate severe damages to U.S. crop yields under climate change. *Proceedings of the National Academy of Sciences USA* 106: 15594–15598.

Schlesinger, W. H., J. F. Reynolds, G. L. Cunningham, L. F. Huenneke, W. M. Jarrell, R. A. Virginia, and W. G. Whitford. 1990. Biological feedbacks in global desertification. *Science* 247: 1043–1048.

Scholz, N. L., E. Fleishman, L. Brown, I. Werner, M. L. Johnson, M. L. Brooks, C. L. Mitchelmore, and D. Schlenk. 2012. A perspective on modern pesticides, pelagic fish declines, and unknown ecological resilience in highly managed ecosystems. *BioScience* 62: 428–434.

Schowalter, T. D. 2011. *Insect Ecology: An Ecosystem Approach,* 3rd ed. San Diego: Elsevier/Academic.

Schowalter, T. D. and P. Turchin. 1993. Southern pine beetle infestation development: Interaction between pine and hardwood basal areas. *Forest Science* 39: 201–210.

Schowalter, T. D., Y. L. Zhang, and J.J . Rykken. 2003. Litter invertebrate responses to variable density thinning in western Washington forest. *Ecological Applications* 13: 1204–1211.

Sears, M. K., R. L. Hellmich, D. E. Stanley-Horn, K. S. Oberhauser, J. M. Pleasants, H. R. Mattila, B. D. Siegfried, and G. P. Dively. 2001. Impact of Bt corn pollen on monarch butterfly populations: A risk assessment. *Proceedings of the National Academy of Sciences USA* 98: 11937–11942.

Şekercioğlu, C. H., P. R. Ehrlich, G. C. Daily, D. Aygen, D. Goehring, and R. F. Sandi. 2002. Disappearance of insectivorous birds from tropical forest fragments. *Proceedings of the National Academy of Sciences USA* 99: 263–267.

Showler, A. T. 2009. Roles of host plants in boll weevil range expansion beyond tropical Mesoamerica. *American Entomologist* 55: 234–242.

Shrewsbury, P. M. and M. J. Raupp. 2006. Do top-down or bottom-up forces determine *Stephanitis pyrioides* abundance in urban landscapes? *Ecological Applications* 16: 262–272.

Sidhu, A. B. S., D. Verdier-Pinard, and D. A. Fidock. 2002. Chloroquine resistance in *Plasmodium falciparum* malaria parasites conferred by pfcrt mutations. *Science* 298: 210–213.

Singh, G. and S. Prakash. 2009. Efficacy of *Bacillus sphaericus* against larvae of malaria and filarial vectors: An analysis of early resistance detection. *Parasitology Research* 104: 763–766.

Sisterson, M. S., Y. Carrière, T. J. Dennehy, and B. E. Tabashnik. 2007. Nontarget effects of transgenic insecticidal crops: Implications of source-sink population dynamics. *Environmental Entomology* 36: 121–127.

Skovmand, O., T. D. A. Ouedraogo, E. Sanogo, H. Samuelsen, L. P. Toé, and T. Baldet. 2009. Impact of slow-release *Bacillus sphaericus* granules on mosquito populations followed in a tropical urban environment. *Journal of Medical Entomology* 46: 67–76.

Smith, J. F., R. G. Luttrell, and J. K. Greene. 2009. Seasonal abundance, species composition, and population dynamics of stink bugs in production fields of early and late soybean in south Arkansas. *Journal of Economic Entomology* 102: 229–236.

Smith, R. H. 2007. *History of the Boll Weevil in Alabama.* Alabama Agricultural Experiment Station Bulletin 670. Auburn, AL: Auburn University.

Smith, S., T. E. Reagan, J. L. Flynn, and G. H. Willis. 1983. Azinphosmethyl and fenvalerate runoff loss from a sugarcane-insect IPM system. *Journal of Environmental Quality* 12: 534–537.

Soderlund, D. M. and J. R. Bloomquist. 1990. Molecular mechanisms of insecticide resistance. In *Pesticide Resistance in Arthropods*, R. T. Roush and B. E. Tabashnik, eds. 58–96. New York: Chapmann & Hall.

Sparks, T. C. 1996. Toxicology of insecticides and acaricides. In *Cotton Insects and Mites: Characterization and Management*, E. G. King, J. R. Phillips, and R. J. Coleman, eds. 283–322. Memphis, TN: The Cotton Foundation.

Spaulding, H. L. and L. K. Rieske. 2010. The aftermath of an invasion: Structure and composition of Central Appalachian hemlock forests following establishment of the hemlock woolly adelgid, *Adelges tsugae. Biological Invasions* 12: 3135–3143.

Spencer, H. J. and G. R. Port. 1988. Effects of roadside conditions on plants and insects. II. Soil conditions. *Journal of Applied Ecology* 25: 709–715.

Spencer, H. J., N. E. Scott, G. R. Port, and A. W. Davison. 1988. Effects of roadside conditions on plants and insects. I. Atmospheric conditions. *Journal of Applied Ecology* 25: 699–707.

Stern, V. M., R. F. Smith, R. van den Bosch, and K. S. Hagen. 1959. The integration of chemical and biological control of the spotted alfalfa aphid. Part 1. The integrated control concept. *Hilgardia* 29: 81–101.

Stewart, M. M. and L. L. Woolbright. 1996. Amphibians. In *The Food Web of a Tropical Rain Forest*, D. P. Reagan and R. B. Waide, eds. 273–320. Chicago: University of Chicago Press.

Stiles, J. H. and R. H. Jones. 1998. Distribution of the red imported fire ant, *Solenopsis invicta*, in road and powerline habitats. *Landscape Ecology* 13: 335–46.

Suarez, A. V., D. T. Bolger, and T. J. Case. 1998. Effects of fragmentation and invasion on native ant communities in coastal southern California. *Ecology* 79: 2041–2056.

Swan, C. M., P. D. Jensen, G. P. Dively, and W. O. Lamp. 2009. Processing of transgenic crop residues in stream ecosystems. *Journal of Applied Ecology* 46: 1304–1313.

Tabashnik, B. E. and Y. Carriére. 2008. Evolution of insect resistance to transgenic plants. In *Specialization, Speciation, and Radiation: The Evolutionary Biology of Herbivorous Insects*, K. J. Tilmon, ed. 267–279. Berkeley: University of California Press.

Tabashnik, B. E., F. R. Groeters, N. Finson, Y. B. Liu, M. W. Johnson, D. G. Heckel, K. Luo, and M. L. Adang. 1996. Resistance to *Bacillus thuringiensis in Plutella xylostella*: the moth heard round the world. In *Molecular Genetics and Evolution of Pesticide Resistance*, American Chemical Society Symposium Series 645, 130–140. Washington, D.C.

Tabashnik, B. E., Y. B. Liu, N. Finson, L. Masson, and D. G. Heckel. 1997. One gene in diamondback moth confers resistance to four *Bacillus thuringiensis toxins*. *Proceedings of the National Academy of Sciences USA* 94: 1640–1644.

Tabashnik, B. E., Y. Carriére, T. J. Dennehy, S. Morin, M. Sisterson, R. T. Roush, A. M. Shelton, and J.-Z. Zhao. 2003. Insect resistance to transgenic Bt crops: Lessons from the laboratory and field. *Journal of Economic Entomology* 96: 1031–1038.

Tabashnik, B. E., A. J. Gassmann, D. W. Crowder, and Y. Carriére. 2008. Insect resistance to Bt crops: Evidence vs. theory. *Nature Biotechnology* 26: 199–202.

Tanner, C. M., F. Kamel, G. W. Ross, J. A. Hoppin, S. M. Goldman, M. Korell, C. Marras, G. S. Bhudhikanok, M. Kasten, A. R. Chade, K. et al. 2011. Rotenone, paraquat, and Parkinson's disease. *Environmental Health Perspectives* 119: 866–872.

Tarkalson, D. D., S. D. Kachman, J. M. N. Knops, J. E. Thies, and C. S. Wortmann. 2008. Decomposition of Bt and non-Bt corn hybrid residues in the field. *Nutrient Cycling in Agroecosystems* 80: 211–222.

Thaler, J. S., M. J. Stout, R. Karban, and S. S. Duffey. 2001. Jasmonate-mediated induced plant resistance affects a community of herbivores. *Ecological Entomology* 26: 312–324.

Traill, L. W., M. L. M. Lim, N. S. Sodhi, and C. J. A. Bradshaw. 2010. Mechanisms driving change: Altered species interactions and ecosystem function through global warming. *Journal of Animal Ecology* 79: 937–947.

Trouvelot, L. 1867. The American silk worm. *American Naturalist* 1: 30–38, 85–94, 145–149.

Trumble, J. T. and P. D. Jensen. 2004. Ovipositional response, developmental effects, and toxicity of hexavalent chromium to *Magaselia scalaris*, a terrestrial detritivore. *Archives of Enviromental Contamination and Toxicology* 46: 372–376.

Tscharntke, T., R. Bommarco, Y. Clough, T. O. Crist, T. Kleijn, T. A. Rand, J. M. Tylianakis, S. van Nouhoys, and S. Vidal. 2007. Conservation biological control and enemy diversity on a landscape scale. *Biological Control* 43: 294–309.

Tscharntke, T., A. M. Klein, A. Kruess, I. Steffan-Dewenter, and C. Thies. 2005. Landscape perspectives on agricultural intensification and biodiversity–ecosystem service management. *Ecology Letters* 8: 857–874.

Valkama, E., J. Koricheva, and E. Oksanen. 2007. Effects of elevated O_3, alone and in combination with elevated CO_2, on tree leaf chemistry and insect herbivore performance: A meta-analysis. *Global Change Biology* 13: 184–201.

van den Berg, H. 2009. Global status of DDT and its alternatives for use in vector control to prevent disease. *Environmental Health Perspectives* 117: 1656–1663.

Van der Putten, W. H., M. Macel, and M. E. Visser. 2010. Predicting species distribution and abundance responses to climate change: Why it is essential to include biotic interactions across trophic levels. *Philosophical Transactions of the Royal Society B* 635: 2025–2034.

Vasconcelos, H. L., E. H. M. Vieira-Neto, and F. M. Mundim. 2006. Roads alter the colonization dynamics of a keystone herbivore in neotropical savannas. *Biotropica* 38: 661–665.

Vezenegho, S. B., B. D. Brooke, R. H. Hunt, M. Coetzee, and L. L. Koekemoer. 2009. Malaria vector composition and insecticide susceptibility status in Guinea Conakry, West Africa. *Medical and Veterinary Entomology* 23: 326–334.

Victor, F. C., D. E. Cohen, and N. A. Soter. 2010. A 20-year analysis of previous and emerging allergens that elicit photoallergic contact dermatitis. *Journal of the American Academy of Dermatology* 62: 605–610.

Vittor, A. Y., R. H. Gilman, J. Tielsch, G. Glass, T. Shields, W. S. Lozano, V. Pinedo-Cancino, and J. A. Patz. 2006. The effect of deforestation on the human-biting rate of *Anopheles darlingi*, the primary vector of falciparum malaria in the Peruvian Amazon. *American Journal of Tropical Medicine and Hygiene* 74: 3–11.

Waring, R. H. and S. W. Running. 1998. *Forest Ecosystems: Analysis at Multiple Scales*, 2nd ed. San Diego: Academic.

Whitford, W. G. 2002. *Ecology of Desert Systems*. San Diego: Academic.

Williams, D. W. and A. M. Liebhold. 1995. Forest defoliators and climatic change: Potential changes in spatial distribution of outbreaks of western spruce budworm (Lepidoptera: Tortricidae) and gypsy moth (Lepidoptera: Lymantriidae). *Environmental Entomology* 24: 1–9.

Williams, D. W. and A. M. Liebhold. 2002. Climate change and the outbreak ranges of two North American bark beetles. *Agricultural and Forest Entomology* 4: 87–99.

Wilson, R. J., D. Gutiérrez, J. Gutiérrez, and V. J. Monserrat. 2007. An elevational shift in butterfly species richness and composition accompanying recent climate change. *Global Change Biology* 13: 1873–1887.

Witcosky, J. J., T. D. Schowalter, and E. M. Hansen. 1986. The influence of time of precommercial thinning on the colonization of Douglas-fir by three species of root-colonizing insects. *Canadian Journal of Forest Research* 16: 745–749.

Xue, Y., K. N. Liou, and A. Kashahara. 1990. Investigation of biophysical feedback on the African climate using a two-dimensional model. *Journal of Climate* 3: 337–352.

Yamanaka, T. and A. M. Liebhold. 2009. Spatially implicit approaches to understand the manipulation of mating success for insect invasion management. *Population Ecology* 51: 427–444.

Yang, C.-C., D. D. Shoemaker, J.-C. Wu, Y.-K. Lin, C.-C. Lin, W.-J. Wu, and C.-J. Shih. 2009. Successful establishment of the invasive fire ant *Solenopsis invicta* in Taiwan: Insights into interactions of alternate social forms. *Diversity and Distributions* 15: 709–719.

Yemshanov, D., F. H. Koch, D. W. McKenney, M. C. Downing, and F. Sapio. 2009. Mapping invasive species risks with stochastic models: A cross-border United States-Canada application for *Sirex noctilio* Fabricius. *Risk Analysis* 29: 868–884.

Yu, G., H. Shen and J. Liu. 2009. Impacts of climate change on historical locust out-breaks in China. *Journal of Geophysical Research* 114: 1–11.

Zangerl, A. R., D. McKenna, C. L. Wraight, M. Carroll, P. Ficarello, R. Warner, and M. R. Berenbaum. 2001. Effects of exposure to event 176 *Bacillus thuringiensis* corn pollen on monarch and black swallowtail caterpillars under field conditions. *Proceedings of the National Academy of Sciences USA* 98: 11908–11912.

Zavala, J. A., C. L. Casteel, E. H. DeLucia, and M. R. Berenbaum. 2008. Anthropogenic increase in carbon dioxide compromises plant defense against invasive insects. *Proceedings of the National Academy of Sciences USA* 105: 5129–5133.

Zettler, J. A., M. D. Taylor, C. R. Allen, and T. P. Spira. 2004. Consequences of forest clear-cuts for native and nonindigenous ants. *Annals of the Entomological Society of America* 97: 513–518.

Zhang, D. D., P. Brecke, H. F. Lee, Y. Q. He, and J. Zhang. 2007. Global climate change, war, and population decline in recent human history. *Proceedings of the National Academy of Sciences USA* 104: 19214–19219.

Zheng, X. and E. A. B. Eltahir. 1998. The role of vegetation in the dynamics of West African monsoons. *Journal of Climate* 11: 2078–2096.

Zimmerman, P. R., J. P. Greenberg, S. O. Wandiga, and P. J. Crutzen. 1982. Termites: A potentially large source of atmospheric methane, carbon dioxide, and molecular hydrogen. *Science* 218: 563–565.

Zvereva, E. L. and M. V. Kozlov 2006. Consequences of simultaneous elevation of carbon dioxide and temperature for plant–herbivore interactions: A metaanalysis. *Global Change Biology* 12: 27–41.

7

Effects of Insects on Ecosystem Services

> Root crops do well, and vegetables of all kinds attain immense propor-
> tions, owing to the freedom from weeds and fertility resulting from the
> dung and bodies of the dead locusts. . . . there are other ways in which
> good may grow out of the locust troubles when they are severe . . . such
> a devastated country is apt to be free from most noxious insects during
> the subsequent two or three years.
>
> **Charles Valentine Riley (1878)**

Insects are not the only factors that affect ecosystem services. Obviously, precipitation is a key factor that affects water yields, and water availability and
soil fertility are key factors affecting net primary production and the services
it supports (Wood et al. 2009; Wright et al. 2011; Lǚ et al. 2012; Pasquini and
Santiago 2012). However, insects affect these and other ecosystem services
in complex, often complementary, and dramatic ways. Pollination services
are highly valued for crop production, but bee stings can be life-threatening
to sensitive individuals. Herbivory conspicuously reduces plant growth and
survival but may stimulate long-term primary production, thereby maintaining net primary productivity (NPP) near carrying capacity (see Chapter 5).
Other insect functional groups (Box 7.1) also affect services in complex ways.
The variety of insect effects on ecosystem services is the focus of this chapter.

BOX 7.1 INSECT FUNCTIONAL GROUPS

A functional group is defined as a group of species that affect ecosystem structure or function in a particular way (Schowalter 2011).
Assignment of insects to functional groups depends on particular objectives because many insects, especially those with complete
metamorphosis, can be members of more than one functional group
during their lives. For example, most butterflies and moths are herbivores as immatures but pollinators as adults. Many mosquito larvae
are important aquatic detritivores as immatures but blood feeders as
adults. Many herbivores and detritivores also disperse plant seeds and
microbial spores. Furthermore, functional groups can be defined on
the basis of resource use (e.g., herbivores, pollinators, or predators), or
role in ecosystem change (e.g., stress-adapted, pioneers, plant mortality
agents, or seed dispersers), etc.

Combining species into functional groups does not imply redundancy because each species has a unique combination of effects on rate, timing, and direction of various ecosystem processes (Schowalter 2011). Loss of any species potentially jeopardizes processes that it affects. Major functional groups defined relative to ecosystem services and human health are described as follows (Romoser and Stoffolano 1998; Schowalter 2011).

Herbivores feed on living plant material and transfer the energy and nutrients from plant biomass into consumer biomass and, ultimately, detritus. Subgroups of herbivores that affect plants differently include *grazers* that chew foliage, stems, flowers, pollen, seeds, and roots; *miners* and *borers* that feed between plant surfaces; *gall-formers* that reside and feed within the plant and induce the production of abnormal growth reactions by plant tissues; *sap-suckers* that siphon plant fluids; and *frugivores* and *seed predators* that consume the reproductive parts of plants. These different modes of consumption affect plants in different ways. For example, *folivores* (species that chew foliage) directly reduce the area of photosynthetic tissue, whereas sap-sucking insects affect the flow of fluids and nutrients within the plant, and *root-feeders* reduce plant capacity to acquire nutrients from the soil or to remain upright.

Pollinators are an important group of herbivores that feed on pollen or nectar and, in the process, transport pollen among individual plants, ensuring adequate fertilization and seed production. A large number of pollinators are *generalists* with respect to plant species. This functional group includes honey bees and many beetles, flies, and thrips that forage on any floral resources available. *Specialist* pollinators exploit particular floral characteristics that may exclude other pollinators. For example, the diversity of orchid flowers represents a variety of colors and structures that attract specific insects that acquire pollen in unique ways to ensure pollination (Corner 1964). *Pollen feeders* feed primarily on pollen (e.g., beetles and thrips) and are likely to transport pollen acquired during feeding, whereas others are primarily *nectar-feeders* (e.g., beetles, butterflies, moths, and flies) that transport pollen more coincidentally. In fact, many nectar feeders avoid the reproductive organs, often by perforating the base of the flower to reach the nectar, for example, *nectar thieves* (Dedej and Delaplane 2004), or, in the case of ants, they may reduce pollen viability (Peakall et al. 1987). Bees, especially *Apis* spp., feed on pollen and nectar. Functional groupings also reflect sensitivity and attraction to floral odors (Chittka and Raine 2006). For example, *dung-, fungus-,* and *carrion-feeding* flies and beetles are the primary pollinators of plants that emit dung or carrion odors (Norman and Clayton 1986; Appanah 1990; Norman et al. 1992).

Seed dispersers include frugivores, seed predators, and detritivores (see next paragraph) that relocate seeds in the commission of their other functions. Seed predator and seed disperser functional groups can be distinguished on the basis of consumption of fruits or seeds versus transport of seeds. *Seed cachers* eat some seeds and move others from their original location to storage locations. Although ants and rodents are best known for caching seeds (Brown et al. 1979), at least one carabid beetle (*Synuchus impunctatous*) caches seeds of *Melampyrum* in hiding places after consuming the caruncle at the end of the seed (Manley 1971). *Seed vectors* primarily include vertebrates that carry seeds adapted to stick to fur or feathers. Insects generally are too small to transport seeds in this way but often transmit spores of microorganisms adapted to adhere to insect exoskeletons or to pass through insect digestive systems. These functional groups also can be subdivided on the basis of *pre-* or *post-dispersal* seed predation, seed size, and so on. Pre-dispersal frugivores and seed predators feed on the concentrated fruits and seeds developing on the parent plant, whereas post-dispersal frugivores and seed predators find scattered fruits and seeds that have fallen to the ground.

Detritivores feed on dead plant or animal material (detritus) and are instrumental in release of energy and nutrients from detritus for uptake into plants or export into aquatic ecosystems. Subgroups are based on their effect on decomposition processes. *Coarse* and *fine comminuters* are instrumental in the initial fragmentation of large litter material. Major taxa in terrestrial ecosystems include millipedes, earthworms, termites, and beetles (coarse) and mites, collembolans, and various other small arthropods (fine). Many species are primarily *fungivores* or *bacteriovores* that fragment substrates while feeding on the surface microflora. A number of species, including dung beetles, millipedes, and termites, are *coprophages*, either feeding on the feces of larger species or reingesting their own feces following microbial decay and enrichment (McBrayer 1975; Coe 1977; Holter 1979; Cambefort 1991; Kohlmann 1991; Dangerfield 1994). In aquatic ecosystems, *scrapers* (including mayflies, caddisflies, chironomid midges, and elmid beetles) that graze or scrape microflora from mineral and organic substrates, and *shredders* (including stoneflies, caddisflies, crane fly larvae, crayfish, and shrimp), that chew or gouge large pieces of decomposing material, represent coarse comminuters; *gatherers* (including stoneflies, mayflies, and copepods) that feed on fine particles of decomposing organic material deposited in streams, and *filterers* (mayflies, caddisflies, and black flies), that have specialized structures for sieving fine suspended organic material, represent fine comminuters (Cummins 1973; Wallace et al. 1992; Wallace and Webster 1996).

Xylophages are a diverse group of detritivores specialized to excavate and fragment woody litter. Aquatic xylophages include crane flies and elmid beetles. Terrestrial taxa include curculionid, buprestid, cerambycid, and lyctid beetles, siricid wasps, carpenter ants, *Camponotus* spp., and termites, with different species often specialized on particular wood species, sizes, or stages of decay. Most of these species either feed on fungal-colonized wood or support mutualistic, internal, or external fungi, bacteria, or protozoa that are necessary to digest cellulose and enhance the nutritional quality of wood (Siepel and de Ruiter-Dijkman 1993; Breznak and Brune 1994).

Carrion feeders represent another specialized group that breaks down animal carcasses. Major taxa include staphylinid, sylphid, scarabaeid, and dermestid beetles, callophorid, muscid and sarcophagid flies, and various ants. Different species typically specialize on particular stages of decay and on particular animal groups, for example, reptiles versus mammals (Watson and Carlton 2003).

Burrowers redistribute large amounts of soil or detritus during nesting, foraging, or feeding activities (Kohlmann 1991). Fossorial functional groups can be distinguished on the basis of their food source and the mechanism and volume of soil/detrital mixing. *Subterranean nesters* burrow primarily for shelter. A variety of vertebrates (e.g., squirrels, wood rats, and coyotes) and invertebrates (including crickets and solitary wasps) excavate tunnels of various sizes, typically depositing soil on the surface and introducing some organic detritus into nests. *Gatherers*, primarily social insects, actively concentrate organic substrates in colonies. Ants and termites redistribute large amounts of soil and organic matter during construction of extensive subterranean, surficial, or arboreal nests (Haines 1978; Anderson 1988). These insects can affect a large volume of substrate (up to 1000 m^3), especially as a result of restructuring and lateral movement of the colony (Moser 1963; Whitford et al. 1976; Hughes 1990). *Fossorial feeders*, such as gophers, moles, earthworms, mole crickets (Gryllotalpidae), and benthic invertebrates feed on subsurface resources (plant, animal, or detrital substrates) as they burrow, constantly mixing mineral substrate and organic material in their wake.

Predators generally are considered to be those animals that kill and consume other animals (prey) (Price 1997), although some plants—such as Venus fly traps, sundews, and pitcher plants—also function as predators of insects. Predators represent important regulators of prey population size and condition and often are incorporated into pest management strategies. Subgroups can be defined on the basis of prey size, for example, *carnivores* (such as wolves and giant water bugs) or *insectivores* (such as many birds, flies, and wasps) or capture strategy, for example, *hunters* (such as many flies, wasps, beetles, and spiders) or *ambushers* (such as ant lions, dragonfly larvae, and trapdoor spiders).

Parasites feed on or in living prey. *Parasitoidism* is unique to insects, especially flies and wasps, and involves the adult parasitoid depositing eggs or larvae on, in, or near multiple hosts, with the larvae subsequently feeding on their living host and eventually killing it. Parasitic interactions are relatively specific associations between coevolved parasites and their particular host species and involve adaptations by the parasite to overcome immune responses of the host. Because of this specificity, parasites and parasitoids tend to be more effective than predators in responding to and controlling population irruptions of their hosts; therefore, they have been primary agents in biological control programs (Hochberg 1989). *Ectoparasites* feed externally, by inserting mouthparts into the host (e.g., lice, fleas, mosquitoes, and ticks), and *endoparasites* feed internally, within the host's body (e.g., bacteria, nematodes, bot flies, and wasps). *Primary parasites* develop on or in a nonparasitic host, whereas *hyperparasites* develop on or in another parasite. Some parasites parasitize other members of the same species (*autoparasitism* or *adelphoparasitism*). *Superparasitism* occurs when more individuals occupy a host than can develop to maturity. *Multiple parasitism* occurs when more than one parasitoid species is present in the host simultaneously.

7.1 Effects on Provisioning Services

Provisioning services include ecosystem products that can be harvested or used to meet human needs. Most crop species originated as members of natural ecosystems and were exploited during their growing seasons by early hunters and gatherers. Later, seeds of desirable species were collected and planted to provide a more reliable supply of food or other products. Over the ensuing millennia, constant selection, and later breeding, for the most desirable qualities has led to crop plant genotypes and phenotypes that are radically different from their ancestral varieties (Diamond 1999). In many cases, these changes improved taste to humans at the expense of distasteful or toxic defenses against insects and pathogens. In other cases, selection for rapid growth led to reduced production of defenses (Michaud and Grant 2009).

Clearly, feeding by insects on plants can reduce plant production of these resources or cause undesirable blemishes that affect cosmetic value (Figure 7.1). Outbreaks of insects typically last only three to four years in natural or seminatural ecosystems, followed by a precipitous decline as food resources

FIGURE 7.1 (SEE COLOR INSERT.)
Extensive defoliation of soybeans by the soybean looper, *Chrysodeixis includens,* can reduce soybean production substantially.

are depleted. Chronically high population sizes are more characteristic of managed monocultures of crop and forest species.

On the other hand, insects are the food base for important fisheries and game animals. Insects falling into streams comprise 30–80% of the diets of young salmon (Kawaguchi and Nakano 2001; Allan et al. 2003; Baxter et al. 2005). Insects falling into nutrient-poor lakes and headwater streams add substantial amounts of carbon, nitrogen, and phosphorus that increase productivity of these aquatic ecosystems (Carlton and Goldman 1984; Mehner et al. 2005; Nowlin et al. 2007; Menninger et al. 2008; Pray et al. 2009). Aquatic insects also are critical to sustaining most freshwater game fish populations. Most gallinaceous birds (including pheasants, quails, chickens, and turkeys), used as food by most human cultures for millennia, feed primarily on insects. Environmental changes or management practices that jeopardize the availability of insects for these animals also threaten these food resources for humans. Therefore, insect effects on fish and wildlife production may be largely positive.

A fresh water supply is one of the most valued ecosystem services. Human communities have been established almost invariably on, or near, sources of fresh water. Constant yield of fresh water remains the primary management goal for municipal watersheds serving urban population centers. Insects and foliage fragments falling into streams during outbreaks may affect water quality (Lovett et al. 2002). Herbivorous insects reduce canopy cover, interception of precipitation, and accumulation of winter snow, thereby increasing the rate and amount of water flowing into streams (Bewley et al. 2010). Soil and litter insects affect soil porosity and decomposition rate, factors that affect the rate of water movement through the substrate (Coleman et al. 2004). Changes in water yield and quality resulting from insect activity may or may not be desirable, depending on needs of downstream users.

Insects or their products also represent important provisioning services (see Chapter 2). Honey has been among our most valuable food resources for millennia (Chapter 2). Insects are an important source of protein in many cultures (Cerritos and Cano-Santana 2008; Ramos-Elorduy 2009; Yen 2009), and maintenance of edible insect populations represents a primary ecosystem management goal in some cultures (Mbata et al. 2002). Insects have provided pharmaceutical compounds or other medical products (Gudger 1925; Epstein and Kligman 1958; Namba et al. 1988; Pemberton 1999; Sherman et al. 2000, 2007; Kerridge et al. 2005), including cantharidin, a defensive alkaloid produced by blister beetles (Meloidae) that is used commercially to remove warts, and surgical silk. Silk production has been practiced at least since 2000–3000 B.C. in China and is among the most widely traded commercial products (Anelli and Prischmann-Voldseth 2009). Scale insects have been important sources of dyes and shellac (Clausen 1954; Greenfield 2005; Anelli and Prischmann-Voldseth 2009). More recently, insect enzymes offer efficient tools for conversion of cellulosic materials into biofuels (Cook and Doran-Peterson 2010).

7.2 Effects on Cultural Services

Ecosystems provide various spiritual, recreational, and other cultural services, including hiking, backpacking, hunting and fishing, and educational and scientific activities. Effects of insects on cultural services can be positive or negative depending on public perceptions.

Insects are the primary food resources for many insectivorous birds and fish that provide popular recreational activities, such as bird-watching and sportfishing. In fact, the creation of artificial lures that resemble various insects is popular for artistic value, as well as attraction of game fish.

Some insects have been important cultural symbols (see Chapter 2). In China, cicadas are symbols of rebirth, and crickets are symbols of good fortune. Both insects often are caged and kept as pets for their songs or fighting ability, but both are considered nuisances in many other cultures (Clausen 1954). Insects also can be objects for entertainment. Butterfly gardens, ant farms, and pet tarantulas, millipedes, and scorpions are available from a variety of biological supply and pet sources. Insects have inspired much artwork (Figure 7.2) and documentary, as well as science fiction, literature, and film.

On the other hand, a number of insects can interfere with outdoor experiences. Biting flies, stinging bees, wasps and spiders, ants and parasitic chiggers, and ticks can be more than just nuisances. Some vector serious disease pathogens, whereas others can cause painful bites or stings. Individuals who are allergic to insect or spider venoms may suffer life-threatening,

FIGURE 7.2 (SEE COLOR INSERT.)
Detail of jade carving with two orthopterans on cabbage, Qing Dynasty, 19th century.

hypersensitive reactions. A trail sign in Taiwan warns hikers to be alert for venomous hornets, centipedes, and snakes (see Figure 2.8).

Plants defoliated or killed by insects reduce shade and can create unsightly conditions or safety hazards in camping or hiking areas. Insect feces and tissues falling on people or eating surfaces are considered a nuisance. Furthermore, some caterpillars are venomous or allergenic (Perlman et al. 1976; Schowalter 2011), exacerbating the nuisance. These detrimental effects can reduce visitation to those recreational sites experiencing outbreaks.

Few studies have evaluated effects of insect outbreaks on cultural values. Downing and Williams (1978) reported that a Douglas-fir tussock moth, *Orgyia pseudotsugata*, outbreak in Oregon did not significantly affect recreational land use but that recreational use appeared to increase as a result of curiosity. Although 75% of visitors were aware of the outbreak, few chose to avoid the area. In fact, the only negative effect mentioned in their study was avoidance of salvage logging operations that were considered unappealing or hazardous. On the other hand, extensive tree mortality during a bark beetle outbreak was viewed as unattractive or hazardous (Michalson 1975). Sheppard and Picard (2006) compiled a number of studies in which subjects were shown pairs of photos, one with insect damage, the other without (Figure 7.3). In general, visual preference depended on the extent of defoliation or tree mortality. Most studies showed a decline in visual preference with increasing defoliation or plant mortality, sometimes showing a threshold of about 10% of visible landscape above which additional defoliation or mortality had less effect. Some studies showed that visual preference was affected by the subject's awareness of the cause. Müller and Job (2009) reported that tourist attitudes toward bark beetle outbreaks in a national park in Germany were largely neutral and against control efforts but were more positive toward noncontrol among tourists who were more knowledgeable about bark beetles and the process of forest recovery following such natural disturbances.

FIGURE 7.3 (SEE COLOR INSERT.)
Series of photos with manipulated variation in extent of pine crown mortality shown randomly to survey participants to evaluate visual preference for forests affected by insects or diseases. Note the progressively greater crown defoliation of the central tree from (a) through (c). (Courtesy of F. Baker. From P. Barbosa, D. K. Letourneau, and A. A. Agrawal, eds., *Insect Outbreaks Revisited*, Wiley/Blackwell, Hoboken, NJ, 2012. With permission.)

7.3 Effects on Supporting Services

Supporting services include primary production, pollination, and soil formation. Primary production represents the energy and matter accumulated by plants (see Chapter 5) and supports all provisioning services through its effects on water and nutrient fluxes and climate. Pollination is essential for fruit and seed production and for reproduction by many plants. Soil formation provides resources necessary for primary production. Disruption of these processes interferes with provisioning and cultural services.

7.3.1 Primary Production

Primary production is the source for all plant-derived ecosystem services, such as fruits, vegetables, wood, and fiber. Primary production also supports the variety of animals and other organisms that must feed on plants to meet their nutritional needs. In addition, primary production supports water yield from ecosystems by controlling rates of water uptake and evapotranspiration, climate moderation by shading and by increasing local precipitation, and carbon sequestration by fixing carbon in biomass.

Herbivory by insects, as well as other animals, removes plant tissues, but plants, including crop species, have considerable capacity to compensate for herbivory, depending on plant condition, timing, and intensity of herbivory and on the availability of water and nutrients (Pedigo et al. 1986; Trumble et al. 1993; Feeley and Terborgh 2005; Dungan et al. 2007; Schowalter 2011). Healthy plants can compensate for lost tissues better than can stressed plants. Larger plants typically have greater carbohydrate storage for reallocation to compensatory growth than do smaller plants. Seedlings are particularly vulnerable to herbivores because of their limited resource storage capacity and limited ability to replace tissues lost to herbivores. Survival of tropical tree seedlings was highly correlated with the percentage of original leaf area present one month after germination and with the number of leaves present at seven months of age (Clark and Clark 1985).

Herbivory typically is focused on less efficient and/or less defended plant tissues that require less investment in search time or detoxification by insects for consumption and digestion. Plants produce more foliage than is required for efficient photosynthesis, and in some cases, excess living or dead foliage can inhibit further primary production (Knapp and Seastedt 1986). Therefore, removal of shaded or senescing plant tissues by herbivores allows plants to reallocate resources to more productive tissues (Knapp and Seastedt 1986; Gutschick and Wiegel 1988; Trumble et al. 1993). In fact, some plants, especially grasses, require pruning or low-to-moderate grazing to maintain production (see Figure 5.9) (Knapp and Seastedt 1986; Williamson et al. 1989; Belovsky and Slade 2000). Short-term growth losses in defoliated conifers can be followed by several years, or decades, of growth rates that exceed predefoliation rates (Wickman 1980; Alfaro and Shepherd 1991), replacing at least some of the short-term losses. Similarly, annual wood production in at least some pine forests reached or exceeded pre-attack levels within 10 to 15 years following outbreaks of mountain pine beetles, *Dendroctonus ponderosae* (Romme et al. 1986).

Plants are best able to compensate for foliage loss in the spring, when environmental conditions favor new or continued growth to replace lost foliage, but they become less able to compensate later in the season. Grasshoppers, *Aulocara elliotti*, did not significantly reduce blue grama grass, *Bouteloua gracilis*, biomass when feeding occurred early in the growing season but significantly reduced grass biomass when feeding occurred late in the growing season (Thompson and Gardner 1996).

Water and nutrient availability are critical to permitting compensatory growth. Lovett and Tobiessen (1993) found that experimental defoliation (80%) of red oak, *Quercus rubra*, seedlings resulted in a significant 50% increase in photosynthetic rates, and seedlings that were provided with elevated nitrogen levels were able to maintain high photosynthetic rates for a longer time than were seedlings at lower nitrogen levels.

Herbivory on aboveground plant parts can reduce root growth and plant ability to acquire sufficient water and nutrients to compensate. Starch concentrations in roots were related inversely to the level of mechanical damage to shoots of a tropical tree, *Cedrela odorata* (Rodgers et al. 1995). Folivory on piñon pines (*Pinus edulis*) adversely affected mycorrhizal fungi, perhaps through reduced carbohydrate supply to roots (Gehring and Whitham 1995). Defoliation of Douglas-fir, *Pseudotsuga menziesii*, seedlings decreased biomass but reduced water stress and increased photosynthesis, compared with undefoliated seedlings, improving seedling survival under drought conditions in greenhouse experiments (Kolb et al. 1999).

Obviously, high levels of herbivory, especially during outbreaks, can overwhelm plants' ability to compensate and may lead to plant stress, decline, and perhaps death. Severe outbreaks by one insect species can weaken plants and increase vulnerability to other insect species. The Douglas-fir beetle, *Dendroctonus pseudotsugae*, and fir engraver beetle, *Scolytus ventralis*, preferentially colonized Douglas-fir trees that had lost more than 90% of foliage to Douglas-fir tussock moths, although larval survival was greater in nondefoliated than in defoliated trees (Wright et al. 1986). However, intense defoliation during drought also can reduce moisture stress that otherwise would limit plant survival (Kolb et al. 1999).

Differential herbivory among host and nonhost plants and plant species influences the rate and direction of change in community structure and explains distributions of many plant species (Louda et al. 1990; Davidson 1993). High intensity of herbivory during outbreaks can dramatically reduce the density of preferred host species and favor replacement by nonhosts (Mattson and Addy 1975; McNaughton 1979; Schowalter 1981; Knapp and Seastedt 1986; Williamson et al. 1989; Belovsky and Slade 2000; Feeley and Terborgh 2005; Dungan et al. 2007; Cairns et al. 2008). Davidson (1993) compiled data indicating that herbivores retard or reverse succession during early seres but advance succession during later seres.

In at least some cases, alteration of vegetation composition by insects may tailor overall plant demand for water and nutrients to prevailing conditions at a site, for example, replacement of nitrogen-rich species by low-nitrogen species (Ritchie et al. 1998; Belovsky and Slade 2000). Succession from pioneer pine forest to late successional fir forest in western North America can be retarded or advanced by insects, depending primarily on moisture availability and condition of the dominant vegetation (Schowalter 2008). When moisture is adequate (e.g., in riparian corridors and at high elevations), the mountain pine beetle advances succession by facilitating the replacement of host pines by more shade-tolerant, fire-intolerant, understory firs. However, limited moisture and short fire return intervals at lower elevations favor pine dominance. In the absence of fire during drought periods, western spruce budworms, *Choristoneura occidentalis*, Douglas-fir tussock moths, and bark beetles concentrate on the understory firs, truncating (or reversing) succession (see Figure 5.11). Fire fueled by fir mortality also leads to eventual regeneration of pine forest.

In one study, canopy openings resulting from spruce budworm, *Choristoneura fumiferana*, outbreaks had a greater diversity of saplings and trees and larger perimeter/area ratios than did canopy openings resulting from clear-cut harvests with protection of advance regeneration practices (Belle-Isle and Kneeshaw 2007). These results suggested that stand recovery and contributions by the surrounding forest should be greater in budworm-generated openings than in harvest-generated openings.

Therefore, insect outbreaks often have less negative effects on primary production (and provisioning and other ecosystem services that it supports) over long time periods than is generally perceived. Biomass reduction by insects may not always be desirable, especially in intensively managed forest or rangeland, but can increase primary production in the same manner as prescribed mowing, pruning, and/or thinning. Improved survival of defoliated plants during drought (Kolb et al. 1999) would mitigate effects of drought, which is a common trigger for outbreaks (Mattson and Haack 1987; Van Bael et al. 2004).

7.3.2 Pollination

Pollination is among the most visible and important ecosystem services (see Figure 2.2). Although many cereal crops and widespread plant species in temperate grasslands and forests are effectively pollinated by wind, pollen limitation is a major factor controlling fruit and seed production for plant species that tend to occur at low densities, especially in deserts and tropical forests and for isolated forbs and understory trees in temperate grasslands and forests (Steffan-Dewenter and Tscharntke 1999; Knight et al. 2005; Vamosi et al. 2006; Ricketts et al. 2008). Consequently, pollination is necessary for reproduction of 60–70% of all plant species and for 35% of global crop production (Figure 7.4) (Losey and Vaughan 2006; Klein et al. 2007; Kremen et al. 2007). The economic importance of insect pollinators, especially honey bees, has led to widespread transport and introduction of honey bees to most regions of the globe. However, many plant species, including crop species in some areas, depend primarily on native specialist pollinators, which are more efficient pollinators than are the more generalist honey bee. For example, native bees are the primary pollinators of squash crops in the United States (Kremen et al. 2004). Native pollinators, where protected, are capable of providing full pollination service and compensating for decline or absence of honey bees (Kremen et al. 2002; Ricketts 2004). Recent dramatic declines in abundances of pollinators have generated concern for maintenance of pollination services (Biesmeijer et al. 2006; Cox-Foster et al. 2007; Genersch 2010). The threat is threefold.

First, a major concern in North America is the mysterious honey bee colony collapse disorder (CCD) that has resulted in the loss of 50–90% of hives in the United States (Cox-Foster et al. 2007). The disorder is indicated by the unexplained absence of live or dead adult workers in or near hives, despite

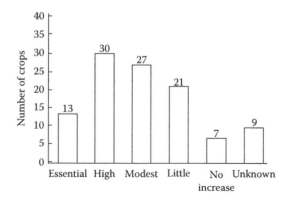

FIGURE 7.4
Level of dependence on animal-mediated pollination by crops that produce fruits or seeds for direct human use as food. Crops for which seeds are only used for breeding or to grow vegetable parts for direct human use or for forage and crops known to be only wind-pollinated, passively self-pollinated, or vegetatively reproduced are not included. Essential = pollinators essential for most varieties (production reduced by >90% in the absence of animal pollinators); high = animal pollinators are extremely beneficial (40 to <90% reduction); modest = animal pollinators are clearly beneficial (10 to <40% reduction); little = some evidence suggests that animal pollinators are beneficial (0 to <10% reduction); no increase = no production increase with animal-mediated pollination; unknown = empirical studies are missing. (From Klein, A.-M, B. E. Vaissière, J. H. Cane, I. Steffan-Dewenter, S. A. Cunningham, C. Kremen, and T. Tscharntke, *Proceedings of the Royal Society B*, 274, 303–313, 2007. With permission.)

abundant brood, honey, and pollen (Cox-Foster et al. 2007; Johnson et al. 2009). Explanations have ranged from increased pesticide exposure to undetected diseases to disorientation and other stresses during transportation of hives among agricultural regions. Claudianos et al. (2006) and Weinstock et al. (2006) reported that the honey bee genome contains substantially fewer genes for immunity or detoxifying enzymes, compared with other known insect genomes, explaining the high sensitivity of honey bees to pesticides. The parasitic mite, *Varroa destructor*, is known to suppress the bee's immune system and may be acting in concert with emerging diseases to cause CCD (Genersch 2010).

Second, abundances of native pollinators have declined in many areas, especially in agricultural regions. Abundances of native pollinators are threatened by a combination of habitat loss (Ricketts 2004; Kremen 2004; Holzschuh et al. 2007; Klein et al. 2007; Taki et al. 2007; Williams and Kremen 2007) and competition from introduced honey bee populations (Aizen and Feinsinger 1994). Declining diversity of pollinators is likely to threaten the yield of many crops requiring specialized pollinators (Hoehn et al. 2008).

Finally, agricultural intensification and habitat fragmentation have isolated pollinator habitats from the crops that must be pollinated. Pollination services by native bees and feral honey bees are positively related to the proximity of natural habitat to agricultural crops (Crane 1999; Steffan-Dewenter and Tscharntke 1999; Kremen et al. 2004; Ricketts 2004; Williams and Kremen 2007). Balvanera et al. (2005) found that increasingly intensive conventional agriculture resulted in the loss of 60% of native bee pollinators, reduced abundances of the most functionally important pollinators, a 60–80% loss of pollination function, and reduced consistency of pollination, compared with organic agriculture within a conserved forest matrix.

7.3.3 Decomposition and Soil Formation

A variety of insects and other animals are instrumental in the decomposition of organic matter that makes nutrients reavailable for plant growth (Coleman et al. 2004; Schowalter 2011). In the absence of detritivorous insects and other arthropods, decomposition may be dramatically reduced or delayed, leading to accumulation of dead organic matter and bottlenecks in nutrient cycling. A number of studies have demonstrated that microarthropods are responsible for up to 80% of the total decay rate, depending on litter quality and ecosystem conditions (Figure 7.5) (Vossbrinck et al. 1979; Seastedt 1984; Heneghan et al. 1999; González and Seastedt 2001; Coleman et al. 2004; Hättenschwiler and Gasser 2005).

Cuffney et al. (1990) and Wallace et al. (1991) reported that a 70% reduction in abundance of shredders from a small headwater stream in the eastern United States reduced leaf litter decay rates by 25–28% and the export of fine particulate organic matter by 56%. As a result, unprocessed leaf litter accumulated (Wallace et al. 1995). Anderson et al. (1984) noted that aquatic xylophagous tipulid larvae fragmented more than 90% of decayed red alder, *Alnus rubra*, wood in a one-year period. Wise and Schaefer (1994) found that

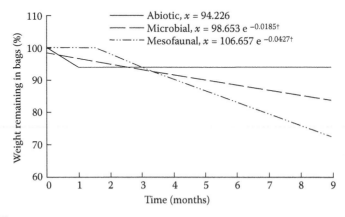

FIGURE 7.5

Decomposition rate of blue grama grass in litter bags treated to permit decomposition by abiotic factors alone, abiotic factors + microbes, and abiotic factors + microbes + mesofauna (microarthropods). Decomposition in the abiotic treatment was insignificant after the first month; decomposition showed a two-month time lag in the treatment including mesofauna. (From Vossbrinck, C. R., D. C. Coleman, and T. A. Woolley, *Ecology*, 60, 265–271, 1979. With permission.)

excluding macroarthropods and earthworms from leaf litter of selected plant species in a beech forest reduced decay rates 36–50% for all litter types except fresh beech litter. When all detritivores were excluded, comparable reduction in decay rate was 36–93%, indicating the prominent role of large arthropods and earthworms in decomposition.

Termites have received considerable attention because of their substantial ecological and economic importance in forest, grassland, desert, and urban ecosystems. Based on laboratory feeding rates, Lee and Butler (1977) estimated that wood consumption by termites in dry sclerophyll forest in South Australia was equivalent to about 25% of annual woody litter increment and 5% of total annual litterfall. Whitford et al. (1982) used termite exclusion plots to demonstrate that termites consumed up to 40% of surficial leaf litter in a warm desert ecosystem in the southwestern United States. Overall, termites in this ecosystem consumed at least 50% of estimated annual litterfall (Johnson and Whitford 1975; Silva et al. 1985). Collins (1981) reported that termites in tropical savannas in West Africa consumed 60% of annual wood fall and 3% of annual leaf fall (24% of total litter production), compared with fire, that removed 0.2% of annual wood fall and 49% of annual leaf fall (31% of total litter production). In that study, fungus-feeding termites were responsible for 95% of the litter removed by termites. Termites apparently consume virtually all litter in tropical savannas in East Africa (Jones 1989, 1990). Termites consume a lower proportion of annual litter inputs in more mesic ecosystems. Collins (1983) reported that termites consumed about 16% of annual litter production in a Malaysian rainforest receiving 2,000 mm precipitation year[-1] and 1–3% of annual litter production in a Malaysian rainforest receiving 5,000 mm precipitation year[-1].

Accumulation of dung from domestic livestock has become a serious problem in many arid and semiarid ecosystems. Termites can remove as much as 100% of cattle dung in three months (Coe 1977; Whitford et al. 1982; Herrick and Lal 1996). On average, termites in the tropics remove 33% of dung in a particular habitat within one month of deposition (Freymann et al. 2008). In the absence of termites, dung would require 25–30 years to disappear (Whitford 1986).

Dung beetles and earthworms also are important consumers of dung in many tropical and subtropical ecosystems (Coe 1977; Holter 1979; Kohlmann 1991). Dung beetles are instrumental in burying vertebrate dung within a few days and contributing to its decomposition (Figure 7.6), but individual species are relatively specific with regard to substrate conditions and host range of colonized dung (Davis 1996).

Native dung beetles in Australia prefer marsupial dung to introduced livestock dung and provided only limited processing of cattle dung for a few weeks in spring and autumn (Tyndale-Biscoe 1994). Ferrar (1975) reported that experimentally marked cattle dung survived at least three months and sometimes more than a year. Accumulating dung smothered pasture vegetation and increased reproductive habitat for two hematophagous flies, the

FIGURE 7.6
Dung beetle rolling ball of dung in which it will lay its eggs. (From Schowalter, T. D., *Insect Ecology: An Ecosystem Approach*, Elsevier, San Diego, 2011. With permission.)

buffalo fly, *Haematobia irritans exigua*, and the bush fly, *Musca vetustissima*, which became serious pests of cattle and humans (Ferrar 1975; Tyndale-Biscoe and Vogt 1996). Beginning in 1967, a number of African dung beetles were introduced into Australia to accelerate dung disintegration and nutrient turnover and to manage fly populations (Hughes et al. 1978). Initial introductions resulted in substantially increased dung disintegration and burial, from less than 7% per week at sites with only one exotic species to 30% at sites with five exotic species, but this rate fluctuated from 0–70% depending on beetle abundance (Tyndale-Biscoe 1994). Suppression of fly reproduction occurred primarily through dung disturbance (Hughes et al. 1978), but the first exotic species were most active during the warm monsoon season and relatively ineffective earlier when bush flies first appeared (Davis 1996; Tyndale-Biscoe and Vogt 1996). Subsequent research identified additional dung beetle species that could be active earlier (Ridsdill-Smith and Kirk 1985) and demonstrated the importance of phoretic mites (e.g., native *Macrocheles glaber* and exotic *M. peregrinus*) that prey on fly larvae in dung pads (Roth et al. 1988). Although we do not know all the consequences of these introductions, mean abundances of native dung beetles have remained similar to their pre-introduction abundances (Tyndale-Biscoe and Vogt 1996). These studies demonstrate the importance of detritivores in preventing accumulation of organic debris.

Decomposition also releases nutrients (mineralization) from organic detritus, making them available for plant uptake and production of new plant tissues. Setälä and Huhta (1991) demonstrated that soil invertebrates increased birch (*Betula pendula*) leaf, stem, and root biomass by 70%, 53%, and 38%, respectively, and increased foliar nitrogen and phosphorus contents threefold and 1.5-fold, respectively, compared with controls with microorganisms only. More recently, Laakso and Setälä (1999) found that experimental removal of microbe- or detritus-feeding soil fauna, especially the microbial-detritivorous enchytraeid worm, *Cognettia sphagnetorum*, reduced plant biomass and uptake of nitrogen.

Outbreaks of herbivorous insects move considerable organic matter from vegetation to the soil surface, affecting the rate and timing of litterfall, decomposition, and soil conditions. Insect feces, in particular, provide an abundant manure (Figure 7.7) and increase the rate and amount of nutrient fluxes (especially nitrogen) to soil, both of which support plant compensatory growth (Chapman et al. 2003; Frost and Hunter 2004, 2007, 2008; Classen et al. 2005; Fonte and Schowalter 2005; Schowalter et al. 2011). Organic matter accumulation, especially where outbreaks increase the amount of coarse woody debris, also affects soil texture, water-holding capacity, and fertility. As organic matter decomposes, its porosity and water retention capacity increase, facilitating aerobic decomposition, carbon loss, and nitrogen-fixation (Harmon et al. 1986; Progar et al. 2000; Coleman et al. 2004). These processes contribute to soil water content and fertility that are necessary for primary production and for those provisioning and cultural services that they support (see earlier). Wood et al. (2009) demonstrated that experimental addition of litter on a plot scale increased foliage production and nitrogen and phosphorus content. These results indicate that biomass and nutrients transferred to the forest floor during outbreaks could be incorporated quickly into new plant tissues. The fertilization effects of herbivores potentially mitigate nutrient limitation and reduce the need for exogenous fertilizers in managed ecosystems.

Fossorial insects and other arthropods alter soil structure by redistributing soil and organic material and increasing soil porosity (Anderson 1988).

FIGURE 7.7 (SEE COLOR INSERT.)
Insect feces collected on a 1 m^2 sheet during an outbreak of cypress leaf roller, *Archips goyerana*, and forest tent caterpillar, *Malacosoma disstria*, in a cypress–tupelo swamp in southern Louisiana.

Porosity determines the depth to which air and water penetrate the substrate. The concentration of organic resources in nests increases soil fertility and moisture around nests. Termite and ant nests frequently support islands of higher primary production and advanced succession in semiarid ecosystems (Parker et al. 1982; Holdo and McDowell 2004; Brody et al. 2010; Fox-Dobbs et al. 2010), potentially contributing to provisioning services in these ecosystems.

Ants and termites are particularly important soil engineers (Dangerfield et al. 1998; MacMahon et al. 2000; Jouquet et al. 2006). Colonies of these insects often occur at high densities and introduce cavities into large volumes of substrate. Eldridge (1993) reported that densities of funnel ant, *Aphaenogaster barbigula*, nest entrances could reach 37 per m^2, equivalent to 9% of the surface area over portions of the eastern Australian landscape. Nests of leaf-cutting ants, *Atta vollenweideri*, reach depths of more than 3 m in pastures in western Paraguay (Jonkman 1978). Moser (2006) excavated a leaf-cutting ant, *Atta texana*, nest in the southern United States and found 97 fungus-garden chambers, 27 dormancy chambers, 45 detritus chambers (for disposal of depleted foliage substrate), and a central cavity at 4 m depth in which the ants and fungus overwinter. The nest extended over an area of 12 m by 17 m on the surface and at least 4 m deep. The bottom of the colony could not be reached, but vertical tunnels extended to at least 7.5 m and might have extended to the water table at 32 m. Whitford et al. (1976) excavated nests of desert harvester ants, *Pogonomyrmex* spp., in the southwestern United States and mapped their three-dimensional architecture. Colony densities were 21–23 per ha at four sites, and each colony consisted of 12 to 15 interconnected galleries (each about 0.035 m^3) within a 1.1 m^3 volume (1.5 m diameter x 2 m deep) of soil, equivalent to about 10 m^3 per ha of cavity space. These colonies frequently penetrated the calcified hardpan (caliche) layer 1.7–1.8 m below the surface, permitting infiltration of water below this hardpan.

The excavation of large galleries and tunnels greatly alters soil structure and chemistry. Termite and ant nests typically represent sites of concentrated organic matter and nutrients (Herzog et al. 1976; Culver and Beattie 1983; Salick et al. 1983; Anderson 1988; Jones 1990; Wagner 1997; Wagner et al. 1997; Lesica and Konnowski 1998; Mahaney et al. 1999; MacMahon et al. 2000; Holdo and McDowell 2004; Wagner and Jones 2004; Risch et al. 2005; Ackerman et al. 2007; Jurgensen et al. 2008). Nests may have concentrations of macronutrients two to three times higher than surrounding soil. Jones (1990) and Salick et al. (1983) noted that soils outside termite nest zones become relatively depleted of organic matter and nutrients. Parker et al. (1982) reported that experimental exclusion of termites for four years increased soil nitrogen concentration by 11%. Ant nests also have been found to have higher rates of microbial activity and carbon and nitrogen mineralization than do surrounding soils (Dauber and Wolters 2000; Lenoir et al. 2001; Wagner and Jones 2004) and represent sites of concentrated carbon dioxide efflux (Risch et al. 2005; Domisch et al. 2006; Jurgensen et al. 2008).

Termites and ants also transport large amounts of soil from lower horizons to the surface and above for construction of nests (Figure 7.8), gallery tunnels, and "carton"—the soil deposited around litter material by termites for protection and to retain moisture during feeding aboveground (Whitford 1986). Whitford et al. (1982) reported that termites brought 10–27 g per m^2 of fine-textured soil material (35% coarse sand, 45% medium fine sand, and 21% very fine sand, clay, and silt) to the surface and deposited 6–20 g of soil carton per gram of litter removed. Herrick and Lal (1996) found that termites deposited an average of 2.0 g of soil at the surface for every gram of dung removed. Mahaney et al. (1999) reported that the termite mound soil contained significantly more (20%) clay than did surrounding soils.

A variety of vertebrate species in Africa have been observed to selectively ingest termite mound soil (Mahaney et al. 1999; Holdo and McDowell 2004). Mahaney et al. (1999) suggested that the higher clay content of termite mounds, along with higher pH and nutrient concentrations, could mitigate gastrointestinal ailments and explain termite soil consumption by chimpanzees. Termite mound soils, as well as surrounding soils, had high concentrations of metahalloysite, used pharmaceutically, and other clay minerals that showed mean binding capacities of 74–95% for four tested alkaloids. Chimpanzees could bind most of the dietary toxins present in 1–10 g of leaves by eating 100 mg of termite mound soil.

Ant and termite nests have particularly important effects on soil moisture because of the large substrate surface areas and volumes affected

FIGURE 7.8 (SEE COLOR INSERT.)
Termite castle formed by transportation of belowground material to build the aboveground structure. (From Schowalter, T. D., *Insect Ecology: An Ecosystem Approach*, Elsevier, San Diego, 2011. With permission.)

(MacMahon et al. 2000). Wagner (1997) reported that soil near ant nests had higher moisture content than did more distant soil. Elkins et al. (1986) compared runoff and water infiltration in plots with termites present or excluded during the previous four years in the southwestern United States. Plots with less than 10% plant cover had higher infiltration rates when termites were present (88 mm per hour) than when termites were absent (51 mm per hour); runoff volumes were twice as high in the termite-free plots with low plant cover (40 mm) as in untreated plots (20 mm). Infiltration and runoff volumes did not differ between shrub-dominated plots (higher vegetation cover) with or without termites.

Eldridge (1993, 1994) measured effects of funnel ants and subterranean harvester termites, *Drepanotermes* spp., on infiltration of water in semiarid eastern Australia. He found that infiltration rates in soils with ant nest entrances were four- to tenfold higher (1,030–1,380 mm per hr) than in soils without nest entrances (120–340 mm per hr). Infiltration rate was correlated positively with nest entrance diameter. However, infiltration rate on the subcircular pavements covering the surface over termite nests was an order of magnitude lower than in the annular zone surrounding the pavement or in interpavement soils. The cemented surface of the pavement redistributed water and nutrients from the pavement to the surrounding annular zone. Ant and termite control of infiltration creates wetter microsites in moisture-limited environments.

Carbon dioxide is the major product of litter decomposition, but incomplete oxidation of organic compounds occurs in some ecosystems, resulting in the release of other trace gases, especially methane (Khalil et al. 1990). Zimmerman et al. (1982) first suggested that termites could contribute up to 35% of global emissions of methane. A number of arthropod species—including most tropical representatives of millipedes, cockroaches, termites, and scarab beetles—are important hosts for methanogenic bacteria and are relatively important sources of biogenic global methane emissions (Hackstein and Stumm 1994). Termites have received the greatest attention as sources of methane because of their high abundance and relatively sealed colonies that are warm and humid, with low oxygen concentrations that favor fermentation and emission of methane or acetate (Brauman et al. 1992; Wheeler et al. 1996). Thirty of 36 temperate and tropical termite species assayed by Brauman et al. (1992), Hackstein and Stumm (1994), and Wheeler et al. (1996) produced methane and/or acetate. Generally, acetogenic bacteria outproduce methanogenic bacteria in wood- and grass-feeding termites, but methanogenic bacteria are much more important in fungus-growing and soil-feeding termites (Brauman et al. 1992). Khalil et al. (1990), Martius et al. (1993), and Sanderson (1996) calculated carbon dioxide and methane fluxes based on global distribution of termite biomass, and concluded that termites contribute about 2% of the total global flux of carbon dioxide (3,500 tg per year) and 4–5% of the global flux of methane (≤20 tg per year). However, emissions of carbon dioxide by termites are 25–50% of the annual emissions from fossil fuel combustion (Khalil et al. 1990). Contributions to atmospheric composition by this ancient insect group may have been more substantial prior to anthropogenic production of carbon dioxide, methane, and other greenhouse gases.

7.4 Effects on Regulating Services

Regulating services control rates of supporting or other services, thereby stabilizing the supply of ecosystem services. Clearly, disruption of regulating services threatens sustainability of necessary ecosystem services. The network of species interactions within ecosystems has considerable capacity to regulate processes and services. Accumulating evidence indicates that insects may serve as important regulators of primary production (Mattson and Addy 1975) and decomposition, which support other ecosystem services. Therefore, the role of insects in regulating services warrants special consideration.

7.4.1 Primary Production

In addition to affecting primary production as a supporting service (see earlier), insects regulate primary production in the same way that predation regulates prey abundance. Accumulating data suggest that insect–plant interactions represent a complex feedback loop that maintains relatively stable populations of plants and insects during normal periods. Despite our biased perspective of insect–plant interactions, insects may be as necessary to maintenance of healthy, stable populations of plants as are wolves to the maintenance of healthy stable populations of elk and moose (Peterson 1999; Wilmers et al. 2006).

During favorable environmental conditions, predation, plant chemical defenses, and optimal spacing among plants of a given species limit population growth by insect herbivores. The resulting balance between plant growth and insect abundance tends to stabilize food webs based on plant or insect resources (Schowalter 2011). In turn, low-to-moderate levels of herbivory often stimulate primary production by removing excess plant material that can interfere with continued growth (Knapp and Seastedt 1986; Gutschick 1999), resulting in compensatory growth by plants (McNaughton 1979; Williamson et al. 1989; Trlica and Rittenhouse 1993; Trumble et al. 1993; Belovsky and Slade 2000; Feeley and Terborgh 2005; Dungan et al. 2007) and relatively stable rates of primary production.

This balance is disrupted by environmental changes that stress plants or lead to increased production or density of particular plant species (White 1969, 1976, 1984; Mattson and Haack 1987; Schowalter and Turchin 1993; Schowalter et al. 1999; Van Bael et al. 2004; Jactel and Brockerhoff 2007). Under these conditions, reduced plant defense and/or easier discovery of nearby host plants by searching insects leads to increased insect population growth (Schowalter 2011). Elevated levels of herbivory decrease host density and primary production, reducing water and nutrient demands and alleviating host stress (Webb 1978; Kolb et al. 1999; Ford and Vose 2007; Cairns et al. 2008) and favoring plant species more tolerant of the altered environmental conditions (Ritchie et al. 1998; Belovsky and Slade 2000). These effects, combined with the tendency of outbreaks to reduce the abundance of the most stressed plant species, should stabilize primary production at intermediate levels that may be most sustainable under prevailing environmental conditions (Schowalter 2011).

Tests of this hypothesis are difficult and depend on one's perspective. For example, the Douglas-fir tussock moth outbreak depicted in Figure 5.11 caused devastating losses in timber supply and increased the risk of wildfire in the short term. However, over longer time periods, reduced fir density returned forest structure to historic conditions that prevailed prior to fire suppression in Sierran forests and that are the recommended management goal for these forests (North et al. 2007). From this perspective, the outbreak improved forest stability and sustainability of ecosystem services from this forest. To the extent that outbreaks function to regulate primary production, suppression may be counterproductive to sustainability of at least some ecosystem services, including timber production.

7.4.2 Biological Control

Control of prey populations has been recognized as a key regulatory function of predation and parasitism (see Chapters 4 and 5). Native predators and parasites provide top-down control of prey populations but cannot always prevent population growth by prey populations released from bottom-up regulation as a result of abundant or stressed hosts. Nevertheless, the importance of predation and parasitism is the basis for introduction of predators or parasites from an invasive species' country of origin as a means of biological control. Insects are both agents and targets of biological control. Although the importance of herbivorous insects in regulating host populations and preventing uncontrolled plant growth generally is not appreciated, plant growth often suffers decline in the absence of herbivores (Knapp and Seastedt 1986). An enormous diversity of predaceous insects and other arthropods, fish, amphibians, reptiles, birds, and mammals, as well as parasitic viruses, bacteria, fungi, and nematodes contribute to the regulation of insect populations in natural ecosystems (Figure 4.12; see also Schowalter 2011).

7.4.3 Disease Cycles

As described earlier for insect-plant interactions, pathogen-host interactions reflect a feedback loop that maintains relatively stable populations of both during normal periods. Host immune systems and distances among hosts limit vector and pathogen abundances. To a large extent, transmission of human diseases is a function of human, as well as vector, density. Our most serious diseases are "crowd diseases" that become epidemic in dense urban populations. In other words, disease transmission functions in a density-dependent manner and operates to reduce dense populations that may have exceeded their carrying capacity, leading to large numbers of susceptible individuals. Again, despite our biased perspective, our interactions with pathogens reflect feedback mechanisms that would normally maintain healthy stable populations of both.

In some cases, interactions among various ecosystem components that affect disease transmission can be quite complex. For example, Jones et al. (1998) found that the abundance of tick vectors of the Lyme disease bacterium, *Borrelia burgdorferi*, in the northeastern United States is related to the abundance of small mammal reservoirs, which reflects acorn production that, in turn, is affected by the extent of defoliation by the gypsy moth, *Lymantria dispar*.

As described earlier for insect-plant interactions, normal balances among pathogens, vectors, and hosts can be disrupted by environmental changes. Stapp et al. (2004) found that local extinction of black-tailed prairie dog, *Cynomys ludovicianus*, colonies in western North America was significantly greater during El Niño years, due to flea-transmitted plague, *Yersinia pestis*, which spreads more rapidly during warmer, wetter conditions (Parmenter et al. 1999). Similarly, Zhou et al. (2002) reported that extremely high populations of sand flies, *Lutzomyia verrucarum*, were associated with El Niño conditions in Peru, resulting in near doubling of human cases of bartonellosis, an emerging, vector-borne, highly fatal infectious disease in the region.

Anthropogenic changes also affect disease transmission. Póvoa et al. (2003) suggested that the reappearance of *Anopheles darlingi* and malaria, caused by several species of *Plasmodium* bacteria, in Belém, Brazil in 1992, after its presumed elimination in 1968, resulted from human encroachment into deforested areas that had become more favorable mosquito habitat. Vittor et al. (2006) explored this hypothesis in northeastern Peru, where malaria also had dropped dramatically during the 1960s as a result of eradication efforts and remained below two cases per 1,000 population until the 1990s. Construction of the Iquitos–Nauta road into the region during the 1980s and 1990s initiated deforestation and allowed rapid settlement and small-scale subsistence agriculture. A sudden increase in the incidence of malaria was observed during the 1990s, reaching more than 120,000 cases (340 per 1,000 population) in 1997. During 2000, Vittor et al. (2006) selected replicate sites along the Iquitos–Nauta road to represent high, medium, or low percentages of deforestation (based on satellite imagery) and human population density (within a 500-m radius around the sample site). Rates of mosquito landing on research personnel were measured at each site between 1800 and 2400 hours (period of peak mosquito activity) and compared among land use and demography treatments. Because mosquito reproduction occurred primarily in ponds and fish farms associated with cleared or naturally open areas, and adult mosquitoes did not fly far from breeding sites, biting rates reflected local populations of mosquitoes. Sites with less than 20% forest and more than 30% grass/crop cover had a 278-fold higher biting rate than did sites with more than 70% forest and less than 10% grass/crop cover. Based on mean percentages of infective *A. darlingi* in the Amazon region (0.5–2.1% infective mosquitoes), Vittor et al. (2006) calculated 38 infective bites per year per km^2 in areas with more than 35% grass/crop cover and 8 to 11 infective bites per year per km^2 in areas with 2–35% grass/crop cover, compared with 0.1 infective bites per year per km^2 in areas with less than 2% grass/crop cover, presenting serious challenges for disease control as deforestation progresses and human population increases. Human contact with novel zoonotic diseases is likely to increase as intrusion into previously unpopulated areas increases (Smith et al. 2007).

7.4.4 Nutrient Cycling

Crossley and Howden (1961) were the first to demonstrate that insect herbivores accelerate nutrient fluxes via consumption of foliage. Subsequent research has demonstrated that insect herbivores affect biogeochemical cycling through changes in vegetation structure and composition and altered rate, seasonal pattern, and quality of throughfall (net precipitation reaching the ground) and litterfall (see earlier).

Outbreaks affect ecosystem sequestration of carbon and nutrients. For example, widespread pine mortality during outbreaks of mountain pine beetle reduced carbon uptake and increased carbon emission from decaying trees (Kurz et al. 2008). A similar outcome was found as a result of defoliation by the gypsy moth (Clark et al. 2010). In both studies, the change in net carbon flux converted the forest from a carbon sink to a carbon source. However, forests recovering from mortality caused by mountain pine beetle can remain growing season carbon sinks as a result of increased photosynthesis by surviving trees and understory vegetation whereas nearby harvested stands may remain carbon sources ten years after harvest (Brown et al. 2010). Brown et al. (2010) recommended deferral of salvage harvest of outbreak sites with substantial surviving trees and understory vegetation, to prevent such sites from being converted from carbon sinks to sources for extended periods. Ritchie et al. (1998) reported that insect herbivory reduced the abundance of nitrogen-rich plant species, leading to replacement by plant species with lower nitrogen concentrations in an oak savanna, potentially reducing competition for limited nitrogen. Invasive insect species may have long-term effects on carbon sequestration through the alteration of species composition, primary productivity, and nutrient fluxes (Peltzer et al. 2010).

Outbreaks affect uptake and use of water and nutrients by vegetation, affecting water quality and yield for downstream uses. Removal of foliage or other plant tissues reduces rates of precipitation interception and evapo-transpiration (Foley et al. 2003). As a result, more precipitation reaches the ground, temporarily increasing soil water content. Leaching of excess water exports nutrients from the system (Swank et al. 1981; Eshleman et al. 1998; Lovett et al. 2002). However, reduced albedo resulting from canopy opening increases evaporation from the soil and reduces cloud formation and local precipitation (Foley et al. 2003; Juang et al. 2007; Janssen et al. 2008).

Outbreaks increase the flux of nutrients from vegetation to soil in several ways. Herbivory increases nutrient flux in throughfall, precipitation enriched with nutrients leached from damaged foliage (Seastedt et al. 1983; Stachurski and Zimka 1984; Schowalter et al. 1991; Reynolds et al. 2003). However, these nutrients may not contribute immediately to primary production. In ecosystems with high annual precipitation, herbivore-induced nutrient fluxes may be masked by greater inputs to soil via precipitation (Schowalter et al. 1991; Fonte and Schowalter 2005), and in ecosystems with high background levels of nitrogen, herbivore-induced nitrogen flux may be immobilized quickly by soil microorganisms (Lovett and Ruesink 1995; Stadler and Müller 1996; Stadler et al. 1998, 2001; Treseder 2008).

Herbivory increases the amount and alters seasonal pattern and form of nutrients in litterfall. In the absence of herbivory, litterfall is highly seasonal (i.e., concentrated at the onset of cold or dry conditions) and has low nutrient concentrations, especially of nitrogen or other nutrients that are re-absorbed from senescing foliage (Marschner 1995; Gutschick 1999; Fonte and Schowalter 2004). Herbivory increases the amount and nutrient content of litterfall during the growing season (as fragmented plant material, insect tissues, and feces), but the nutritional quality of litter for detritivores and decomposers is affected by herbivore-induced defenses that may retard decomposition.

Insect tissues and feces have particularly high concentrations of nutrients, especially nitrogen, compared with plant material, and increase the rate of nutrient flux to soil (Frost and Hunter 2004; Schowalter et al. 2011). Hollinger (1986) reported that an outbreak of the California oak moth, *Phryganidia californica*, increased fluxes of nitrogen and phosphorus from trees to litter by twofold, and feces and insect remains accounted for 60–70% of the total fluxes. Deposition of folivore feces can explain 62% of the variation in soil nitrate availability (Reynolds et al. 2003). Christenson et al. (2002) and Frost and Hunter (2007) demonstrated, using ^{15}N, that early season herbivore feces were rapidly decomposed, whereas leaf litter nitrogen remained in litter, and some feces nitrogen was incorporated into foliage and, subsequently, into late-season defoliators during the same growing season. To the extent that nutrients, especially nitrogen, often are immobilized in plant tissues and limited in availability for plant use (Gutschick 1999), such herbivore-induced turnover may be an important mechanism for maintaining nutrient supply for new plant tissues.

Outbreaks also affect litter decomposition and mineralization rates through alteration of the soil/litter environment. Decomposition is strongly affected by litter moisture (Meentemeyer 1978; Whitford et al. 1981), a factor affected by canopy opening (Foley et al. 2003; Classen et al. 2005). Experimental addition of herbivore feces or throughfall increased abundances of Collembola and fungal and bacterial feeding nematodes (Reynolds et al. 2003). Although long-term canopy opening may result in evaporation of soil moisture (Foley et al. 2003), outbreak-induced canopy opening and litter deposition increase soil moisture and decomposition (Classen et al. 2005; see section 7.4.5).

7.4.5 Climate and Disturbance

Vegetation has considerable capacity to modify climate and mitigate disturbances, depending on vegetation height and density. Several studies have demonstrated the importance of vegetation to shading and protecting the soil surface, abating wind speed, and controlling water fluxes (Foley et al. 2003; Classen et al. 2005). Vegetation cover reduces albedo and diurnal soil surface temperatures (Foley et al. 2003). Evapotranspiration contributes to canopy cooling and to convection-generated condensation above the canopy, thereby increasing local precipitation (Meher-Homji 1991; Foley et al. 2003; Juang et al. 2007; Janssen et al. 2008). Vegetation removal results in evaporation of soil moisture and loss of control of soil temperature. Exposed soil surfaces can reach midday temperatures lethal to most organisms (Seastedt and Crossley 1981). Unimpeded wind and precipitation erode and degrade soils.

Although effects of outbreaks on climate have not been studied directly, reduced vegetation cover over large areas during outbreaks could have similar effects on regional climate. However, unlike most anthropogenic vegetation removal, defoliation or tree mortality due to insect outbreaks retains some shade and adds water-retaining litter to the soil surface. Schowalter et al. (1991) reported that 20% defoliation of experimental Douglas-fir saplings doubled the amount of water and litterfall at the soil surface, compared with undefoliated saplings. Classen et al. (2005) reported that canopy opening by manipulated abundances of scale insects, *Matsucoccus acalyptus*, increased soil temperature and moisture by 26% and 35%, respectively, similar to global change scenarios and sufficient to alter ecosystem processes (Foley et al. 2003).

Outbreaks affect the probability or severity of future disturbances, especially fire or storms. Increased fuel accumulation generally has been considered to increase the likelihood and severity of fire (McCullough et al. 1998), but this is not necessarily the case. Bebi et al. (2003) concluded that spruce, *Picea engelmannii*, mortality to spruce beetle, *Dendroctonus rufipennis*, did not increase the occurrence of subsequent fires. The probability of fire resulting from outbreaks depends on the amount and decomposition rate of increased litter. Grasshopper outbreaks that reduce grass biomass should reduce the severity of subsequent grassland fire (Knapp and Seastedt 1986). Outbreaks that increase only fine litter material (e.g., foliage fragments) may increase the probability and spread of low intensity fire, whereas outbreaks that cause tree mortality (and increase abundance of ladder fuels) are more likely to increase the risk of catastrophic fire (Jenkins et al. 2008). Insect outbreaks that open the canopy increase penetration of high wind speeds and the probability of tree fall but also reduce wind resistance of defoliated trees. Pruning at least 80% of the canopy can reduce wind stress significantly (Moore and Maguire 2005). Wind-related tree mortality following spruce budworm defoliation in eastern Canada was related to outbreak severity (Taylor and MacLean 2009). Tree mortality during storms peaked 11 to 15 years after outbreak, due to greater exposure of surviving trees to wind. Obviously, insect-induced disturbances can interfere with the supply of provisioning services and likely reduce cultural values, but they also may prevent undesirable changes in ecosystem structure or composition (e.g., succession from grassland to forest or from pine forest to fir forest in the absence of fire).

7.5 Summary

Insects affect ecosystem services in a variety of positive and negative ways. Pollinators are generally considered to benefit crop, and other plant, production, but herbivores, detritivores, and disease vectors often conflict with human interests by destroying crops, undermining wooden structures, and spreading human diseases. However, many of these insects also provide food, medicinal and industrial products, and cultural icons and can provide critical supporting and regulating services.

Most insect species have multiple, often opposite, effects. For example, honey bees and other bees and wasps are important plant pollinators, necessary for the production of fruit and vegetable crops, but they also pose substantial health hazards and diminish recreational values of ecosystems.

The destructive potential of herbivorous insects is well known, but without insects to reduce plant growth in excess of carrying capacity, continued primary production would be threatened. Overproduction by plants can depress individual growth and lead to widespread competitive stress or fuel fire that would severely disturb the plant community, requiring longer recovery time than when insects provide regulation of plant growth. Plant populations in the absence of insects could fluctuate dramatically between high densities exceeding carrying capacity and low densities incapable of providing the resources on which we depend. Herbivores and predators provide an essential service for host or prey populations, reducing abundance and preferentially removing weak and diseased individuals, thereby maintaining relatively more constant, sustainable populations of healthier individuals. If we were successful in eliminating the insects that annoy us, insectivorous birds (including most songbirds) and freshwater game fish would become rare or absent. Humans would need to find other food resources.

Similarly, termite destruction of wood in homes is undesirable, but the loss of termites or other detritivores from natural ecosystems would leave vital nutrients locked in undecomposed detritus. Livestock dung would accumulate, foul pastures, and discourage grazing without termites and dung beetles to remove it.

In short, insects have complex roles in ecosystems that include a variety of positive and negative effects on ecosystem services. These roles can interfere with human interests, but effective solutions require that management decisions weigh the complementary values of insects in order to protect human interests as well as the sustainability of ecosystem services.

References

Ackerman, I. L., W. G. Teixeira, S. J. Riha, J. Lehmann, and E. C. M. Fernandes. 2007. The impact of mound-building termites on surface soil properties in a secondary forest of central Amazonia. *Applied Soil Ecology* 37: 267–276.

Aizen, M. A. and P. Feinsinger. 1994. Habitat fragmentation, native insect pollinators, and feral honey bees in Argentine "Chaco Serano." *Ecological Applications* 4: 378–392.

Alfaro, R. I. and R. F. Shepherd. 1991. Tree-ring growth of interior Douglas-fir after one year's defoliation by Douglas-fir tussock moth. *Forest Science* 37: 959–964.

Allan, J. D., M. S. Wipfli, J. P. Caouette, A. Prussian, and J. Rodgers. 2003. Influence of streamside vegetation on inputs of terrestrial invertebrates to salmonid food webs. *Canadian Journal of Fisheries and Aquatic Sciences* 60: 309–320.

Anderson, J. M. 1988. Invertebrate-mediated transport processes in soils. *Agriculture, Ecosystems and Environment* 24: 5–19.

Anderson, N. H., R. J. Steedman, and T. Dudley. 1984. Patterns of exploitation by stream invertebrates of wood debris (xylophagy). *Verhandlungen der Internationalen Vereinigung für Theoretische und Angewandte Limnologie* 22: 1847–1852.

Anelli, C. M. and D. A. Prischmann-Voldseth. 2009. Silk batik using beeswax and cochineal dye: An interdisciplinary approach to teaching entomology. *American Entomologist* 55: 95–105.

Appanah, S. 1990. Plant-pollinator interactions in Malaysian rain forests. In *Reproductive Ecology of Tropical Forest Plants*, K. Bawa and M. Hadley, eds. 85–100. Paris: UNESCO/Parthenon.

Balvanera, P., C. Kremen, and M. Martínez-Ramos. 2005. Applying community structure analysis to ecosystem function: Examples from pollination and carbon storage.*Ecological Applications* 15: 360–375.

Baxter C. V., K. D. Fausch, and W. C. Saunders. 2005. Tangled webs: Reciprocal flows of invertebrate prey link streams and riparian zones. *Freshwater Biology* 50: 201–220.

Bebi, P., D. Kilakowski, and T. T. Veblen. 2003. Interactions between fire and spruce beetles in a subalpine Rocky Mountain forest landscape. *Ecology* 84: 362–371.

Belle-Isle, J. and D. Kneeshaw. 2007. A stand and landscape comparison of the effects of a spruce budworm (*Choristoneura fumiferana* [Clem.]) outbreak to the combined effects of harvesting and thinning on forest structure. *Forest Ecology and Management* 246: 163–174.

Belovsky, G. E. and J. B. Slade. 2000. Insect herbivory accelerates nutrient cycling and increases plant production. *Proceedings of the National Academy of Sciences USA* 97: 14412–14417.

Bewley, D., Y. Alila, and A. Varhola. 2010. Variability of snow water equivalent and snow energetics across a large catchment subject to Mountain Pine Beetle infestation and rapid salvage logging. *Journal of Hydrology* 388: 464–479.

Biesmeijer, J.C., S. P. M. Roberts, M. Reemer, R. Ohlemüller, M. Edwards, T. Peeters, A. P. Schaffers, S. G. Potts, R. Kleukers, C. D. Thomas, et al. 2006. Parallel declines in pollinators and insect-pollinated plants in Britain and The Netherlands. *Science* 313: 351–354.

Brauman, A., M. D. Kane, M. Labat, and J. A. Breznak. 1992. Genesis of acetate and methane by gut bacteria of nutritionally diverse termites. *Science* 257: 1384–1387.

Breznak, J. A. and A. Brune. 1994. Role of microorganisms in the digestion of lignocellulose by termites. *Annual Review of Entomology* 39: 453–487.

Brody, A. K., T. M. Palmer, K. Fox-Dobbs, and D. F. Doak. 2010. Termites, vertebrate herbivores, and the fruiting success of *Acacia drepanolobium*. *Ecology* 91: 399–407.

Brown, J. H., O. J. Reichman, and D. W. Davidson. 1979. Granivory in desert ecosystems. *Annual Review of Ecology and Systematics* 10: 201–227.

Brown, M., T. A. Black, Z. Nesic, V. N. Foord, D. L. Spittlehouse, A. L. Fredeen, N. J. Grant, P. J. Burton, and J. A. Trofymow. 2010. Impact of mountain pine beetle on the net ecosystem production of lodgepole pine stands in British Columbia. *Agricultural and Forest Meteorology* 150: 254–264.

Cairns, D. M., C. L. Lafon, J. D. Waldron, M. Tchakerian, R. N. Coulson, K. D. Klepzig, A. G. Birt, and W. Xi. 2008. Simulating the reciprocal interaction of forest landscape structure and southern pine beetle herbivory using LANDIS. *Landscape Ecology* 23: 403–415.

Cambefort, Y. 1991. From saprophagy to coprophagy. In *Dung Beetle Ecology*, I. Hanski and Y. Cambefort, eds. 22–35. Princeton, NJ: Princeton University Press.

Carlton R. G. and C. R. Goldman. 1984. Effects of a massive swarm of ants on ammonium concentrations in a subalpine lake. *Hydrobiologia* 111: 113–117.

Cerritos, R. and Z. Cano-Santana. 2008. Harvesting grasshoppers *Sphenarium purpurascens* in Mexico for human consumption: A comparison with insecticidal control for managing pest outbreaks. *Crop Protection* 27: 473–480.

Chapman, S. K., S. C. Hart, N. S. Cobb, T. G. Whitham, and G. W. Koch. 2003. Insect herbivory increases litter quality and decomposition: An extension of the acceleration hypothesis. *Ecology* 84: 2867–2876.

Chittka, L. and N. E. Raine. 2006. Recognition of flowers by pollinators. *Current Opinion in Plant Biology* 9: 428–435.

Christenson, L. M., G. M. Lovett, M. J. Mitchell, and P. M. Groffman. 2002. The fate of nitrogen in gypsy moth frass deposited to an oak forest floor. *Oecologia* 131: 444–452.

Clark, D. B. and D. A. Clark. 1985. Seedling dynamics of a tropical tree: Impacts of herbivory and meristem damage. *Ecology* 66: 1884–1892.

Clark, K. L., N. Skowronski, and J. Hom. 2010. Invasive insects impact forest carbon dynamics. *Global Change Biology* 16: 88–101.

Classen, A. T., S. C. Hart, T. G. Whitham, N. S. Cobb, and G. W. Koch. 2005. Insect infestations linked to changes in microclimate: Important climate change implications. *Soil Science Society of America Journal* 69: 2049–2057.

Claudianos, C., H. Ranson, R. M. Johnson, S. Biswas, M. A. Schuler, M. R. Berenbaum, R. Feyereisen, and J. G. Oakeshott. 2006. A deficit of detoxification enzymes: Pesticide sensitivity and environmental response in the honey bee. *Insect Molecular Biology* 15: 615–636.

Clausen, L. W. 1954. *Insect Fact and Folklore*. New York: MacMillan.

Coe, M. 1977. The role of termites in the removal of elephant dung in the Tsavo (East) National Park Kenya. *East African Wildlife Journal* 15: 49–55.

Coleman, D. C., D. A. Crossley, Jr., and P. F. Hendrix. 2004. *Fundamentals of Soil Ecology*, 2nd ed. Amsterdam: Elsevier.

Collins, N. M. 1981. The role of termites in the decomposition of wood and leaf litter in the southern Guinea savanna of Nigeria. *Oecologia* 51: 389–399.

Collins, N. M. 1983. Termite populations and their role in litter removal in Malaysian rain forests. In *Tropical Rain Forest: Ecology and Management*, S. L. Sutton, T. C. Whitmore, and A. C. Chadwick, eds. 311–325. London: Blackwell.

Cook, D. M. and J. Doran-Peterson. 2010. Mining diversity of the natural biorefinery housed within *Tipula abdominalis* larvae for use in an industrial biorefinery for production of lignocellulosic ethanol. *Insect Science* 13: 303–312.

Corner, E. J. H. 1964. *The Life of Plants*. Cleveland, OH: World Publishing.

Cox-Foster, D. L., S. Conlan, E. C. Holmes, G. Palacios, J. D. Evans, N. A. Moran, P.-L. Quan, T. Briese, M. Hornig, D. M. Geiser, et al. 2007. A metagenic survey of microbes in honey bee colony collapse disorder. *Science* 318: 283–287.

Crane, E. 1999. *The World History of Beekeeping and Honey Hunting*. New York: Routledge.

Crossley, D. A., Jr., and H. F. Howden. 1961. Insect-vegetation relationships in an area contaminated by radioactive wastes. *Ecology* 42: 302–317.

Cuffney, T. F., J. B. Wallace, and G. J. Lugthart. 1990. Experimental evidence quantifying the role of benthic invertebrates in organic matter dynamics in headwater streams. *Freshwater Biology* 23: 281–299.

Culver, D. C. and A. J. Beattie. 1983. Effects of ant mounds on soil chemistry and vegetation patterns in a Colorado montane meadow. *Ecology* 64: 485–492.

Cummins, K. W. 1973. Trophic relations of aquatic insects. *Annual Review of Entomology* 18: 183–206.

Dangerfield, J. M. 1994. Ingestion of leaf litter by millipedes: The accuracy of laboratory estimates for predicting litter turnover in the field. *Pedobiologia* 38: 262–265.

Dangerfield. J. M., T. S. McCarthy, and W. N. Ellery. 1998. The mound-building termite *Macrotermes michaelseni* as an ecosystem engineer. *Journal of Tropical Ecology* 14: 507–520.

Dauber, J. and V. Wolters. 2000. Microbial activity and functional diversity in the mounds of three different ant species. *Soil Biology and Biochemistry* 32: 93–99.

Davidson, D. W. 1993. The effects of herbivory and granivory on terrestrial plant succession. *Oikos* 68: 23–35.

Davis, A. L. V. 1996. Seasonal dung beetle activity and dung dispersal in selected South African habitats: Implications for pasture improvement in Australia. *Agriculture, Ecosystems and Environment* 58: 157–169.

Dedej, S. and K. S. Delaplane. 2004. Nectar-robbing carpenter bees reduce seed-setting capability of honey bees (Hymenoptera: Apidae) in rabbiteye blueberry, *Vaccinium ashei*, "Climax." *Environmental Entomology* 33: 100–106.

Diamond, J. 1999. *Guns, Germs, and Steel: The Fates of Human Societies*. New York: W.W. Norton.

Domisch, T., L. Finér, M. Ohashi, A. C. Risch, L. Sundström, P. Niemelä, and M. F. Jurgensen. 2006. Contribution of red wood ant mounds to forest floor CO_2 efflux in boreal coniferous forests. *Soil Biology and Biochemistry* 38: 2425–2433.

Downing, K. B. and W. R. Williams. 1978. Douglas-fir tussock moth: Did it affect private recreational businesses in northeastern Oregon? *Journal of Forestry* 76: 29–30.

Dungan, R. J., M. H. Turnbull, and D. Kelly. 2007. The carbon costs for host trees of a phloem-feeding herbivore. *Journal of Ecology* 95: 603–613.

Eldridge, D. J. 1993. Effect of ants on sandy soils in semi-arid eastern Australia: Local distribution of nest entrances and their effect on infiltration of water. *Australian Journal of Soil Research* 31: 509–518.

Eldridge, D. J. 1994. Nests of ants and termites influence infiltration in a semi-arid woodland. *Pedobiologia* 38: 481–492.

Elkins, N.Z., G. V. Sabol, T. J. Ward, and W. G. Whitford. 1986. The influence of subterranean termites on the hydrological characteristics of a Chihuahuan Desert ecosystem. *Oecologia* 68: 521–528.

Epstein W. L. and A. M. Kligman. 1958. Treatment of warts with cantharidin. *American Medical Association Archives of Dermatology* 77: 508–511.

Eshleman, K. N., R. P. Morgan II, J. R. Webb, F. A. Deviney, and J. N. Galloway. 1998. Temporal patterns of nitrogen leakage from mid-Appalachian forested watersheds: Role of insect defoliation. *Water Resources Research* 34: 2005–2116.

Feeley, K. J. and J. W. Terborgh. 2005. The effects of herbivore density on soil nutrients and tree growth in tropical forest fragments. *Ecology* 86: 116–124.

Ferrar, P. 1975. Disintegration of dung pads in north Queensland before the introduction of exotic dung beetles. *Australian Journal of Experimental Agriculture and Animal Husbandry* 15: 325–329.

Foley, J. A., M. H. Costa, C. Delire, N. Ramankutty, and P. Snyder. 2003. Green surprise? How terrestrial ecosystems could affect earth's climate. *Frontiers in Ecology and the Environment* 1: 38–44.

Fonte, S. J. and T. D. Schowalter. 2004. Decomposition of greenfall vs. senescent foliage in a tropical forest ecosystem in Puerto Rico. *Biotropica* 36: 474–482.

Fonte, S. J. and T. D. Schowalter 2005. The influence of a neotropical herbivore (*Lamponius portoricensis*) on nutrient cycling and soil processes. *Oecologia* 146: 423–431.

Ford, C. R. and J. M. Vose. 2007. *Tsuga canadensis* (L.) Carr. mortality will impact hydrologic processes in southern Appalachian forest ecosystems. *Ecological Applications* 17: 1156–1167.

Fox-Dobbs, K., D. F. Doak, A. K. Brody, and T. M. Palmer. 2010. Termites create spatial structure and govern ecosystem function by affecting N_2 fixation in an east African savanna. *Ecology* 91: 1296–1307.

Freymann, B. P., R. Buitenwerf, O. Desouza, and H. Olff. 2008. The importance of termites (Isoptera) for the recycling of herbivore dung in tropical ecosystems: A review. *European Journal of Entomology* 105: 165–173.

Frost, C. J. and M. D. Hunter. 2004. Insect canopy herbivory and frass deposition affect soil nutrient dynamics and export in oak mesocosms. *Ecology* 85: 3335–3347.

Frost, C. J. and M. D. Hunter. 2007. Recycling of nitrogen in herbivore feces: Plant recovery, herbivore assimilation, soil retention, and leaching losses. *Oecologia* 151: 42–53.

Frost, C. J. and M. D. Hunter. 2008. Insect herbivores and their frass affect *Quercus rubra* leaf quality and initial stages of subsequent decomposition. *Oikos* 117: 13–22.

Gehring, C. A. and T. G. Whitham. 1995. Duration of herbivore removal and environmental stress affect the ectomycorrhizae of pinyon pine. *Ecology* 76: 2118–2123.

Genersch, E. 2010. Honey bee pathology: Current threats to honey bees and beekeeping. *Applied Microbiology and Biotechnology* 87: 87–97.

González, G. and T. R. Seastedt. 2001. Soil fauna and plant litter decomposition in tropical and subalpine forests. *Ecology* 82: 955–964.

Greenfield, A. B. 2005. *A Perfect Red: Empire, Espionage, and the Quest for the Color of Desire*. New York: Harper Collins.

Gudger, E. W. 1925. Stitching wounds with the mandibles of ants and beetles. *Journal of the American Medical Association* 84: 1861–1864.

Gutschick, V. P. 1999. Biotic and abiotic consequences of differences in leaf structure. *New Phytologist* 143: 4–18.

Gutschick, V. P. and F. W. Wiegel. 1988. Optimizing the canopy photosynthetic rate by patterns of investment in specific leaf mass. *American Naturalist* 132: 67–86.

Hackstein, J. H. P. and C. K. Stumm. 1994. Methane production in terrestrial arthropods. *Proceedings of the National Academy of Sciences USA* 91: 5441–5445.

Haines, B. L. 1978. Element and energy flows through colonies of the leaf-cutting ant, *Atta columbica*, in Panama. *Biotropica* 10: 270–277.

Harmon, M. E., J. F. Franklin, F. J. Swanson, P. Sollins, S. V. Gregory, J. D. Lattin, N. H. Anderson, S. P. Cline, N. G. Aumen, J. R. Sedell, et al. 1986. Ecology of coarse woody debris in temperate ecosystems. *Advances in Ecological Research* 15: 133–302.

Hättenschwiler, S. and P. Gasser. 2005. Soil animals alter plant litter diversity effects on decomposition. *Proceedings of the National Academy of Sciences USA* 102: 1519–1524.

Heneghan, L., D. C. Coleman, X. Zou, D. A. Crossley, Jr., and B. L. Haines. 1999. Soil microarthropod contributions to decomposition dynamics: Tropical-temperate comparisons of a single substrate. *Ecology* 80: 1873–1882.

Herrick, J. E. and R. Lal. 1996. Dung decomposition and pedoturbation in a seasonally dry tropical pasture. *Biology and Fertility of Soils* 23: 177–181.

Herzog, D. C., T. E. Reagan, D. C. Sheppard, K. M. Hyde, S. S. Nilakhe, M. Y. B. Hussein, M. L. McMahan, R. C. Thomas, and L. D. Newsom. 1976. *Solenopsis invicta* Buren: Influence on Louisiana pasture soil chemistry. *Environmental Entomology* 5: 160–162.

Hochberg, M. E. 1989. The potential role of pathogens in biological control. *Nature* 337: 262–265.

Hoehn, P., T. Tscharntke, J. M. Tylianakis, and I. Steffan-Dewenter. 2008. Functional group diversity of bee pollinators increases crop yield. *Proceedings of the Royal Society B* 275: 2283–2291.

Holdo, R. M. and L. R. McDowell. 2004. Termite mounds as nutrient-rich food patches for elephants. *Biotropica* 36: 231–239.

Hollinger, D. Y. 1986. Herbivory and the cycling of nitrogen and phosphorus in isolated California oak trees. *Oecologia* 70: 291–297.

Holter, P. 1979. Effect of dung-beetles (*Aphodius* spp.) and earthworms on the disappearance of cattle dung. *Oikos* 32: 393–402.

Holzschuh, A., I. Steffan-Dewenter, D. Kleijn, and T. Tscharntke. 2007. Diversity of flower-visiting bees in cereal fields: Effects of farming system, landscape composition, and regional context. *Journal of Applied Ecology* 44: 41–49.

Hughes, L. 1990. The relocation of ant nest entrances: Potential consequences for ant-dispersed seeds. *Australian Journal of Ecology* 16: 207–214.

Hughes, R. D., M. Tyndale-Biscoe, and J. Walker. 1978. Effects of introduced dung beetles (Coleoptera: Scarabaeidae) on the breeding and abundance of the Australian bushfly, *Musca vetustissima* Walker (Diptera: Muscidae). *Bulletin of Entomological Research* 68: 361–372.

Jactel, H. and E. G. Brockerhoff. 2007. Tree diversity reduces herbivory by forest insects. *Ecology Letters* 10: 835–848.

Janssen, R. H. H., M. B. J. Meinders, E. H. van Nes, and M. Scheffer. 2008. Microscale vegetation-soil feedback boosts hysteresis in a regional vegetation-climate system. *Global Change Biology* 14: 1104–1112.

Jenkins, M. J., E. Herbertson, W. Page, and C. A. Jorgensen. 2008. Bark beetles, fuels, fires, and implications for forest management in the intermountain West. *Forest Ecology and Management* 254: 16–34.

Johnson, K. A. and W. G. Whitford. 1975. Foraging ecology and relative importance of subterranean termites in Chihuahuan Desert ecosystems. *Environmental Entomology* 4: 66–70.

Johnson, R. M., J. D. Evans, G. E. Robinson, and M. R. Berenbaum. 2009. Changes in transcript abundance relating to colony collapse disorder in honey bees (*Apis mellifera*). *Proceedings of the National Academy of Sciences USA* 106: 14790–14795.

Jones, C. G., R. S. Ostfeld, M. P. Richard, E. M. Schauber, and J. O. Wolff. 1998. Chain reactions linking acorns to gypsy moth outbreaks and Lyme disease risk. *Science* 279: 1023–1026.

Jones, J. A. 1989. Environmental influences on soil chemistry in central semiarid Tanzania. *Soil Science Society of America Journal* 53: 1748–1758.

Jones, J. A. 1990. Termites, soil fertility, and carbon cycling in dry tropical Africa: A hypothesis. *Journal of Tropical Ecology* 6: 291–305.

Jonkman, J. C. M. 1978. Nests of the leaf-cutting ant *Atta vollenweideri* as accelerators of succession in pastures. *Zeitschrift für angewandte Entomologie* 86: 25–34.

Jouquet, P., J. Dauber, J. Lagerlöf, P. Lavelle, and M. Lapage. 2006. Soil invertebrates as ecosystem engineers: Effects on soil and feedback loops. *Applied Soil Ecology* 32: 153–164.

Juang, J.-Y., G. G. Katul, A. Porporato, P. C. Stoy, M. S. Sequeira, M. Detto, H.-S. Kim, and R. Oren. 2007. Eco-hydrological controls on summertime convective rainfall triggers. *Global Change Biology* 13: 887–896.

Jurgensen, M. F., L. Finér, T. Domisch, J. Kilpeläinen, P. Punttila, M. Ohashi, P. Niemelä, L. Sundström, S. Neuvonen, and A. C. Risch. 2008. Organic mound-building ants: Their impact on soil properties in temperate and boreal forests. *Journal of Applied Entomology* 132: 266–275.

Kawaguchi Y. and S. Nakano. 2001. Contribution of terrestrial invertebrates to the annual resource budget for salmonids in forest and grassland reaches of a headwater stream. *Freshwater Biology* 46: 303–316.

Kerridge, A., H. Lappin-Scott, and J.R. Stevens. 2005. Antibacterial properties of larval secretions of the blowfly, *Lucilia sericata*. *Medical and Veterinary Entomology* 19: 333–337.

Khalil, M. A. K., R. A. Rasmussen, J. R. J. French, and J. A. Holt. 1990. The influence of termites on atmospheric trace gases: CH_4, CO_2, $CHCl_3$, N_2O, CO, H_2, and light hydrocarbons. *Journal of Geophysical Research* 95: 3619–3634.

Klein, A.-M, B. E. Vaissière, J. H. Cane, I. Steffan-Dewenter, S. A. Cunningham, C. Kremen, and T. Tscharntke. 2007. Importance of pollinators in changing landscapes for world crops. *Proceedings of the Royal Society B* 274: 303–313.

Knapp, A. K. and T. R. Seastedt. 1986. Detritus accumulation limits productivity of tallgrass prairie. *BioScience* 36: 662–668.

Knight, T. M., J. A. Steets, J. A. Vamosi, S. J. Mazer, M. Burd, D. R. Campbell, M. R. Dudash, M. O. Johnston, R. J. Mitchell, and T.-L. Ashman. 2005. Pollen limitation of plant reproduction: Pattern and process. *Annual Review of Ecology, Evolution and Systematics* 36: 467–497.

Kohlmann, B. 1991. Dung beetles in subtropical North America. In *Dung Beetle Ecology*, I. Hanski and Y. Cambefort, eds. 116–132. Princeton, NJ: Princeton University Press.

Kolb, T. E., K. A. Dodds, and K. M. Clancy. 1999. Effect of western spruce budworm defoliation on the physiology and growth of potted Douglas-fir seedlings. *Forest Science* 45: 280–291.

Kremen, C., N. M. Williams, M. A. Aizen, B. Gemmill-Herren, G. LeBuhn, R. Minckley, L. Packer, S. G. Potts, T. Roulston, I. Steffan-Dewenter, et al. 2007. Pollination and other ecosystem services produced by mobile organisms: A conceptual framework for the effects of land-use change. *Ecology Letters* 10: 299–314.

Kremen, C., N. M. Williams, R. L. Bugg, J. P. Fay, and R. W. Thorp. 2004. The area requirements of an ecosystem service: Crop pollination by native bee communities in California. *Ecology Letters* 7: 1109–1119.

Kremen, C., N. M. Williams, and R. W. Thorp. 2002. Crop pollination from native bees as risk from agricultural intensification. *Proceedings of the National Academy of Sciences USA* 99: 16812–16816.

Kurz, W. A., C. C. Dymond, G. Stinson, G. J. Rampley, E. T. Neilson, A. L. Carroll, T. Ebata, and L. Safranyik. 2008. Mountain pine beetle and forest carbon feedback to climate change. *Nature* 452: 987–990.

Laakso, J. and H. Setälä. 1999. Sensitivity of primary production to changes in the architecture of belowground food webs. *Oikos* 87: 57–64.

Lee, K. E. and J. H. A. Butler. 1977. Termites, soil organic matter decomposition, and nutrient cycling. *Ecological Bulletin (Stockholm)* 25: 544–548.

Lenoir, L., T. Persson, and J. Bengtsson. 2001. Wood ant nests as potential hot spots for carbon and nitrogen mineralization. *Biology and Fertility of Soils* 34: 235–240.

Lesica, P. and P. B. Kannowski. 1998. Ants create hummocks and alter structure and vegetation of a Montana fen. *American Midland Naturalist* 139: 58–68.

Losey, J. E. and M. Vaughn. 2006. The economic value of ecological services provided by insects. *BioScience* 56: 311–323.

Louda, S. M., K. H. Keeler, and R. D. Holt. 1990. Herbivore influences on plant performance and competitive interactions. In *Perspectives on Plant Competition*, J. B. Grace and D. Tilman, eds. 413–444. San Diego: Academic.

Lovett, G. and P. Tobiessen. 1993. Carbon and nitrogen assimilation in red oaks (*Quercus rubra* L.) subject to defoliation and nitrogen stress. *Tree Physiology* 12: 259–269.

Lovett, G. M., L. M. Christenson, P. M. Groffman, C. G. Jones, J. E. Hart, and M. J. Mitchell. 2002. Insect defoliation and nitrogen cycling in forests. *BioScience* 52: 335–341.

Lovett, G. M. and A. E. Ruesink. 1995. Carbon and nitrogen mineralization from decomposing gypsy moth frass. *Oecologia* 104: 133–138.

Lű, X.-T., D.-L. Kong, Q.-M. Pan, M. E. Simmons, and X.-G. Han. 2012. Nitrogen and water availability interact to affect leaf stoichiometry in a semi-arid grassland. *Oecologia* 168: 301–310.

MacMahon, J. A., J. F. Mull, and T. O. Crist. 2000. Harvester ants (*Pogonomyrmex* spp.): Their community and ecosystem influences. *Annual Review of Ecology and Systematics* 31: 265–291.

Mahaney, W. C., J. Zippin, M. W. Milner, K. Sanmugadas, R. G. V. Hancock, S. Aufreiter, S. Campbell, M. A. Huffman, M. Wink, D. Malloch, et al. 1999. Chemistry, mineralogy, and microbiology of termite mound soil eaten by the chimpanzees of the Mahale Mountains, western Tanzania. *Journal of Tropical Ecology* 15: 565–588.

Manley, G. V. 1971. A seed-cacheing carabid (Coleoptera). *Annals of the Entomological Society of America* 64: 1474–1475.

Marschner, H. 1995. *The Mineral Nutrition of Higher Plants*, 2nd ed. San Diego: Academic.

Martius, C., R. Wassmann, U. Thein, A. Bandeira, H. Rennenberg, W. Junk, and W. Seiler. 1993. Methane emission from wood-feeding termites in Amazonia. *Chemosphere* 26: 623–632.

Mattson, W. J. and N. D. Addy. 1975. Phytophagous insects as regulators of forest primary production. *Science* 190: 515–522.

Mattson, W. J. and R. A. Haack. 1987. The role of drought in outbreaks of plant-eating insects. *BioScience* 37: 110–118.

Mbata, K. J., E. N. Chidumayo, and C. M. Lwatula. 2002. Traditional regulation of edible caterpillar exploitation in the Kopa area of Mpika district in northern Zambia. *Journal of Insect Conservation* 6: 115–130.

McBrayer, J. F. 1975. Exploitation of deciduous leaf litter by *Apheloria montana* (Diplopoda: Eurydesmidae). *Pedobiologia* 13: 90–98.

McCullough, D. G., R. A. Werner, and D. Neumann. 1998. Fire and insects in northern and boreal forest ecosystems of North America. *Annual Review of Entomology* 43: 107–127.

McNaughton, S. J. 1979. Grazing as an optimization process: Grass-ungulate relationships in the Serengeti. *American Naturalist* 113: 691–703.

Meentemeyer, V. 1978. Macroclimate and lignin control of litter decomposition rates. *Ecology* 59: 465–472.

Meher-Homji, V. M. 1991. Probable impact of deforestation on hydrological processes. *Climate Change* 19: 163–173.

Mehner T., J. Ihlau, H. Dörner, M. Hupfer, and F. Hölker. 2005. Can feeding of fish on terrestrial insects subsidize the nutrient pool of lakes? *Limnology and Oceanography* 50: 2022–2031.

Menninger, H. L., M. A. Palmer, L. S. Craig, and D. C. Richardson. 2008. Periodical cicada detritus impacts stream ecosystem metabolism. *Ecosystems* 11: 1306–1317.

Michalson, E. L. 1975. Economic impact of mountain pine beetle on outdoor recreation. *Southern Journal of Agricultural Economics* 7(2): 43–50.

Michaud, J. P. and A. K. Grant. 2009. The nature of resistance to *Dectes texanus* (Col., Cerambycidae) in wild sunflower, *Helianthus annuus. Journal of Applied Entomology* 133: 518–523.

Moore, J. R. and D. A. Maguire. 2005. Natural sway frequencies and damping ratios of trees: Influence of crown structure. *Trees* 19: 363–373.

Moser, J. C. 1963. Contents and structure of *Atta texana* nest in summer. *Annals of the Entomological Society of America* 56: 286–291.

Moser, J. C. 2006. Complete excavation and mapping of a Texas leafcutting ant nest. *Annals of the Entomological Society of America* 99: 891–897.

Müller, M. and H. Job. 2009. Managing natural disturbance in protected areas: Tourists' attitude towards the bark beetle in a German national park. *Biological Conservation* 142: 375–383.

Namba, T., Y. H. Ma, and K. Inagaki. 1988. Insect-derived crude drugs in the Chinese Song Dynasty. *Journal of Ethnopharmacology* 24: 247–285.

Norman, E. M. and D. Clayton. 1986. Reproductive biology of two Florida pawpaws: *Asimina obovata* and *A. pygmaea* (Annonaceae). *Bulletin of the Torrey Botanical Club* 113: 16–22.

Norman, E. M., K. Rice, and S. Cochran. 1992. Reproductive biology of *Asimina parviflora* (Annonaceae). *Bulletin of the Torrey Botanical Club* 119: 1–5.

North, M., J. Innes, and H. Zald. 2007. Comparison of thinning and prescribed fire restoration treatments to Sierran mixed-conifer historic conditions. *Canadian Journal of Forest Research* 37: 331–342.

Nowlin, W. H., M. J. González, M. J. Vanni, M. H. H. Stevens, M. W. Fields, and J. J. Valenti. 2007. Allochthonous subsidy of periodical cicadas affects the dynamics and stability of pond communities. *Ecology* 88: 2174–2186.

Parker, L. W., H. G. Fowler, G. Ettershank, and W. G. Whitford. 1982. The effects of subterranean termite removal on desert soil nitrogen and ephemeral flora. *Journal of Arid Environments* 5: 53–59.

Parmenter, R. R., E. P. Yadav, C. A. Parmenter, P. Ettestad, and K. L. Gage. 1999. Incidence of plague associated with increased winter-spring precipitation in New Mexico. *American Journal of Tropical Medicine and Hygiene* 61: 814–821.

Pasquini, S. C. and L. S. Santiago. 2012. Nutrients limit photosynthesis in seedlings of a lowland tropical tree species. *Oecologia* 168: 311–319.

Peakall, R., A. J. Beattie, and S. H. James. 1987. Pseudocopulation of an orchid by male ants: A test of two hypotheses accounting for the rarity of ant pollination. *Oecologia* 73: 522–524.

Pedigo, L. P., S. H. Hutchins, and L. G. Higley. 1986. Economic injury levels in theory and practice. *Annual Review of Entomology* 31: 341–368.

Peltzer, D. A., R. B. Allen, G. M. Lovett, D. Whitehead, and D. A. Wardle. 2010. Effects of biological invasions on forest carbon sequestration. *Global Change Biology* 16: 732–746.

Pemberton, R. W. 1999. Insects and other arthropods used as drugs by Korean traditional medicine. *Journal of Ethnopharmacology* 65: 207–216.

Perlman, F., E. Press, J. A. Googins, A. Malley, and H. Poarea. 1976. Tussockosis: Reactions to Douglas fir tussock moth. *Annals of Allergy* 36: 302–307.

Peterson, R. O. 1999. Wolf–moose interaction on Isle Royale: The end of natural regulation? *Ecological Applications* 9: 10–16.

Póvoa, M. M., J. E. Conn, C. D. Schlichting, J. C. O. F. Amaral, M. N. O. Segura, A. N. M. da Silva, C. C. B. dos Santos, R. N. L. Lacerda, R. T. L. de Souza, D. Galiza, et al. 2003. Malaria vectors, epidemiology, and the re-emergence of *Anopheles darlingi* in Belém, Pará, Brazil. *Journal of Medical Entomology* 40: 379–386.

Pray, C. L, W. H. Nowlin, and M. J. Vanni. 2009. Deposition and decomposition of periodical cicadas (Homoptera: Cicadidae: Magicicada) in woodland aquatic ecosystems. *Journal of the North American Benthological Society* 28: 181–195.

Price, P. W. 1997. *Insect Ecology*, 3rd ed. New York: John Wiley & Sons.

Progar, R. A., T. D. Schowalter, C. M. Freitag, and J. J. Morrell. 2000. Respiration from coarse woody debris as affected by moisture and saprotroph functional diversity in western Oregon. *Oecologia* 124: 426–431.

Ramos-Elorduy, J. 2009. Anthro-entomophagy: Cultures, evolution, and sustainability. *Entomological Research* 39: 271–288.

Reynolds, B. C., D. A. Crossley, Jr., and M. D. Hunter. 2003. Response of soil invertebrates to forest canopy inputs along a productivity gradient. *Pedobiologia* 47: 127–139.

Ricketts, T. H. 2004. Tropical forest fragments enhance pollinator activity in nearby coffee crops. *Conservation Biology* 18: 1262–1271.

Ricketts, T. H., J. Regetz, I. Steffan-Dewenter, S. A. Cunningham, C. Kremen, A. Bogdanski, B. Gemmill-Herren, S. S. Greenleaf, A. M. Klein, M. M. Mayfield, et al. 2008. Landscape effects on crop pollinator services: Are there general patterns? *Ecology Letters* 11: 499–515.

Ridsdill-Smith, T. J. and A. A. Kirk. 1985. Selecting dung beetles (Scarabaeinae) from Spain for bushfly control in south-western Australia. *Entomophaga* 30: 217–223.

Riley, C. V. 1878. *First Annual Report of the United States Entomological Commission for the Year 1877 Relating to the Rocky Mountain Locust and the Best Methods of Preventing Its Injuries and of Guarding Against Its Invasions, in Pursuance of an Appropriation Made by Congress for This Purpose.* Washington, DC: U.S. Department of Agriculture.

Risch, A. C., M. F. Jurgensen, M. Schütz, and D. S. Page-Dumbroese. 2005. The contribution of red wood ants to soil C and N pools and CO_2 emissions in subalpine forests. *Ecology* 85: 419–430.

Ritchie, M. E., D. Tilman, and J. M. H. Knops. 1998. Herbivore effects on plant and nitrogen dynamics in oak savanna. *Ecology* 79: 165–177.

Rodgers, H. L., M. P. Brakke, and J. J. Ewel. 1995. Shoot damage effects on starch reserves of *Cedrela odorata*. *Biotropica* 27: 71–77.

Romme, W. H., D. H. Knight, and J. B. Yavitt. 1986. Mountain pine beetle outbreaks in the Rocky Mountains: Regulators of primary productivity? *American Naturalist* 127: 484–494.

Romoser, W. S. and J. G. Stoffolano, Jr. 1998. *The Science of Entomology*, 4th ed. Boston; McGraw-Hill.

Roth, J. P., A. MacQueen, and D. E. Bay. 1988. Predation by the introduced phoretic mite, *Macrocheles peregrinus* (Acari: Macrochelidae), on the buffalo fly, *Haematobia irritans exigua* (Diptera: Muscidae), in Australia. *Environmental Entomology* 17: 603–607.

Salick, J., R. Herrera, and C. F. Jordan. 1983. Termitaria: Nutrient patchiness in nutrient-deficient rain forests. *Biotropica* 15: 1–7.

Sanderson, M. G. 1996. Biomass of termites and their emissions of methane and carbon dioxide: A global database. *Global Biogeochemical Cycles* 10: 543–557.

Schowalter, T. D. 1981. Insect herbivore relationship to the state of the host plant: Biotic regulation of ecosystem nutrient cycling through ecological succession. *Oikos* 37: 126–130.

Schowalter, T. D. 2008. Insect herbivore responses to management practices in conifer forests in North America. *Journal of Sustainable Forestry* 26: 204–222.

Schowalter, T. D. 2011. *Insect Ecology: An Ecosystem Approach*, 3rd ed. San Diego: Elsevier/Academic.

Schowalter, T. D. 2012. Insect outbreak effects on ecosystem services. In *Insect Outbreaks Revisited*, P. Barbosa, D. K. Letourneau, and A. A. Agrawal, eds. 246–265. Hoboken, NJ: Wiley/Blackwell.

Schowalter, T. D., S. J. Fonte, J. Geagan, and J. Wang. 2011. Effects of manipulated herbivore inputs on nutrient flux and decomposition in a tropical rainforest in Puerto Rico. *Oecologia* 167: 1141–1149.

Schowalter, T. D., D. C. Lightfoot, and W. G. Whitford. 1999. Diversity of arthropod responses to host-plant water stress in a desert ecosystem in southern New Mexico. *American Midland Naturalist* 142: 281–290.

Schowalter, T. D., T. E. Sabin, S. G. Stafford, and J. M. Sexton. 1991. Phytophage effects on primary production, nutrient turnover, and litter decomposition of young Douglas-fir in western Oregon. *Forest Ecology and Management* 42: 229–243.

Schowalter, T. D. and P. Turchin. 1993. Southern pine beetle infestation development: Interaction between pine and hardwood basal areas. *Forest Science* 39: 201–210.

Schowalter, T. D., Y. L. Zhang, and T. E. Sabin. 1998. Decomposition and nutrient dynamics of oak *Quercus* spp. logs after five years of decomposition. *Ecography* 21: 3–10.

Seastedt, T. R. 1984. The role of microarthropods in decomposition and mineralization processes. *Annual Review of Entomology* 29: 25–46.

Seastedt, T. R. and D. A. Crossley, Jr. 1981. Microarthropod response following cable logging and clear-cutting in the southern Appalachians. *Ecology* 62: 126–135.

Seastedt, T. R., D. A. Crossley, Jr., and W. W. Hargrove. 1983. The effects of low-level consumption by canopy arthropods on the growth and nutrient dynamics of black locust and red maple trees in the southern Appalachians. *Ecology* 64: 1040–1048.

Setälä, H. and V. Huhta. 1991. Soil fauna increase *Betula pendula* growth: Laboratory experiments with coniferous forest floor. *Ecology* 72: 665–671.

Sheppard, S. and P. Picard. 2006. Visual-quality impact of forest pest activity at the landscape level: A synthesis of published knowledge and research needs. *Landscape and Urban Planning* 77: 321–342.

Sherman, R. A., M. J. R. Hall, and S. Thomas. 2000. Medical maggots: An ancient remedy for some contemporary afflictions. *Annual Review of Entomology* 45: 55–81.

Sherman, R. A., H. Stevens, D. Ng, and E. Iversen. 2007. Treating wounds in small animals with maggot debridement therapy: A survey of practitioners. *Veterinary Journal* 173: 138–143.

Siepel, H. and E. M. de Ruiter-Dijkman. 1993. Feeding guilds of oribatid mites based on their carbohydrase activities. *Soil Biology and Biochemistry* 25: 1491–1497.

Silva, S. I., W. P. MacKay, and W. G. Whitford. 1985. The relative contributions of termites and microarthropods to fluff grass litter disappearance in the Chihuahuan Desert. *Oecologia* 67: 31–34.

Smith, K. F., D. F. Sax, S. D. Gaines, V. Guernier, and J.-F. Guégan. 2007. Globalization of human infectious disease. *Ecology* 88: 1903–1910.

Stachurski, A. and J. R. Zimka. 1984. The budget of nitrogen dissolved in rainfall during its passing through the crown canopy in forest ecosystems. *Ekologia Polska* 32: 191–218.

Stadler, B., B. Michalzik, and T. Müller. 1998. Linking aphid ecology with nutrient fluxes in a coniferous forest. *Ecology* 79: 1514–1525.

Stadler, B. and T. Müller. 1996. Aphid honeydew and its effect on the phyllosphere microflora of *Picea abies* (L.) Karst. *Oecologia* 108: 771–776.

Stadler, B., S. Solinger, and B. Michalzik. 2001. Insect herbivores and the nutrient flow from the canopy to the soil in coniferous and deciduous forests. *Oecologia* 126: 104–113.

Stapp, P., M. F. Antolin, and M. Ball. 2004. Patterns of extinction in prairie dog meta-populations: Plague outbreaks follow El Niño events. *Frontiers in Ecology and the Environment* 2: 235–240.

Steffan-Dewenter, I. and T. Tscharntke. 1999. Effects of habitat isolation on pollinator communities and seed set. *Oecologia* 121: 432–440.

Swank, W. T., J. B. Waide, D. A. Crossley, Jr., and R. L. Todd. 1981. Insect defoliation enhances nitrate export from forest ecosystems. *Oecologia* 51: 297–299.

Taki, H., P. G. Kevan, and J. S. Ascher. 2007. Landscape effects of forest loss in a pollination system. *Landscape Ecology* 22: 1575–1587.

Taylor, S. L. and D. A. MacLean. 2009. Legacy of insect defoliators: Increased wind-related mortality two decades after a spruce budworm outbreak. *Forest Science* 55: 256–267.

Thompson, D. C. and K. T. Gardner. 1996. Importance of grasshopper defoliation period on southwestern blue grama-dominated rangeland. *Journal of Range Management* 49: 494–498.

Treseder, K. K. 2008. Nitrogen additions and microbial biomass: A meta-analysis of ecosystem studies. *Ecology Letters* 11: 1111–1120.

Trlica, M. J. and L. R. Rittenhouse. 1993. Grazing and plant performance. *Ecological Applications* 3: 21–23.

Trumble, J. T., D. M. Kolodny-Hirsch, and I. P. Ting. 1993. Plant compensation for arthropod herbivory. *Annual Review of Entomology* 38: 93–119.

Tyndale-Biscoe, M. 1994. Dung burial by native and introduced dung beetles (Scarabaeidae). *Australian Journal of Agricultural Research* 45: 1799–1808.

Tyndale-Biscoe, M. and W. G. Vogt. 1996. Population status of the bush fly, *Musca vetustissima* (Diptera: Muscidae), and native dung beetles (Coleoptera: Scarabaeinae) in south-eastern Australia in relation to establishment of exotic dung beetles. *Bulletin of Entomological Research* 86: 183–192.

Vamosi, J. C., T. M. Knight, J. A. Steets, S. J. Mazer, M. Burd, and T.-L. Ashman. 2006. Pollination decays in biodiversity hotspots. *Proceedings of the National Academy of Sciences USA* 103: 956–961.

Van Bael, S. A., A. Aiello, A. Valderrama, E. Medianero, M. Samaniego, and S. J. Wright. 2004. General herbivore outbreak following an El Niño-related drought in a lowland Panamanian forest. *Journal of Tropical Ecology* 20: 625–633.

Vittor, A. Y., R. H. Gilman, J. Tielsch, G. Glass, T. Shields, W. S. Lozano, V. Pinedo-Cancino, and J. A. Patz. 2006. The effect of deforestation on the human-biting rate of *Anopheles darlingi*, the primary vector of falciparum malaria in the Peruvian Amazon. *American Journal of Tropical Medicine and Hygiene* 74: 3–11.

Vossbrinck, C. R., D. C. Coleman, and T. A. Woolley. 1979. Abiotic and biotic factors in litter decomposition in a semiarid grassland. *Ecology* 60: 265–271.

Wagner, D. 1997. The influence of ant nests on Acacia seed production, herbivory, and soil nutrients. *Journal of Ecology* 85: 83–93.

Wagner, D., M. J. F. Brown, and D. M. Gordon. 1997. Harvest ant nests, soil biota, and soil chemistry. *Oecologia* 112: 232–236.

Wagner, D. and J. B. Jones. 2004. The contribution of harvester ant nests, *Pogonomyrmex rugosus* (Hymenoptera, Formicidae), to soil nutrient stocks and microbial biomass in the Mojave Desert. *Environmental Entomology* 33: 599–607.

Wallace, J. B., T. F. Cuffney, J. R. Webster, G. J. Lugthart, K. Chung, and G. S. Goldwitz. 1991. Export of fine organic particles from headwater streams: Effects of season, extreme discharges, and invertebrate manipulation. *Limnology and Oceanography* 36: 670–682.

Wallace, J. B. and J. R. Webster. 1996. The role of macroinvertebrates in stream ecosystem function. *Annual Review of Entomology* 41: 115–139.

Wallace, J. B., J. R. Webster, and R. L. Lowe. 1992. High-gradient streams of the Appalachians. In *Biodiversity of Southeastern United States: Aquatic Communities*, C. T. Hackney, S. M. Adams, and W. A. Martin, eds. 133–191. New York: John Wiley.

Wallace, J. B., M. R. Whiles, S. Eggert, T. F. Cuffney, G. J. Lugthart, and K. Chung. 1995. Long-term dynamics of coarse particulate organic matter in three Appalachian Mountain streams. *Journal of the North American Benthological Society* 14: 217–232.

Watson, E. J. and C. E. Carlton. 2003. Spring succession of necrophilous insects on wildlife carcasses in Louisiana. *Journal of Medical Entomology* 40: 338–347.

Webb, W. L. 1978. Effects of defoliation and tree energetics. In *The Douglas-fir Tussock Moth: A Synthesis*, M. H. Brookes, R. W. Stark, and R. W. Campbell, eds. 77–81. USDA Forest Service Tech. Bull. 1585. Washington, DC: USDA Forest Service.

Weinstock, G. M., G. E. Robinson, and members of the Honeybee Genome Sequencing Consortium. 2006. Insights into social insects from the genome of the honeybee *Apis mellifera*. *Nature* 443: 931–949.

Wheeler, G. S., M. Tokoro, R. H. Scheffrahn, and N. Y. Su. 1996. Comparative respiration and methane production rates in Nearctic termites. *Journal of Insect Physiology* 42: 799–806.

White T. C. R. 1969. An index to measure weather-induced stress of trees associated with outbreaks of psyllids in Australia. *Ecology* 50: 905–909.

White, T. C. R. 1976. Weather, food, and plagues of locusts. *Oecologia* 22: 119–134.

White, T. C. R. 1984. The abundance of invertebrate herbivores in relation to the availability of nitrogen in stressed food plants. *Oecologia* 63: 90–105.

Whitford, W. G. 1986. Decomposition and nutrient cycling in deserts. In *Pattern and Process in Desert Ecosystems*, W. G. Whitford, ed. 93–117. Albuquerque: University of New Mexico Press.

Whitford, W. G., P. Johnson, and J. Ramirez. 1976. Comparative ecology of the harvester ants *Pogonomyrmex barbatus* (F. Smith) and *Pogonomyrmex rugosus* (Emery). *Insectes Sociaux* 23: 117–132.

Whitford, W. G., V. Meentemeyer, T. R. Seastedt, K. Cromack, Jr., D. A. Crossley, Jr., P. Santos, R. L. Todd, and J. B. Waide. 1981. Exceptions to the AET model: Deserts and clear-cut forest. *Ecology* 62: 275–277.

Whitford, W. G., Y. Steinberger, and G. Ettershank. 1982. Contributions of subterranean termites to the "economy" of Chihuahuan Desert ecosystems. *Oecologia* 55: 298–302.

Wickman, B. E., 1980. Increased growth of white fir after a Douglas-fir tussock moth outbreak. *Journal of Forestry* 78: 31–33.

Williams, N. M. and C. Kremen. 2007. Resource distributions among habitats determine solitary bee offspring production in a mosaic landscape. *Ecological Applications* 17: 910–921.

Williamson, S. C., J. K. Detling, J. L. Dodd, and M. I. Dyer. 1989. Experimental evaluation of the grazing optimization hypothesis. *Journal of Range Management* 42: 149–152.

Wilmers, C. C., E. Post. R. O. Peterson, and J. A. Vucetich. 2006. Predator disease out-break modulates top-down, bottom-up, and climatic effects on herbivore population dynamics. *Ecology Letters* 9: 383–389.

Wise, D. H. and M. Schaefer. 1994. Decomposition of leaf litter in a mull beech forest: Comparison between canopy and herbaceous species. *Pedobiologia* 38: 269–288.

Wood, T. E., D. Lawrence, D. A. Clark, and R. L. Chazdon. 2009. Rain forest nutrient cycling and productivity in response to large-scale litter manipulation. *Ecology* 90: 109–121.

Wright, L. C., A. A. Berryman, and B. E. Wickman. 1986. Abundance of the fir engraver, *Scolytus ventralis*, and the Douglas-fir beetle, *Dendroctonus pseudotsugae*, following tree defoliation by the Douglas-fir tussock moth, *Orgyia pseudotsugata*. *Canadian Entomologist* 116: 293–305.

Wright, S. J., J. B. Yavitt, N. Wurzburger, B. L. Turner, E. V. J. Tanner, E. J. Sayer, L. S. Santiago, M. Kaspari, L. O. Hedin, K. E. Harms, et al. 2011. Potassium, phosphorus, or nitrogen limit root allocation, tree growth, or litter production in a lowland tropical forest. *Ecology* 92: 1616–1625.

Yen, A. L. 2009. Entomophagy and insect conservation: Some thoughts for digestion. *Journal of Insect Conservation* 13: 667–670.

Zhou, J., W. K.-M. Lau, P. M. Masuoka, R. G. Andre, J. Chamberlin, P. Lawyer, and L. W. Laughlin. 2002. El Niño helps spread Bartonellosis epidemics in Peru. *Eos, Transactions, American Geophysical Union* 83: 157, 160–161.

Zimmerman, P. R., J. P. Greenberg, S. O. Wandiga, and P. J. Crutzen. 1982. Termites: A potentially large source of atmospheric methane, carbon dioxide, and molecular hydrogen. *Science* 218: 563–565.

8

Valuation of Insect and Management Effects

> Not too many years ago the blotches caused by lygus bugs feeding on an occasional lima bean were of little concern, and lygus bugs were considered a minor pest on this crop. However, with the emphasis on product appearance in the frozen-food industry, a demand was created for a near-perfect bean. For this reason . . . lygus bugs are now considered serious pests of lima beans.
>
> **Stern et al. (1959)**

Ecosystem services are produced free of charge, as outputs of ecosystem processes described in Chapter 5. Although humans, unlike other organisms, are capable of transporting services great distances, we still ultimately depend on food and water supplies that result from ecosystem processes. The only costs associated with our use of ecosystem services are those required for extraction and transportation, costs that are passed along from land managers and utilities to individual consumers. The costs of maintaining artificially high densities of crop species in agroecosystems or of replacing ecosystem services that are lost as a result of ecosystem degradation, are generally ignored (Christensen et al. 2000).

Traditional value systems have emphasized provisioning services, such as food, fiber, wood, fresh water, medicinal and industrial products, and cultural and recreational services. These values are easily computed from market values and can be used to establish action thresholds for preventing damage by insects (Pedigo et al. 1986). For example, the value of fresh water and timber are based on their selling price and accounting for costs of replacing these from other sources if these ecosystem services were not available (Christensen et al. 2000). Similarly, additional expenses required for protection of services or unscheduled harvest to prevent loss can be computed. Johnson et al. (2006) reported that more than US$194 million was spent on monitoring and control of gypsy moths in the United States during 1985–2004. However, in many cases, the marginal benefits of insect control in forests and grasslands may warrant action only for targeted sites with very high outbreak populations or very high resource values (Shewchuk and Kerr 1993; Zimmerman et al. 2004). Costs of insect-vectored diseases to human health and social stability are more problematic. Prevention of diseases such as plague, yellow fever, and malaria seems far preferable to the costs of treatment or deaths following infection, but are we willing to invest sufficient resources for inoculation and economic development in impoverished reservoirs of these diseases? Interruption of transmission by treating vector habitats has obvious benefits but at what cost in terms of long-term sustainability of ecosystem services that are disrupted by such control efforts?

Only recently have the potential values of supporting and regulating services, necessary to sustain extractive and cultural services, been recognized. These values are difficult to calculate because these services do not have market values (Dasgupta et al. 2000). Nevertheless, we would be wise to compare the economic and social costs of crashes in supply of services were our maximum rates of use to continue versus reducing these rates to more sustainable levels. Costanza et al. (1997) estimated regional and global values of all ecosystem services, using market values and user fees where available, at a total global value of US $33 trillion annually, but Jørgensen (2010) concluded that the full value of ecosystem services is much higher.

The importance of biodiversity in maintaining ecosystem services is clear (Ewel 1986; Ewel et al. 1991; Tilman and Downing 1994; Hooper et al. 2005; Hoehn et al. 2008; Allan et al. 2009; Duffy 2009; see also Chapter 5) but the value of protecting individual species is not. Biodiversity is analogous to the mechanical parts of an airplane or clock (Leopold 1949; Ehrlich and Ehrlich 1981). Some parts may be removed with little apparent effect, but as more parts are removed, a point will be reached at which the mechanism no longer works. Similarly, as species disappear, ecosystems may reach thresholds beyond which they will no longer support human needs. We do not know enough about the mechanisms that underlie ecosystem integrity to know which parts are critical. However, the importance of biodiversity in maintaining pollination and biological control of pests in agricultural landscapes already has become sufficiently evident to promote policy changes in the United States and European Union to offset biodiversity loss through agri-environmental programs that subsidize producers for conservation or restoration practices (Donald and Evans 2006).

This chapter summarizes available data on economic or social values of insect and management effects on various ecosystem services. For nonprovisioning services, these values are not available, and even for provisioning services, the net values that include long-term benefits from compensatory growth, regulation of nutrient fluxes, and soil development have not been calculated. Therefore, an explicit objective of this chapter is to draw attention to the need for more complete accounting of the multiple, often opposing, social and economic benefits and costs of insect effects on ecosystem services.

8.1 Valuation of Insect Effects on Provisioning Services

Ecosystems are the source of all food, fresh water, fiber, wood, and biofuel for human use (see Chapter 1). Plants and some animals, including insects, remain sources of many important medical and industrial products (Zenk and Juenger 2007). Some of these products are difficult to synthesize, and bioprospecting for sources of new products continues in many ecosystems (Helson et al. 2009). Which species may be the source of a future remedy? Clearly, insect outbreaks that reduce the supply of important plant or animal-derived resources will increase costs associated with their loss or replacement, but low-to-moderate levels of herbivory might actually increase production of plant products (see Chapter 7). Additional supply costs may be incurred if insect-killed trees must be salvaged to avoid deterioration, thereby disrupting planned harvest schedules in managed forests. Costs of plant growth loss, fruit or seed loss, mortality, and/or unscheduled salvage harvest are easy to calculate. The value lost at various insect densities has been used to develop economic thresholds for pest control (Pedigo et al. 1986).

However, some short-term costs of reduced resource production or unscheduled salvage harvest may be reduced if harvest can be delayed until compensatory growth can replace lost products (McNaughton 1979; Wickman 1980; Knapp and Seastedt 1986; Romme et al. 1986; Alfaro and Shepherd 1991). Furthermore, some costs might have been avoided by pruning, thinning, or employing other management practices to prevent outbreaks (see Chapters 4, 5, and 9).

Ecosystems also are important sources of fish and wildlife, many of which depend on insects as food resources. A variety of insectivorous songbirds, an attraction for birdwatchers, would decline or disappear in the absence of abundant insect prey, including mosquitoes, and often congregate to sites of abundant insects (Koenig and Liebhold 2005). Bats and many small mammals, lizards, and amphibians also are supported primarily by insects and, in turn, serve as prey to larger game or fur-bearing animals. Salmonids are among the most important global fisheries that are supported for much of their life cycles by insect prey, with up to half of their in-stream diets consisting of terrestrial insects falling into the water (Kawaguchi and Nakano 2001; Allan et al. 2003; Baxter et al. 2005). Therefore, insect effects on fish and wildlife production are largely positive.

Ecosystems are valued sources of fresh water, and adequate water supply often is the primary management goal for municipal watersheds. Insects falling into nutrient-poor lakes and headwater streams affect water quality (Carlton and Goldman 1984; Mehner et al. 2005; Nowlin et al. 2007; Menninger et al. 2008; Pray et al. 2009) but probably not substantially (Lovett et al. 2002). Herbivorous insects reduce canopy cover and increase the volume of precipitation reaching the ground and flowing into streams. Increased water yield could be beneficial to downstream communities during droughts (a typical trigger for defoliator outbreaks) that otherwise would reduce water yields. However, increased yields could contribute to flooding if water levels downstream were already high. Soil and litter insects affect soil porosity and decomposition rates that, in turn, control the rate of water movement through the substrate. Changes in water yield and quality resulting from insect herbivory may or may not be desirable, depending on the needs of downstream users.

The food value of insects or their products is substantial in many areas (Clausen 1954; DeFoliart 1999; Cerritos and Cano-Santana 2008; Ramos-Elorduy 2009; Yen 2009). Currently, commercial honey production amounts to about US$2 billion globally, with China accounting for 20% of the total production and 25% of world honey exports (Parker 2003). Substantial amounts of honey are produced by feral colonies, primarily in forests, and honey production remains a major use of forest ecosystems (Bradbear 2009).

The value of insects as a food source directs ecosystem management practices in some regions. Mbata et al. (2002) described the process by which local governments manage harvest of edible caterpillars (primarily two saturniids, *Gynanisa maja* and *Gonimbrasia zambesina*) in Zambian forests. In one unique study, Cerritos and Cano-Santana (2008) reported that harvesting grasshoppers for food during an outbreak in Mexico provided US$3,000 per family and substantially reduced grasshopper abundance and reproduction (Figure 8.1), compared with US$150 for insecticide treatment had control tactics been implemented.

Control of insects often is demanded by the general public that, especially in Western societies, is intolerant of blemished produce or increased food prices. Acceptance of control costs by producers and the general public varies with the degree of visual damage or food shortage (Torrell et al. 1989; Sheppard and Picard 2006). When control costs are substantially subsidized by the government, crop producers or resource managers are inclined to control insects at lower densities than would be acceptable to taxpayers. For example, an individual rancher's 50% share of the US$6.18 ha^{-1} cost for control at an economic injury level of 18 grasshoppers ha^{-1} is equivalent to a rancher's economic injury level of less than three grasshoppers ha^{-1} (Torrell et al. 1989). Gatto et al. (2009) conducted an economic analysis of pest management for processionary moth, *Thaumetopoea pityocampa*, in Portugal and concluded that pest management costs outweighed market revenues for maritime pine, *Pinus pinaster*, plantations—at least in the short term—making control undesirable for private landowners. Taxpayer support also would be unwise, based on provisioning service values alone, but could be justified by potential benefits to the public through other types of ecosystem services, such as improved carbon sequestration, recreation, and public health, that is, nonprovisioning ecosystem services.

Pest management often is more detrimental to other ecosystem services than would be justified by its value for the provisioning service. For example, insecticides have documented toxicity for many nontarget species, especially fish and pollinators, threatening sustainability of provisioning and supporting services derived from these species (Smith et al. 1983; Claudianos et al. 2006; Baldwin et al. 2009). Furthermore, nontarget effects could undermine important ecosystem functions that contribute to the long-term sustainability of other ecosystem services (Downing and Leibold 2002; Hättenschwiler and Gasser 2005).

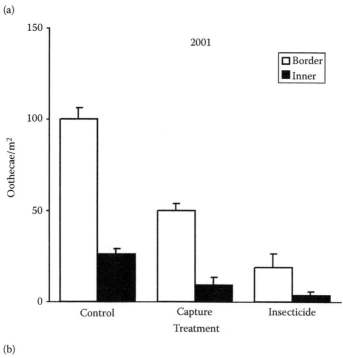

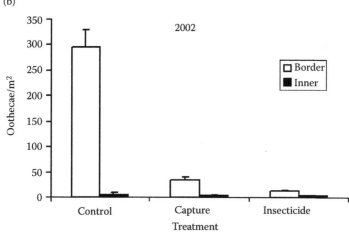

FIGURE 8.1

Density of grasshopper, *Sphenarium purpurascens* egg cases (mean + SE) by treatment in the Puebla–Tlaxcala Valley of Mexico in 2001 (a) and 2002 (b). Treatments were grasshopper capture for food or insecticide treatment. Data were collected along borders or in the interior of alfalfa fields. (From Cerritos, R. and Z. Cano-Santana, *Crop Protection*, 27, 473–480, 2008. With permission.)

8.2 Valuation of Insect Effects on Cultural Services

Insect effects on cultural services can be positive or negative depending on public perceptions. Insects are valued as cultural icons and objects of art or natural beauty. The global value of recreational services alone (which often can be calculated from usage fees) has been estimated at US$815 billion by Costanza et al. (1997), and insects influence these through public use or perceptions.

Birdwatching is a popular recreational use of ecosystems, and many of the most attractive songbirds are insectivorous species whose abundance depends on the availability of insect prey (Losey and Vaughn 2006). Similarly, sportfishing provides billions of dollars in equipment sales and tourist revenues. Most game fish are primarily or exclusively insectivorous species (see Chapter 7) and provide exciting action when caught with artificial lures based on insects.

Some insects themselves provide tourism destinations. Huntly et al. (2005) reported that 95% of tourists to Hluhluwe-Imfolozi Park in South Africa expressed interest in information on invertebrates. Tours can be arranged to view monarch butterfly migrations, morpho butterflies, dragonflies, or jewel beetles (Huntly et al. 2005). Butterfly farmers rear a variety of temperate and tropical butterflies for butterfly houses (Figure 8.2) and for release at weddings and other special events.

Insects also can be objects for entertainment. Butterfly gardens, ant farms, and pet tarantulas, millipedes, and scorpions are available from a variety of biological supply and pet sources. Insects have inspired much lucrative artwork (see Figure 7.2) and documentary, as well as science fiction, literature, and film.

FIGURE 8.2 (SEE COLOR INSERT.)
Brightly colored morpho butterflies, *Morpho* spp., delight visitors to a commercial butterfly house. Butterfly houses have become popular tourist destinations in many cities.

Plants defoliated or killed by insects reduce shade and can create unsightly conditions or safety hazards in camping, hiking, or other recreational areas. Insect feces and tissues falling on people or eating surfaces are considered a nuisance. Furthermore, some caterpillars are venomous or allergenic (Perlman et al. 1976; Schowalter 2011; see Figure 3.6c), exacerbating the nuisance. These detrimental effects can reduce visitation to recreational sites experiencing outbreaks.

Few studies have evaluated effects of insect outbreaks on cultural values. Downing and Williams (1978) reported that a Douglas-fir tussock moth outbreak in Oregon did not significantly affect recreational land use but that recreational use appeared to increase as a result of curiosity. Although 75% of visitors were aware of the outbreak, few chose to avoid the area. In fact, the only negative effect mentioned in their study was avoidance of salvage logging operations that were considered unappealing or hazardous. On the other hand, extensive defoliation or plant mortality may be viewed as unattractive or hazardous (Michalson 1975). Sheppard and Picard (2006) reported that visual preference depended on the extent of defoliation or tree mortality (see Figure 7.3). Some, but not all, studies showed that visual preference was affected by the subject's awareness of the cause (Müller and Job 2009). Because visitors' perceptions of ecosystem changes resulting from insect outbreaks determine willingness to stay and spend money on recreational activities or in nearby towns, negative perceptions can reduce the economic value of cultural services regardless of longer-term costs or benefits of outbreaks (see Section 8.3). However, even when tourist perceptions of insect damage do not warrant control efforts (Müller and Job 2009), recreational values versus timber value represents a trade-off. The net value of insect effects on cultural services has not been calculated.

8.3　Valuation of Insect Effects on Supporting Services

Changes in values resulting from insect effects on supporting services can be positive or negative, depending on management goals and time frame. However, it is important to remember that, regardless of our perspectives, insects have been among the factors that supported services that humans have valued since before we were able to manipulate them to our apparent advantage. Therefore, valuation requires consideration of trade-offs among effects of insects on supporting services.

Primary production is among the most important supporting services in that it provides or controls supply of all provisioning services. Plant growth losses, mortality, or species replacement due to herbivory by insects may not always be desirable, but plants, including crop species, often are able to compensate for short-term growth losses following herbivore outbreaks and some require pruning for sustained growth (Knapp and Seastedt 1986; Pedigo et al. 1986; Williamson et al. 1989; Alfaro and Shepherd 1991; Trumble et al. 1993; Gutschick 1999; see Chapter 7). Herbivory contributes to primary production in the same manner as prescribed mowing, pruning, and/or thinning that are often used in managed ecosystems to improve primary production. Compensatory growth following loss of plant tissues reflects selective removal of less efficient or less defended plant parts, permitting plant allocation of carbon and nutrients to new, more productive plant tissues (Gutschick 1999). Compensatory growth over the long term can largely replace short-term reductions in primary production, offsetting costs of short-term losses (Figure 8.3). Improved survival of defoliated plants during drought would mitigate effects of drought, which is a common trigger for outbreaks (Mattson and Haack 1987; Van Bael et al. 2004). Furthermore, suppression of the most abundant plant species increases biodiversity and often tailors species composition to the prevailing climate and resource supply capacity of the ecosystem by replacing intolerant species with more tolerant species (Ritchie et al. 1998; Schowalter 2008, 2011).

Therefore, insect outbreaks often have less negative effects on primary production (and provisioning and other ecosystem services that it supports) over long time periods than is generally perceived. In fact, the effect of the Douglas-fir tussock moth on Sierran forests shown in Figure 5.11 is the recommended forest structure for sustainable management (North et al. 2007). If harvest of ecosystem products (e.g., timber) can be delayed, the cost of

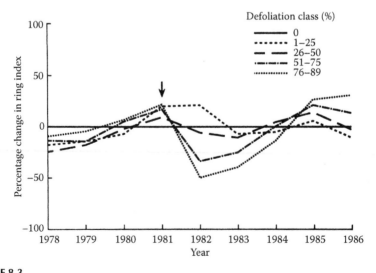

FIGURE 8.3
Changes in ring width indices for Douglas-fir defoliated at different intensities by the Douglas-fir tussock moth in 1981 (arrow). The horizontal line at 0% represents the ring width index for nondefoliated trees. (From Alfaro, R. I. and R. F. Shepherd, *Forest Science*, 37, 959–964, 1991. With permission.)

unscheduled salvage harvest may be offset by long-term replacement of lost products (McNaughton 1979; Wickman 1980; Knapp and Seastedt 1986; Romme et al. 1986; Alfaro and Shepherd 1991). In other words, although the costs of plant growth loss, mortality, and/or unscheduled salvage harvest are easy to calculate, and have contributed to largely negative perceptions of herbivorous insects, these costs might have been avoided by pruning, thinning, or other management practices recommended to prevent outbreaks.

Pollination of agricultural crops by honey bees, *Apis mellifera*, and other insects is necessary for 35% of global fruit and vegetable production for human consumption (Klein et al. 2007), clearly a positive effect of insects on supporting services. Bommarco et al. (2012) reported that insect pollinators (primarily honey bees but including hover flies, bumble bees, and other insects) increased oilseed rape seed weight by 18% and market value by 20% per plant. Insect pollination of plants used as livestock or wildlife feed, such as clover and alfalfa, provides additional indirect value (Berenbaum et al. 2007). Pollination service is worth about US$120 billion per year globally (Costanza et al. 1997) and $5–$10 billion per year in the United States (Losey and Vaughn 2006; Berenbaum et al. 2007; Isaacs et al. 2009). Economic costs of food production can increase when availability of mobile honey bee colonies becomes limited (Allen-Wardell et al. 1998). However, fruit abortion can complicate the calculation of benefits of pollination services (Bos et al. 2007)

Many crops require more efficient pollination by native pollinators than is provided by generalist honey bees. For example, squash and melons are pollinated primarily by native bee species that decline in intensively managed or fragmented landscapes (Figure 8.4) (Kremen et al. 2002, 2004). Such plants could not be grown in areas from which native pollinators have disappeared, thereby eliminating a crop option for farmers or increasing the cost of producing manually pollinated crops.

Decomposition is necessary for release of nutrients from dead organic matter and for preventing accumulation of detritus. Insects, along with millipedes, oribatid mites, and earthworms, are instrumental in both processes, as described in Chapter 7 (Seastedt 1984; Zhong and Schowalter 1989; Setälä and Huhta 1991). A number of studies have demonstrated that arthropod detritivores are responsible for up to 80% of the total decay rate, depending on litter quality and ecosystem conditions, especially litter moisture (see Figure 7.5) (Vossbrinck et al. 1979; Seastedt 1984; Coleman et al. 2004). In the absence of this natural process, more expensive methods for organic matter removal and fertilization would be required.

For example, the economic value of dung beetles that reduce cattle dung in pastures and rangelands is estimated at more than $400 million in the United States (Losey and Vaughn 2006). In the absence of detritivores that can process livestock dung effectively (as described in Chapter 7 for Australia following the introduction of livestock), dung accumulates, fouling pasture

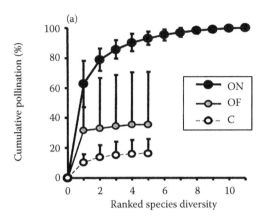

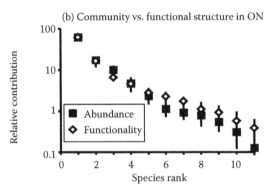

FIGURE 8.4

Watermelon pollination function by native bees in California under three management regimes: ON = organic farm near native habitat, OF = organic farm isolated from native habitat, C = conventional farm. (a) Relationship between cumulative pollination function (as a percentage of total in the ON treatment) and species richness (in rank order of decreasing contribution to pollination). (b) Comparison between community and functional structures for ON treatment. (c) Functional structure for pollination for each management treatment. In panels (b) and (c), values are means with 95% confidence intervals over four (OF) or five (ON, C) replicates per management treatment. (From Balvanera, P., C. Kremen, and M. Martínez-Ramos, *Ecological Applications*, 15, 360–375, 2005. With permission.)

grasses and discouraging grazing (Figure 8.5) as well as providing breeding material for nuisance flies (Ferrar 1975; Tyndale-Biscoe 1994; Tyndale-Biscoe and Vogt 1996). Specialized dung beetles, along with phoretic predatory mites, were introduced into Australia from South Africa at considerable cost to provide this ecosystem service. Although the beetles and mites eliminated dung quickly and reduced fly populations (Tyndale-Biscoe 1994), reducing costs of livestock production and fly control, the full costs of such introductions for other ecosystem services remain unknown (Tyndale-Biscoe and Vogt 1996).

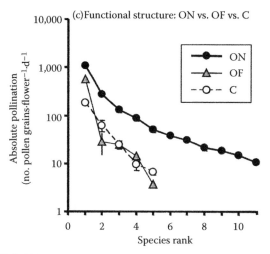

FIGURE 8.4 (CONTINUED)
Watermelon pollination function by native bees in California under three management regimes: ON = organic farm near native habitat, OF = organic farm isolated from native habitat, C = conventional farm. (a) Relationship between cumulative pollination function (as a percentage of total in the ON treatment) and species richness (in rank order of decreasing contribution to pollination). (b) Comparison between community and functional structures for ON treatment. (c) Functional structure for pollination for each management treatment. In panels (b) and (c), values are means with 95% confidence intervals over four (OF) or five (ON, C) replicates per management treatment. (From Balvanera, P., C. Kremen, and M. Martínez-Ramos, *Ecological Applications*, 15, 360–375, 2005. With permission.)

FIGURE 8.5 (SEE COLOR INSERT.)
Fouling of pasture grasses by cattle dung in the absence of effective processing by dung beetles. In addition to covering forage grasses, dung also may adhere to grass blades, thereby reducing palatability (inset).

8.4 Valuation of Insect Effects on Regulating Services

In contrast to supporting services, regulating services provide feedback that maintains more consistent supply (i.e., sustainability) of other services. For example, control of climate and biogeochemical cycling rates by vegetation cover and interactions among other organisms maintains a more consistent supply of water and nutrients and thereby maintains more consistent primary production and the services it supports than would occur in the absence of such regulation (Foley et al. 2003; Schowalter 2011). Insects themselves represent a regulatory mechanism that can maintain more consistent primary production and community composition than occurs in their absence (see Chapter 7). Obviously, insect-induced disturbances can interfere with harvest of provisioning services and can reduce cultural values, as described earlier, but they also may prevent undesirable changes in ecosystem structure or composition (e.g., succession from grassland to forest or from pine forest to fir forest in the absence of fire).

The value of natural regulatory effects is difficult to calculate. The value of biological control of native crop pests by native or introduced insects, a single regulatory mechanism among many, has been estimated at $5.4 billion per year in the United States (Losey and Vaughn 2006). However, this estimate did not include the value of controlling disease vectors, invasive crop pests, or weeds. Landis et al. (2008) estimated the value of native predators and parasites in controlling a single invasive pest, the soybean aphid (*Aphis glycines*), in four U.S. states at more than US$239 million per year.

Clearly, although our economic system encourages maximum exploitation of ecosystem services, a more consistent supply of services would prevent catastrophic oscillations that can destabilize society and lead to population displacement and epidemics of disease, as seen during famines of the past (Riley 1878; Bray 1996; Diamond 1999; Acuña-Soto et al. 2002; Smith et al. 2007; Zhang et al. 2007; Perry et al. 2008; Bora et al. 2010; Hsiang et al. 2011). What is such natural regulation of ecosystem services worth? Once this value is established, the value of regulation by insects and other ecosystem components could be calculated as the difference in long-term resource production, cultural use, and futures market values accruing or discounted from changes in ecosystem conditions resulting from the presence or absence of insect activity.

8.5 Valuation of Insect Management Strategies

Efforts to control insects also have measurable costs and benefits. These costs and benefits must be compared with the full values of insect effects (see earlier) to evaluate net value of control efforts.

Short-term values of insect management strategies are relatively simple to calculate. They include costs of time, labor, and materials to implement control and benefits in terms of disease prevention, lives saved, or improved resource production. For example, the per hectare costs of insecticide application to reduce disease vector populations or crop damage can be compared with the costs of medical treatment or reduced commodity production or with alternative strategies such as crop residue manipulation or crop rotation that could reduce insect population growth. Current annual expenditures for malaria control amount to nearly US$1 billion, although the funds needed for adequate prevention could be twice that (Snow et al. 2008).

The relatively low short-term costs of insecticide application have appeared to justify their selection over alternative, but potentially more sustainable, methods for pest management (see Chapter 6). However, Reay-Jones et al. (2003) estimated that planting resistant varieties of sugarcane could reduce crop losses to the Mexican rice borer, *Eoreuma loftini*, by 24%, and irrigation with 30 cm of water would reduce losses by 29%, substantially reducing the need for other pest control tactics. Mixing of different crops can reduce pest populations and crop losses substantially (Risch 1980, 1981; Zhou et al. 2009). Even mixing resistant and susceptible varieties of the same crop in the same field can provide significant benefit for pest management (Garrett and Mundt 1999). The ecological and market costs of (potentially) reduced yields from planting resistant crop varieties versus use of insecticides to protect vulnerable varieties should be provided so that consumers can evaluate the trade-offs between these two pest management strategies.

Biological control also has measurable costs and benefits. The labor and travel costs of foreign exploration, importation, quarantine, rearing, and field release of prospective biological control agents represent costs that can be compared with benefits achieved in reducing crop losses to target pests. However, introduced biological control agents may have additional effects on nontarget species that also should be considered in such cost-benefit analysis.

Introducing biological control agents from the pest's region of origin requires consideration of the agent's ability to become established in the new community and its effects on nontarget species (Symondson et al. 2002; Louda et al. 2003; Delfosse 2005; van Lenteren et al. 2006; McCoy and Frank 2010). Refinement of quarantine and testing procedures for potential biocontrol agents and selection for more specialized agents have minimized problems, but it is difficult to anticipate all consequences of introductions, including prey switching from invasive to native species (Louda et al. 2003; Delfosse 2005; Hokkanen et al. 2007; McCoy and Frank 2010). Introduced species do not necessarily stop where first introduced. The cactus moth, *Cactoblastis cactorum*, was introduced from South America into Australia and other countries to control invasive prickly pear cacti, *Opuntia* spp. (introduced to support a cochineal dye industry [see Chapter 2]), but recently invaded the southeastern United States (perhaps via nursery stock from the Dominican Republic) and now threatens native prickly pear cacti (Pemberton and Cordo 2001; Simonsen et al. 2008). Introduction of biological control agents is discouraged in this case because these likely would attack native *Cactoblastis* species that help regulate native prickly pear cacti (Pemberton and Cordo 2001).

Longer-term values are more difficult to calculate because of the complexity of ecosystem processes leading to changes in ecosystem services.

However, some costs have become evident and, in some cases, are preventable. For example, DDT and other broad-spectrum toxins introduced into ecosystems for insect control have disrupted food webs and threatened ecosystem processes that support ecosystem services (Carson 1962; Marquis and Whelan 1994; Letourneau and Dyer 1998; Dyer and Letourneau 1999a, b; Knight et al. 2005; Mooney 2007; Letourneau et al. 2009). In particular, detrimental effects on predaceous arthropods and insectivorous birds may undermine natural regulation of prey populations and lead to population increases of target and nontarget pests, requiring more intensive control efforts (Bajwa and Aliniazee 2001; Baldwin et al. 2009). Reduced pollinator, fish, and aquaculture productivity represent more direct negative effects on ecosystem services (Smith et al. 1983; Claudianos et al. 2006; Barbee and Stout 2009). Many pesticides also have detrimental effects on human health that at least partially offset the benefits of reduced vector abundance (Tanner et al. 2011). These costs must be considered in assessment of the need for insecticide application.

Some economic analyses have demonstrated that suppressing insects may provide no benefit or even be counterproductive (Torrell et al. 1989; Gatto et al. 2009). Cerritos and Cano-Santana (2008) reported that harvesting grasshoppers for food instead of applying insecticides for control provided adequate control and a net benefit of US$3150 per Mexican family. In other cases, the net benefits of insect control are not justified by protected resource values. These examples illustrate the dangers of implementing insect control without adequate consideration of the net benefits and costs for long-term sustainability of ecosystem services and human health, irrespective of the feasibility of long-term control of insects.

8.6 Summary

Insects have a variety of positive and negative effects on ecosystem services. Insects are beneficial for fish and wildlife production and have generally neutral, or compensatory, effects on water yield and cultural services. Plant growth loss and mortality may remain undesirable in ecosystems managed for production of plant products, but many plants are able to compensate over the long term for short-term losses to insects, given adequate availability of water and nutrients. Costs of insect damage to plant resources may be tolerable in public ecosystems managed for multiple uses and long periods between harvests. Insect outbreaks can cause serious losses in provisioning services in the short term but also provide regulation of long-term primary productivity that improves the sustainability of ecosystem services. We have only recently recognized that native insect outbreaks can maintain primary production near carrying capacity in the same way that predators maintain more stable and healthy populations of their prey. In situations where insects represent regulating services, control of outbreaks could be counterproductive.

On the other hand, efforts to control insects may cause serious nontarget effects that interfere with sustainability of ecosystem services. Broad-spectrum insecticides are known to disrupt food webs and interfere with predation and other natural regulatory factors, fish and aquaculture productivity, and pollination services. Insecticides also may have negative effects on human health that offset their benefits. When control is necessary, noninsecticidal options may be nearly as effective for pest reduction and far less disruptive to ecosystem services. These options should be considered on the basis of all costs and benefits relative to human health and the sustainability of ecosystem services. New perspectives and approaches to management of insects is the topic of the next, and concluding, chapter.

References

Acuña-Soto, R., D. W. Stahle, M. K. Cleaveland, and M. D. Therrell. 2002. Megadrought and megadeath in 16th century Mexico. *Emerging Infectious Diseases* 8: 360–362.

Alfaro, R. I. and R. F. Shepherd. 1991. Tree-ring growth of interior Douglas-fir after one year's defoliation by Douglas-fir tussock moth. *Forest Science* 37: 959–964.

Allan, B. F., R. B. Langerhans, W. A. Ryberg, W. J. Landesman, N. W. Griffin, R. S. Katz, B. J. Oberle, M. R. Schutzenhofer, K. N. Smyth, A. de St. Maurice, et al. 2009. Ecological correlates of risk and incidence of West Nile virus in the United States. *Oecologia* 158: 699–708.

Allan, J. D., M. S. Wipfli, J. P. Caouette, A. Prussian, and J. Rodgers. 2003. Influence of streamside vegetation on inputs of terrestrial invertebrates to salmonid food webs. *Canadian Journal of Fisheries and Aquatic Sciences* 60: 309–320.

Allen-Wardell, G., P. Bernhardt, R. Bitner, A. Burquez, S. Buchmann, J. Cane, P. A. Cox, V. Dalton, P. Feinsinger, M. Ingram, et al. 1998. The potential consequences of pollinator declines on the conservation of biodiversity and stability of food crop yields. *Conservation Biology* 12: 8–17.

Bajwa, W. I. and M. T. Aliniazee. 2001. Spider fauna in apple ecosystem of western Oregon and its field susceptibility to chemical and microbial insecticides. *Journal of Economic Entomology* 94: 68–75.

Baldwin, D. H., J. A. Spromberg, T. K. Collie, and N. L. Scholz. 2009. A fish of many scales: Extrapolating sublethal pesticide exposures to the productivity of wild salmon populations. *Ecological Applications* 19: 2004–2015.

Balvanera, P., C. Kremen, and M. Martínez-Ramos. 2005. Applying community structure analysis to ecosystem function: Examples from pollination and carbon storage. *Ecological Applications* 15: 360–375.

Barbee, G. C. and M. J. Stout. 2009. Comparative acute toxicity of neonicotinoid and pyrethroid insecticides to non-target crayfish (*Procambarus clarkii*) associated with rice–crayfish crop rotations. *Pest Management Science* 65: 1250–1256.

Baxter C. V., K. D. Fausch, and W. C. Saunders. 2005. Tangled webs: Reciprocal flows of invertebrate prey link streams and riparian zones. *Freshwater Biology* 50: 201–220.

Berenbaum, M., P. Bernhardt, S. Buchmann, N. W. Calderone, P. Goldstein, D. W. Inouye, P. Kevan, C. Kremen, R. A. Medellín, T. Ricketts, et al. 2007. *Status of Pollinators in North America*. Washington, DC: National Academies Press.

Bommarco, R., L. Marini, and B. E. Vassiére. 2012. Insect pollination enhances seed yield, quality, and market value in oilseed rape. *Oecologia* 169: 1025–1032.

Bora, S., I. Ceccacci, C. Delgado, and R. Townsend. 2010. *World Development Report 2011: Food Security and Conflict*. Washington, DC: Agriculture and Rural Development Department, World Bank.

Bos, M. M., D. Veddeler, A. K. Bogdanski, A.-M. Klein, T. Tscharntke, I. Steffen-Dewenter, and J. M. Tylianakis. 2007. Caveats to quantifying ecosystem services: Fruit abortion blurs benefits from crop pollination. *Ecological Applications* 17: 1841–1849.

Bradbear, N. 2009. *Bees and Their Role in Forest Livelihoods: A Guide to the Services Provided by Bees and the Sustainable Harvesting, Processing and Marketing of Their Products*. Rome: FAO.

Bray, R. S. 1996. *Armies of Pestilence: The Impact of Disease on History*. New York: Barnes and Noble.

Carlton R. G. and C. R. Goldman. 1984. Effects of a massive swarm of ants on ammonium concentrations in a subalpine lake. *Hydrobiologia* 111: 113–117.

Carson, R. 1962. *Silent Spring*. New York: Houghton-Mifflin.

Cerritos, R. and Z. Cano-Santana. 2008. Harvesting grasshoppers *Sphenarium purpurascens* in Mexico for human consumption: A comparison with insecticidal control for managing pest outbreaks. *Crop Protection* 27: 473–480.

Christensen, N. L., Jr., S. V. Gregory, P. R. Hagenstein, T. A. Heberlein, J. C. Hendee, J. T. Olson, J. M. Peek, D. A. Perry, T. D. Schowalter, K. Sullivan, et al. 2000. *Environmental Issues in Pacific Northwest Forest Management*. Washington, DC: National Academy Press.

Claudianos, C., H. Ranson, R. M. Johnson, S. Biswas, M. A. Schuler, M. R. Berenbaum, R. Feyereisen, and J. G. Oakeshott. 2006. A deficit of detoxification enzymes: Pesticide sensitivity and environmental response in the honey bee. *Insect Molecular Biology* 15: 615–636.

Clausen, L. W. 1954. *Insect Fact and Folklore*. New York: MacMillan.

Coleman, D. C., D. A. Crossley, Jr., and P. F. Hendrix. 2004. *Fundamentals of Soil Ecology*, 2nd ed. Amsterdam: Elsevier.

Costanza, R., R. d'Arge, R. de Groot, S. Farger, M. Grasso, B. Hannon, K. Limburg, S. Naeem, R. V. O'Neill, J. Paruelo, et al. 1997. The value of the world's ecosystem services and natural capital. *Nature* 387: 253–260.

Dasgupta, P., S. Levin, and J. Lubchenco. 2000. Economic pathways to ecological sustainability. *BioScience* 50: 339–345.

DeFoliart, G. R. 1999. Insects as food: Why the Western attitude is important. *Annual Review of Entomology* 44: 21–50.

Delfosse, E. S. 2005. Risk and ethics in biological control. *Biological Control* 35: 319–329.

Diamond, J. 1999. *Guns, Germs, and Steel: The Fates of Human Societies*. New York: W. W. Norton.

Donald, P. F. and A. D. Evans. 2006. Habitat connectivity and matrix restoration: The wider implications of agri-environmental schemes. *Journal of Applied Ecology* 43: 209–218.

Downing, A. L. and M. A. Leibold. 2002. Ecosystem consequences of species richness and composition in pond food webs. *Nature* 416: 837–841.

Downing, K. B. and W. R. Williams. 1978. Douglas-fir tussock moth: Did it affect private recreational businesses in northeastern Oregon? *Journal of Forestry* 76: 29–30.

Duffy, J. E. 2009. Why biodiversity is important to the functioning of real-world ecosystems. *Frontiers in Ecology and the Environment* 7: 437–444.

Dyer, L. A. and D. K. Letourneau. 1999a. Relative strengths of top-down and bottom-up forces in a tropical forest community. *Oecologia* 119: 265–274.

Dyer, L. A. and D. K. Letourneau. 1999b. Trophic cascades in a complex terrestrial community. *Proceedings of the National Academy of Sciences USA* 96: 5072–5076.

Ehrlich, P. and A. Ehrlich. 1981. *Extinction: The Causes and Consequences of the Disappearance of Species*. New York: Random House.

Ewel, J. J. 1986. Designing agricultural ecosystems for the humid tropics. *Annual Review of Ecology and Systematics* 17: 245–271.

Ewel, J. J., M. J. Mazzarino, and C. W. Berish. 1991. Tropical soil fertility changes under monocultures and successional communities of different structure. *Ecological Applications* 1: 289–302.

Ferrar, P. 1975. Disintegration of dung pads in north Queensland before the introduction of exotic dung beetles. *Australian Journal of Experimental Agriculture and Animal Husbandry* 15: 325–329.

Foley, J. A., M. H. Costa, C. Delire, N. Ramankutty, and P. Snyder. 2003. Green surprise? How terrestrial ecosystems could affect earth's climate. *Frontiers in Ecology and the Environment* 1: 38–44.

Garrett, K. A. and C. C. Mundt. 1999. Epidemiology in mixed host populations. *Phytopathology* 89: 984–990.

Gatto, P., A. Zocca, A. Battisti, M. J. Barrento, M. Branco, and M. R. Paiva. 2009. Economic assessment of managing processionary moth in pine forests: A case study in Portugal. *Journal of Environmental Management* 90: 683–691.

Gutschick, V. P. 1999. Biotic and abiotic consequences of differences in leaf structure. *New Phytologist* 143: 4–18.

Hättenschwiler, S. and P. Gasser. 2005. Soil animals alter plant litter diversity effects on decomposition. *Proceedings of the National Academy of Sciences USA* 102: 1519–1524.

Helson, J. E., T. L. Capson, T. Johns, A. Aiello, and D. M. Windsor. 2009. Ecological and evolutionary bioprospecting: Using aposematic insects as guides to rainforest plants active against disease. *Frontiers in Ecology and the Environment* 7: 130–134.

Hoehn, P., T. Tscharntke, J. M. Tylianakis, and I. Steffan-Dewenter. 2008. Functional group diversity of bee pollinators increases crop yield. *Proceedings of the Royal Society B* 275: 2283–2291.

Hokkanen, H. M. T., J. C. van Lenteren, and I. Menzler-Hokkanen. 2007. Ecological risks of biological control agents: Impacts on IPM. In *Perspectives in Ecological Theory and Integrated Pest Management*, M. Kogan and P. Jepson, eds. 246–268, Cambridge, UK: Cambridge University Press.

Hooper, D. U., F. S. Chapin, III, J. J. Ewel, A. Hector, P. Inchausti, S. Lavorel, J. H. Lawton, D. M. Lodge, M. Loreau, S. Naeem, et al. 2005. Effects of biodiversity on ecosystem functioning: A concensus of current knowledge. *Ecological Monographs* 75: 3–35.

Hsiang, S. M., K. C. Meng, and M. A. Cane. 2011. Civil conflicts are associated with the global climate. *Nature* 476: 438–441.

Huntly, P. M., S. van Noort, and M. Hamer. 2005. Giving increased value to invertebrates through ecotourism. *South African Journal of Wildlife Research* 35: 53–62.

Isaacs, R., J. Tuell, A. Fiedler, M. Gardiner, and D. Landis. 2009. Maximizing arthropod-mediated ecosystem services in agricultural landscapes: The role of native plants. *Frontiers in Ecology and the Environment* 7: 196–203.

Johnson, D. M., A. M. Liebhold, P. C. Tobin, and O. N. Bjørnstad. 2006. Allee effects and pulsed invasion by the gypsy moth. *Nature* 444: 361–363.

Jørgensen, S. E. 2010. Ecosystem services, sustainability, and thermodynamic indicators. *Ecological Complexity* 7: 311–313.

Kawaguchi Y. and S. Nakano. 2001. Contribution of terrestrial invertebrates to the annual resource budget for salmonids in forest and grassland reaches of a headwater stream. *Freshwater Biology* 46: 303–316.

Klein, A.-M, B. E. Vaissière, J. H. Cane, I. Steffan-Dewenter, S. A. Cunningham, C. Kremen, and T. Tscharntke. 2007. Importance of pollinators in changing landscapes for world crops. *Proceedings of the Royal Society B* 274: 303–313.

Knapp, A. K. and T. R. Seastedt. 1986. Detritus accumulation limits productivity of tallgrass prairie. *BioScience* 36: 662–668.

Knight, T. M., M. W. McCoy, J. M. Chase, K. A. McCoy, and R. D. Holt. 2005. Trophic cascades across ecosystems. *Nature* 437: 880–883.

Koenig, W. D. and A. M. Liebhold. 2005. Effects of periodical cicada emergences on abundances and synchrony of avian populations. *Ecology* 86: 1873–1882.

Kremen, C., N. M. Williams, R. L. Bugg, J. P. Fay, and R. W. Thorp. 2004. The area requirements of an ecosystem service: Crop pollination by native bee communities in California. *Ecology Letters* 7: 1109–1119.

Kremen, C., N. M. Williams, and R. W. Thorp. 2002. Crop pollination from native bees at risk from agricultural intensification. *Proceedings of the National Academy of Sciences USA* 99: 16812–16816.

Landis, D. A., M. M. Gardiner, W. van der Werf, and S. M. Swinton. 2008. Increasing corn for biofuel production reduces biocontrol services in agricultural landscapes. *Proceedings of the National Academy of Sciences USA* 105: 20552–20557.

Leopold, A. 1949. *Sand County Almanac.* Oxford: Oxford University Press.

Letourneau, D. K. and L. A. Dyer. 1998. Density patterns of *Piper* ant-plants and associated arthropods: Top-predator trophic cascades in a terrestrial system? *Biotropica* 30: 162–169.

Letourneau, D. K., J. A. Jedlicka, S. G. Bothwell, and C. R. Moreno. 2009. Effects of natural enemy biodiversity on the suppression of arthropod herbivores in terrestrial ecosystems. *Annual Review of Ecology, Evolution and Systematics* 40: 573–592.

Losey, J. E. and M. Vaughn. 2006. The economic value of ecological services provided by insects. *BioScience* 56: 311–323.

Louda, S. M., R. W. Pemberton, M. T. Johnson, and P. A. Follett. 2003. Non-target effects—the Achilles' heel of biocontrol? Retrospective analyses to assess risk associated with biocontrol introductions. *Annual Review of Entomology* 48: 365–396.

Lovett, G. M., L. M. Christenson, P. M. Groffman, C. G. Jones, J. E. Hart, and M. J. Mitchell. 2002. Insect defoliation and nitrogen cycling in forests. *BioScience* 52: 335–341.

Marquis, R. J. and C. J. Whelan. 1994. Insectivorous birds increase growth of white oak through consumption of leaf-chewing insects. *Ecology* 75: 2007–2014.

Mattson, W. J. and R. A. Haack. 1987. The role of drought in outbreaks of plant-eating insects. *BioScience* 37: 110–118.

Mbata, K. J., E. N. Chidumayo, and C. M. Lwatula. 2002. Traditional regulation of edible caterpillar exploitation in the Kopa area of Mpika district in northern Zambia. *Journal of Insect Conservation* 6: 115–130.

McCoy, E. D. and J. H. Frank. 2010. How should the risk associated with the introduction of biological control agents be estimated? *Agricultural and Forest Entomology* 12: 1–8.

McNaughton, S. J. 1979. Grazing as an optimization process: Grass-ungulate relationships in the Serengeti. *American Naturalist* 113: 691–703.

Mehner T., J. Ihlau, H. Dörner, M. Hupfer, and F. Hölker. 2005. Can feeding of fish on terrestrial insects subsidize the nutrient pool of lakes? *Limnology and Oceanography* 50: 2022–2031.

Menninger, H. L., M. A. Palmer, L. S. Craig, and D. C. Richardson. 2008. Periodical cicada detritus impacts stream ecosystem metabolism. *Ecosystems* 11: 1306–1317.

Michalson, E. L. 1975. Economic impact of mountain pine beetle on outdoor recreation. *Southern Journal of Agricultural Economics* 7(2): 43–50.

Mooney, K. A. 2007. Tritrophic effects of birds and ants on a canopy food web, tree growth, and phytochemistry. *Ecology* 88: 2005–2014.

Müller, M. and H. Job. 2009. Managing natural disturbance in protected areas: Tourists' attitude towards the bark beetle in a German national park. *Biological Conservation* 142: 375–383.

North, M., J. Innes, and H. Zald. 2007. Comparison of thinning and prescribed fire restoration treatments to Sierran mixed-conifer historic conditions. *Canadian Journal of Forest Research* 37: 331–342.

Nowlin, W. H., M. J. González, M. J. Vanni, M. H. H. Stevens, M. W. Fields, and J. J. Valenti. 2007. Allochthonous subsidy of periodical cicadas affects the dynamics and stability of pond communities. *Ecology* 88: 2174–2186.

Parker, J. 2003. World honey prices bolstered by smaller 2002 production, U.S. antidumping tariffs, and fears about contaminated Chinese honey. *American Bee Journal* 143: 523–525.

Pedigo, L. P., S. H. Hutchins, and L. G. Higley. 1986. Economic injury levels in theory and practice. *Annual Review of Entomology* 31: 341–368.

Pemberton, R. W. and H. A. Cordo. 2001. Potential and risks of biological control of *Cactobastis cactorum* (Lepidoptera: Pyralidae) in North America. *Florida Entomologist* 84: 513–526.

Perlman, F., E. Press, J. A. Googins, A. Malley, and H. Poarea. 1976. Tussockosis: Reactions to Douglas fir tussock moth. *Annals of Allergy* 36: 302–307.

Perry, D. A., R. Oren, and S. C. Hart. 2008. *Forest Ecosystems*, 2nd ed. Baltimore, MD: Johns Hopkins University Press.

Pray, C. L, W. H. Nowlin, and M. J. Vanni. 2009. Deposition and decomposition of periodical cicadas (Homoptera: Cicadidae: Magicicada) in woodland aquatic ecosystems. *Journal of the North American Benthological Society* 28: 181–195.

Ramos-Elorduy, J. 2009. Anthro-entomophagy: Cultures, evolution, and sustainability. *Entomological Research* 39: 271–288.

Reay-Jones, F. P. F., M. O. Way, M. Sétamou, B. L. Legendre, and T. E. Reagan. 2003. Resistance to the Mexican rice borer (Lepidoptera: Crambidae) among Louisiana and Texas sugarcane cultivars. *Journal of Economic Entomology* 96: 1929–1934.

Riley, C. V. 1878. *First Annual Report of the United States Entomological Commission for the Year 1877 Relating to the Rocky Mountain Locust and the Best Methods of Preventing Its Injuries and of Guarding Against Its Invasions, in Pursuance of an Appropriation Made by Congress for This Purpose*. Washington, DC: U.S. Department of Agriculture.

Risch, S. 1980. The population dynamics of several herbivorous beetles in a tropical agroecosystem: The effect of intercropping corn, beans, and squash in Costa Rica. *Journal of Applied Ecology* 17: 593–612.

Risch, S. J. 1981. Insect herbivore abundance in tropical monocultures and polycultures: An experimental test of two hypotheses. *Ecology* 62: 1325–1340.

Ritchie, M. E., D. Tilman, and J. M. H. Knops. 1998. Herbivore effects on plant and nitrogen dynamics in oak savanna. *Ecology* 79: 165–177.

Romme, W. H., D. H. Knight, and J. B. Yavitt. 1986. Mountain pine beetle outbreaks in the Rocky Mountains: Regulators of primary productivity? *American Naturalist* 127: 484–494.

Schowalter, T. D. 2008. Insect herbivore responses to management practices in conifer forests in North America. *Journal of Sustainable Forestry* 26: 204–222.

Schowalter, T. D. 2011. *Insect Ecology: An Ecosystem Approach*, 3rd ed. San Diego: Elsevier/Academic.

Seastedt, T. R. 1984. The role of microarthropods in decomposition and mineralization processes. *Annual Review of Entomology* 29: 25–46.

Setälä, H. and V. Huhta. 1991. Soil fauna increase *Betula pendula* growth: Laboratory experiments with coniferous forest floor. *Ecology* 72: 665–671.

Sheppard, S. and P. Picard. 2006. Visual-quality impact of forest pest activity at the landscape level: A synthesis of published knowledge and research needs. *Landscape and Urban Planning* 77: 321–342.

Shewchuk, B. A. and W. A. Kerr. 1993. Returns to grasshopper control on rangelands in southern Alberta. *Journal of Range Management* 46: 458–462.

Simonsen, T. J., R. L. Brown, and F. A. H. Sperling. 2008. Tracing an invasion: Phylogeography of *Cactoblastis cactorum* (Lepidoptera: Pyralidae) in the United States based on mitochondrial DNA. *Annals of the Entomological Society of America* 101: 899–905.

Smith, K. F., D. F. Sax, S. D. Gaines, V. Guernier, and J.-F. Guégan. 2007. Globalization of human infectious disease. *Ecology* 88: 1903–1910.

Smith, S., T. E. Reagan, J. L. Flynn, and G. H. Willis. 1983. Azinphosmethyl and fenvalerate runoff loss from a sugarcane-insect IPM system. *Journal of Environmental Quality* 12: 534–537.

Snow, R. W., C. A. Guerra, J. J. Mutheu, and S. I. Hay. 2008. International funding for malaria control in relation to populations at risk of stable *Plasmodium falciparum* transmission. *PLoS Medicine* 5(7): e142.

Stern, V. M., R. F. Smith, R. van den Bosch, and K. S. Hagen. 1959. The integration of chemical and biological control of the spotted alfalfa aphid. Part 1. The integrated control concept. *Hilgardia* 29: 81–101.

Symondson, W. O. C., K. D. Sunderland, and M. H. Greenstone. 2002. Can generalist predators be effective biocontrol agents? *Annual Review of Entomology* 47: 561–594.

Tanner, C. M., F. Kamel, G. W. Ross, J. A. Hoppin, S. M. Goldman, M. Korell, C. Marras, G. S. Bhudhikanok, M. Kasten, A. R. Chade, et al. 2011. Rotenone, paraquat, and Parkinson's disease. *Environmental Health Perspectives* 119: 866–872.

Tilman, D. and J. A. Downing. 1994. Biodiversity and stability in grasslands. *Nature* 367: 363–365

Torrell, L. A., J. H. Davis, E. W. Huddleston, and D. C. Thompson. 1989. Economic injury levels for interseasonal control of rangeland insects. *Journal of Economic Entomology* 82: 1289–1294.

Trumble, J. T., D. M. Kolodny-Hirsch, and I. P. Ting. 1993. Plant compensation for arthropod herbivory. *Annual Review of Entomology* 38: 93–119.

Tyndale-Biscoe, M. 1994. Dung burial by native and introduced dung beetles (Scarabaeidae). *Australian Journal of Agricultural Research* 45: 1799–1808.

Tyndale-Biscoe, M. and W. G. Vogt. 1996. Population status of the bush fly, *Musca vetustissima* (Diptera: Muscidae), and native dung beetles (Coleoptera: Scarabaeinae) in south-eastern Australia in relation to establishment of exotic dung beetles. *Bulletin of Entomological Research* 86: 183–192.

Van Bael, S. A., A. Aiello, A. Valderrama, E. Medianero, M. Samaniego, and S. J. Wright. 2004. General herbivore outbreak following an El Niño-related drought in a lowland Panamanian forest. *Journal of Tropical Ecology* 20: 625–633.

van Lenteren, J. C., J. Bale, F. Bigler, H. M. T. Hokkanen, and A. J. M. Loomans. 2006. Assessing risks of releasing exotic biological control agents of arthropod pests. *Annual Review of Entomology* 51: 609–634.

Vossbrinck, C. R., D. C. Coleman, and T. A. Woolley. 1979. Abiotic and biotic factors in litter decomposition in a semiarid grassland. *Ecology* 60: 265–271.

Wickman, B. E. 1980. Increased growth of white fir after a Douglas-fir tussock moth outbreak. *Journal of Forestry* 78: 31–33.

Williamson, S. C., J. K. Detling, J. L. Dodd, and M. I. Dyer. 1989. Experimental evaluation of the grazing optimization hypothesis. *Journal of Range Management* 42: 149–152.

Yen, A. L. 2009. Entomophagy and insect conservation: Some thoughts for digestion. *Journal of Insect Conservation* 13: 667–670.

Zenk, M. H. and M. Juenger. 2007. Evolution and current status of the phytochemistry of nitrogenous compounds. *Phytochemistry* 68: 2757–2772.

Zhang, D. D., P. Brecke, H. F. Lee, Y. Q. He, and J. Zhang. 2007. Global climate change, war and population decline in recent human history. *Proceedings of the National Academy of Sciences USA* 104: 19214–19219.

Zhong, H. and T. D. Schowalter. 1989. Conifer bole utilization by wood-boring beetles in western Oregon. *Canadian Journal of Forest Research* 19: 943–947.

Zhou, H.-B., J.-L. Chen, D.-F. Cheng, Y. Liu, and J.-R. Sun. 2009. Effects of wheat-pea intercropping on the population dynamics of *Sitobion avenae* (Homoptera: Aphididae) and its main natural enemies. *Acta Entomologica Sinica* 52: 775–782.

Zimmerman, K. M., J. A. Lockwood, and A. V. Latchininsky. 2004. A spatial, Markovian model of rangeland grasshopper (Orthoptera: Acrididae) population dynamics: Do long-term benefits justify suppression of infestations? *Environmental Entomology* 33: 257–266.

9

Conclusions and Recommendations

> Conceived as a mere exercise in technology, pest control amounts to hardly more than bulldozing nature without thought to consequences, and frequently creates more problems than it solves.

Geier (1966)

Sustaining ecosystem services is not optional; it is absolutely necessary to sustain human life on this planet. As described in Chapters 5 and 6, the collapse of civilizations in Mesopotamia and Mesoamerica can be linked to anthropogenic deforestation that exacerbated the severity of drought effects on ecosystem services (Xue et al. 1990; Zheng and Eltahir 1998; Janssen et al. 2008; Lentz and Hockaday 2009; Briant et al. 2010; Cook et al. 2012). Human history is filled with examples of social unrest, population displacement, and war resulting from food or water shortages that often are caused by human alteration of environmental conditions, as well as by insects or other natural causes (Riley 1878; Bray 1996; Diamond 1999; Acuña-Soto et al. 2002; Smith 2007; Zhang et al. 2007; Perry et al. 2008; Bora et al. 2010; Hsiang et al. 2011; see Chapter 6). In the wake of social unrest or displacement come epidemics of crowd diseases, often vectored by insects (Bray 1996; Diamond 1999; Acuña-Soto et al. 2002; Therrell et al. 2004; Brouqui 2011).

The concept that earth's resources are not unlimited and could be depleted is not new. Cooper (1823) foresaw a time when the wasteful cutting of frontier forests and slaughter of passenger pigeons would limit these resources. Unfortunately, wasteful use and unsuccessful attempts to engineer solutions to declining ecosystem services continue, and policies are driven more by short-term profit than by sustainability. The Millennium Ecosystem Assessment (2005) provided data indicating that 60% of our global ecosystem services are either being degraded or being used unsustainably. An estimated half of the historic forest cover of the globe has been removed (Myers 1999; Perry et al. 2008), with consequences for global climate (Foley et al. 2003a, b; Juang et al. 2007; Janssen et al. 2008). Climate change represents a pervasive challenge to ecosystem services. Many species will be forced to disperse to new areas in response to rising temperatures and altered precipitation patterns and some, along with their associated ecosystem services, may fail to survive. Yields of our major crops decline precipitously at temperatures above 30°C (86°F), as they have in recent years (Fedoroff et al. 2010), with predicted reductions of 10–20% for corn, soybeans, and cotton between 2020 and 2049 and

20–80% for these crops between 2070 and 2099 (Schlenker and Roberts 2009). Obviously, this does not bode well for supporting an increasing human population.

As outlined in preceding chapters, insects are highly adaptive to changes in their environment, including the introduction of new toxins, making them difficult to control. Current agricultural practices, urban crowding, and encroachment into previously uninhabited ecosystems provide concentrations of resources that can sustain elevated populations of insect species that become pests. Disruption of food webs, especially management practices (such as habitat fragmentation and insecticides) that reduce predator abundance or predation rate, allows target and nontarget pest populations to increase quickly in the absence of regulation by predators. Furthermore, disruption of food webs alters ecosystem processes, including primary production, biogeochemical cycling, decomposition, and soil processes, in ways that threaten the sustainability of the ecosystem services on which we depend. Efforts to control insects with insecticides have largely failed to provide long-term solutions because target insects quickly become resistant. Clearly, the following changes in the way we manage insects and ecosystems are necessary to ensure a continued supply of food, water, and other ecosystem services.

9.1 Why New Pest Management Approaches Are Needed

What are reasonable expectations for insect control, and under what circumstances is control warranted? Is control feasible and at what cost? How should our perspective and approach to insect control be changed to manage our ecosystems for sustained services and human health?

The preceding chapters make clear that any single factor that causes high mortality will favor rapid adaptation for resistance. Individuals that survive predation or insecticides reproduce and pass on their adaptive genetic traits to their offspring. Over successive generations, surviving individuals increase the frequency of adaptive genes in the population, making the population less vulnerable. Predators must adapt by becoming faster, stealthier, or more powerful. Similarly, we are forced to expend ever increasing effort and cost to kill increasingly resistant insects while simultaneously trying to prevent the undesirable effects of these efforts on ecosystem services, such as disruption of natural biological control or fisheries (Scholz et al. 2012). Alternatively, we must invest ever increasing effort and cost on continued discovery of new tactics or compounds with new modes of action to which insects have not been exposed. The effort and expense of discovering and registering new chemicals every five to ten years make long-term control with insecticides an uncertain goal.

Target insects can be killed by insecticides only if they are exposed to a lethal dose. This dose can be derived from laboratory experiments but must be validated by field studies that evaluate the effectiveness of application rate, timing, and techniques to ensure adequate exposure of insects protected by vegetation. High pressure application of a fine mist is more effective than a lower pressure application of larger drops that are less likely to penetrate dense foliage. Aerial application is effective for insects exposed on plant surfaces in short vegetation, but lateral application is more effective for insects that are more likely to occur in taller or more complex vegetation. Systemic insecticides that are absorbed and moved into various plant tissues are necessary for insects that feed internally in plants. Some insects are virtually impossible to control using insecticides. Bark beetles, for example, are largely protected under bark. The entire bark surface must be soaked with insecticide to penetrate it, but this method is expensive and used only for individual high value trees around homes or in parks (Schowalter 2012).

Given these challenges to ensuring insect exposure to lethal concentrations, many insects survive and pass genes for resistance to their offspring. Resistance to insecticides among target species has become such a serious problem that entomological research is increasingly focused on tactics to delay genetic selection for this trait. In some cases, resistance can reduce the efficacy of insecticides from 99% mortality to 60% (i.e., the number of insects surviving to cause injury increases 40-fold) within five years (Felland et al. 1990). This brings into question the wisdom of expecting to control insects with insecticides over the long term. Crop producers, land managers, and policymakers should avoid the temptation to take the quickest, short-term approach to pest management and make decisions that will protect the sustainability of ecosystem services.

More effective management of insects, consistent with sustaining ecosystem services, can be achieved by adhering to the principles of integrated pest management (IPM). Integrated pest management addresses pest management within the context of complex ecosystems and sustainability of ecosystem services. This concept is not new. The roots of IPM lie in the application of ecological principles (e.g., host abundance and predation) to control crop pests in the late 1800s (Riley 1878; Howard 1896). These principles were largely supplanted during the 1940s when powerful and inexpensive synthetic pesticides became available, and attention shifted from noninsecticidal methods of control to developing and testing chemical insecticides (Kogan 1998). IPM as an explicit strategy gained traction during the 1950s as a result of observed resistance to DDT among target insects and detrimental nontarget effects (Clausen 1954; Roussel and Clower 1957; Stern et al. 1959; Carson 1962; Kogan 1998). During the next two decades, entomologists developed this scientific approach to pest management, supported by federal and Food and Agriculture Organization of the United Nations (UNFAO) policies that promoted application of IPM principles to reduce insecticide use (Kogan 1998). The following principles guide IPM strategy.

9.2 Management Goals

In deciding how to manage insects, it is important to recognize that *sustainability of ecosystem services and human health, not killing insects, must be the focus of management efforts.* Sustainability means a relatively consistent rate of production or supply of ecosystem products or services on which we depend. Efforts to maximize the rate of resource production or harvest that cannot be sustained under prevailing environmental conditions is futile and ultimately self-destructive, because the collapse of ecosystems leads to shortages in food and water supply that can drive (in fact have driven) social unrest, disease epidemics, and the collapse of society (Zhang et al. 2007; Hsiang et al. 2011). Our goal as a society must be protection of ecosystem conditions that are fundamental to a sustainable supply of ecosystem services. Our approach and efforts to manage insects must be consistent with this goal.

In this regard, *it is important to distinguish the different effects of native and exotic insects.* Populations of native insects are kept at low numbers by natural regulatory factors that often are compromised in managed ecosystems. Concentration of resources, such as in agricultural crops or plantation forests, removes a major regulatory factor from population growth of insect species adapted to searching for that host species (see Chapter 4). Crop or forestry breeding programs that favor plant growth over production of natural defenses remove a second major regulator factor (Chapter 4). Finally, agricultural or forestry practices that reduce the abundance of predators and parasites (such as insecticide application and removal of natural habitats that serve as reservoirs for predator and parasite populations) remove the third major regulatory factor (Chapter 4). By contrast, invasive species generally have escaped regulatory factors that controlled their populations in their region of origin. Although many exotic species do not find favorable conditions in new habitats, those species that are not controlled effectively by host defenses or predators and parasites in their new habitat are free to reproduce and spread unchecked, becoming invasive and potentially destructive.

Furthermore, many, if not most, native insect species contribute to essential ecosystem services in their native habitats (see Chapters 7 and 8). Insect pollinators are the best-known supporters of ecosystem services by ensuring pollination and the production of fruit and vegetable crops (see Chapter 7). However, the diversity of non-honey bee pollinators also needs to be protected in order to ensure adequate reproduction of many plant species, including some crop species and species of conservation concern. Restoration of critical ecosystems and associated plant species depends on attention to adequate population sizes of associated pollinators (Archer and Pyke 1991; Norman et al. 1992; Corbet 1997; Knight et al. 2005; Ricketts et al. 2008).

Insect species contributing to decomposition, including cockroaches and termites, are generally maligned as household pests but are critical to the turnover of organic detritus and the return of essential nutrients to the soil pool for plant uptake. The Australian experience with dung accumulation and nuisance flies in the absence of dung beetles to clear pastures of livestock dung demonstrates the value for detritivores to ecosystem services.

9.3 Deciding When to Control

Second, *we need to refine our perspective of insects as pests in deciding when to control.* Most insects have little, if any, direct interaction with humans, but they fill important roles as herbivores, pollinators, and decomposers in natural and managed ecosystems (see Chapters 6–8). Even when insects cause visible damage to agricultural crops or forest products, control efforts do not necessarily improve yields. Application of insecticides is expensive in terms of labor and materials, as well as nontarget consequences (Scholz et al. 2012), and these costs can be saved if insecticide use does not improve human health or ecosystem services. Control efforts should be restricted to species that are serious threats to human or animal health or food production. Native insect species on native hosts should be viewed as natural regulatory mechanisms that maintain plant production near carrying capacity in the same way that predators are necessary to maintain healthy prey populations near their carrying capacity. Interfering with such regulatory mechanisms generally has not improved ecosystem services in the long term.

Insecticides should be used as little as possible and, when necessary, the most effective insecticide should be used as efficiently as possible. Although insecticides have a role in reducing pest numbers quickly, we should avoid using insecticides as a quick and convenient solution. Insecticides are most justified for reducing abundances of vectors of human diseases or insects that can devastate crops and cause famine. Even when insecticides must be used, a number of options are available for reducing the amount used and thereby minimizing or delaying development of resistance.

Identification of action thresholds can be used to delay insecticide application until use is warranted to prevent unacceptable losses (Pedigo et al. 1986). For example, soybean aphids (*Aphis glycines*) cause little economic injury to soybean yields below 675 aphids per plant (Ragsdale et al. 2011). However, control should be initiated at 250 aphids per plant to prevent growing populations from reaching the economic injury level. Controlling aphids at lower densities is unnecessary and undermines natural biological control, as well as being costly (Ragsdale et al. 2011). Tables 9.1 and 9.2 provide examples of ways in which various factors can be evaluated to calculate economic thresholds.

Identification of the most critical stages of plant growth or site conditions can be used to target application more effectively and efficiently, thereby avoiding broadcast or preventive applications. For example, in some cases, only damage to young plants translates into reduced yield (Clark and Clark 1985); older plants

TABLE 9.1

Benefit/Cost Analysis of Processionary Moth (*Thaumetopoea pityocampa*) Effects on Ecosystem Services in Portuguese Pine Forests

Total Benefits (+) and Costs (−)	Scenario		Net Benefit (With—Without)
	Management	No Management	
Financial analysis			
Timber revenue (+)	27,562	22,785	+4,777
Forest management cost (−)	22,266	22,266	0
Pest management cost (−)	6,044	0	−6,044
Cumulative NPV (2%)	−748	519	−1,267
Conventional economic analysis			
Estate value (+)	33,079	28,049	+5,030
Cumulative NPV (2%)	32,331	28,568	+3,763
Extended economic analysis			
Carbon sequestration (+)	9,878	7,902	+1,976
Risk of dermatitis (−)	0	4,058	−4,058
Cumulative NPV (2%)	42,209	32,412	+9,797
Extended economic analysis			
Recreation (+)	11,160	8,928	+2,232
Cumulative NPV (2%)	53,369	41,340	+12,029

Source: From Gatto, P., A. Zocca, A. Battisti, M. J. Barrento, M. Branco, and M. R. Paiva, *Journal of Environmental Management*, 90, 683–691, 2009. With permission.
Net present value (NPV) in thousand euros.

are less susceptible to damage (Ragsdale et al. 2011). In other cases, young plants have more time to compensate for insect feeding; older plants have less time for compensation (Thompson and Gardner 1996). Clearly, the particular insect and crop must be evaluated for likelihood of yield reduction. Incorporation of Bt toxin genes in crop plants targets the insects that feed on the plant and reduces the need for broadcast application of insecticides (Carriére et al. 2001; Cattaneo et al. 2006).

In some cases, *insect behavior permits treatment only of select habitats or resources for effective control.* Baits treated with toxins or growth regulators are effective for treatment of ants and termites around living quarters, gardens, or livestock pens. The location of colonies need not be known; foraging insects will carry the bait back to the colony and spread the toxin or growth regulator throughout the colony during food sharing (trophallaxis). Similarly, pheromones can be used to repel bark beetles from high value trees, disrupt mating, or attract sufficient numbers to baited traps to reduce the population sufficiently to prevent damage (Schowalter 2012).

Protection of individual human or livestock targets from biting flies is sufficient to prevent injury or disease transmission; broadcast application of toxins is unnecessary. Toxin-impregnated bed nets or clothing are effective in preventing

TABLE 9.2

Economic Injury Levels for Grasshopper Control Programs

Model Assumption	Economic Threshold (Number of Grasshoppers m^{-2})	
	Fourth Instar	Adult
Average conditions[1]	18	10
Treatment cost (US$ ha^{-1})		
3.09	<4	<3
4.94	11	7
6.18	18	10
7.41	28	14
8.65	36	17
Forage value (US$ AUM^{-1})		
4	68	26
6	33	16
8	18	10
10	10	6
12	5	4
14	<4	<3
Discount rate (%)		
4	17	8
7	18	10
10	19	10
13	20	11
16	22	12
Population growth rate for untreated grasshoppers		
Decrease by 0.5 yr^{-1} ($r_1 = -0.695$)	49	21
Remain constant ($r_1 = 0$)	18	10
Double each year ($r_1 = 0.695$)	<4	<3
Population growth rate for treated grasshoppers		
Remain constant ($r_2 = 0$)	7	5
Double each year ($r_2 = 0.695$)	18	10
Triple each year ($r_2 = 1.10$)	32	15
Treatment efficacy rate (%)		
99	5	4
95	8	6
90	18	10
85	29	14
80	42	19
75	65	25

continued

TABLE 9.2 (CONTINUED)

Economic Injury Levels for Grasshopper Control Programs

	Economic Threshold (Number of Grasshoppers m^{-2})	
Model Assumption	Fourth Instar	Adult
Grasshopper life stage		
Before fourth instar	18	10
Before fifth instar	20	11
Before adult	24	13

Source: Torell, L. A., J. H. Davis, E. W. Huddleston, and D. C. Thompson, *Journal of Economic Entomology*, 82, 1289–1294, 1989. With permission.

Based on varying levels of treatment cost, forage value, discount rate, grasshopper population growth rate, and treatment efficacy.

[1] Average conditions: treatment cost = US$6.18 ha^{-1}; forage value = US$8 animal unit month (AUM)$^{-1}$; discount rate = 7%; untreated grasshopper density remains constant; treated grasshopper density doubles each year; treatment efficacy = 90%, and treatment occurs at the beginning of the fourth instar. These conditions were altered to compute the economic injury level for alternative model assumptions.

mosquito bites (Goodman and Mills 1999) and employ less insecticide than aerial spraying. Because many biting flies do not fly far from breeding sites in ponds or rodent dens, protection of human habitations can be accomplished by treating only a buffer area representing the flight distance from human habitations. Mascari and Foil (2009) and Mascari et al. (2007) pioneered a feed-through method to control sand fly vectors of leishmaniasis within such buffer zones around villages or towns. Larvae of the flies feed on rodent feces in burrows, whereas adults feed on the blood of rodents and humans, transmitting the leishmaniansis parasite in the process. In lieu of broadcast application of insecticides or rodenticides that disrupt the food chains of surrounding ecosystems, rodent bait treated with insect growth regulators (IGRs) is applied only in the buffer area around human communities and agricultural fields. The bait is nontoxic to rodents, but the insecticide passes through the rodent gut into feces at a sufficient concentration to kill fly larvae and is assimilated by the rodents at sufficient concentrations to kill blood-feeding adult flies.

Incipient outbreaks and invasive species should be targeted earlier, before their control becomes much more expensive and uncertain. Early detection of potential threats is a necessary precursor to appropriate action. Detection may be difficult when access to potential populations is restricted by isolation in rugged terrain, by natural disasters that destroy or block roads and alter other map features, or by warfare or poor law enforcement. For example, desert locust population refuges can be scattered across the entire 13 million km^2 area (10% of the earth's land area) of the Sahara and Arabian deserts where resources for detection are extremely limited. In some cases, remote sensing can reveal areas of green vegetation, plant stress, or other suitable conditions for insect population growth in isolated areas, thereby improving our ability to target incipient outbreaks (Carter and Knapp 2001; Nansen et al. 2009, 2010). In some cases, minimal effort may be sufficient to prevent population growth to sizes that cause problems. For example, biopesticides or barrier treatments that reduce crowding among solitary locusts can prevent phase shift to the gregarious, plague phase and thereby prevent outbreaks at lower cost than can the suppression of widespread plagues (Showler 2002).

9.4 Greater Emphasis on Ecological Principles

Ecological principles should be given greater weight in policies regulating pest management and ecosystem services. As noted earlier (see also Chapters 4 and 5), ecosystem services result from the complex interactions of species in food webs that create ecosystem structure and processes. Each species has a role that we may or may not understand but should not undervalue. Ecologists have identified general factors that promote or limit population growth and supply of ecosystem services. These include the quality and quantity of food and other resources and the variety of predators affecting various life stages. A diversity of native species that contributes to key ecosystem processes is fundamental to the sustainability of ecosystem services. These factors must be protected to prevent reduction in ecosystem services. The loss of any native species, no matter how seemingly irrelevant, or the addition of invasive species could cascade through ecosystem processes to affect sustainability of ecosystem services, as described by the butterfly effect (Bradbury 1952; Lorenz 1993). Unfortunately, policies and management decisions are based more often on perceived need or public demand than on scientific evidence (Carpenter et al. 2009). However, *sustainability is impossible if social or economic benefits are maximized at the expense of ecosystem processes that produce the services.*

We need to protect biodiversity as insurance against the loss of critical ecosystem services. Although extinctions occur naturally as part of the evolutionary process, the current rate of anthropogenic extinction exceeds that of the major extinction events of the past and exceeds the rate at which new species develop (Ehrlich and Ehrlich 1981). We do not yet know enough about the mechanisms underlying ecosystem integrity to anticipate which species' disappearance will cascade through ecosystem processes to jeopardize future ecosystem services. In tinkering with complex ecosystems that we do not understand completely, we should work to keep all the components until we know which can be discarded safely.

As noted in Chapter 8, biodiversity can be compared with the mechanical parts of an airplane or clock (Leopold 1949; Ehrlich and Ehrlich 1981). Some parts may be more critical to the function of the mechanism than are others, but as more parts are removed, a point will be reached at which the mechanism no longer works. Consider how many rivets could be lost from a plane before you would refuse to fly in it. Similarly, as species disappear, or are replaced by nonindigenous species, disruption of ecosystem processes may reach a threshold beyond which ecosystem services will no longer be adequate to support human needs (Ehrlich and Mooney 1983), as found by early civilizations in Mesopotamia and Mesoamerica (Xue et al. 1990; Zheng and Eltahir 1998; Janssen et al. 2008; Lentz and Hockaday 2009; Cook et al. 2012).

We do know that much of the diversity in ecosystems represents species that regulate populations and the effects of other species or that replace less tolerant species to maintain ecosystem processes when environmental conditions favor a different set of species (Tilman and Downing 1994; Naeem 1998; Mooney 2010). For example, primary production and the ecosystem services it supports remained higher during drought in more diverse ecosystems than in less diverse ecosystems (Tilman and Downing 1994).

Science can help policymakers and the general public to understand, evaluate, and balance the trade-offs and risks necessary among multiple competing, and often contradictory, interests in ecosystem services (Maser 2005; Abt et al. 2010; Bradford and D'Amato 2012). Despite the global stature of science in the United States, biologists and physical scientists have envied their European and Asian counterparts whose advice and recommendations seem to be appreciated and sought by their governments. The difference is that many governments consider investment in scientific research to be for the public good, whereas science is widely distrusted and ignored by the public in the United States. For example, among developed nations, the United States is unique in its low public acceptance of such fundamental scientific principles as evolution and anthropogenic effects on climate (Miller et al. 2006; Ayala et al. 2008; Doran and Zimmerman 2009). Unfortunately, many reputable scientists also are unwilling to engage in policy development (Gray and Campbell 2008).

Admittedly, scientific models and predictions always reflect state-of-the-art knowledge that can change with new information. In other words, predicted consequences of management activities for complex ecosystems are subject to the butterfly effect (see Chapter 1). Nevertheless, predictions of anthropogenic effects on ecosystem services, as testable hypotheses, have been largely confirmed over several decades, such as in the direction of climate change, and approaching the magnitude (e.g., effects of deforestation, urban growth, and fossil fuel combustion on climate warming and ecosystem processes; see Chapter 6), warranting confidence in predicted trends and concern for the future (Arndt et al. 2010; Soboll et al. 2011).

We need to involve multiple competing interests in management decisions that affect ecosystem services (Christensen et al. 2000; Maser and Pollio 2011). For example, defoliation of forests may be detrimental to timber interests in the short term but beneficial to fish and wildlife interests and neutral or positive for recreational values (Müller and Job 2009). Compensatory timber production and greater consistency of primary production also may benefit timber interests in the long term. We need to manage insects in ways that minimize loss of critical ecosystem services without jeopardizing long-term sustainability of all services. Bringing all stakeholders together and employing socioeconomic-ecological decision models can clarify objectives, illuminate consequences of various management options, and assist in reaching consensus on an appropriate balance among objectives that accomplishes sustainability. Again, although maximizing socioeconomic returns may be tempting, *decisions that provide inadequate protection of ecosystem structure and processes will undermine sustainability of ecosystem services.*

Christensen et al. (2000) and Maser and Pollio (2011) describe examples of stakeholders working together to improve resource and land use plans that balance various interests. Sustainability cannot be achieved as long as competing interests in ecosystem services strive for dominance but rather will require that all stakeholders accept trade-offs that best balance various interests with a sustainable supply of ecosystem services, perhaps with assistance from a facilitator (Christensen et al. 2000; Maser 2005, 2010; Maser and Pollio 2011). In this respect, watershed councils in Oregon have brought together urban, forestry, agricultural, fisheries, wildlife, and environmental interest groups to participate in identifying relevant data, models, and decision-making criteria that express their objectives and preferences for management of ecosystem services (Lamy et al. 2002). This process led to the identification of five main objectives, 28 subobjectives, and 20 restoration options (Figure 9.1).

9.5 Use of a Broader Range of Tactics

Pest management should employ the broadest spectrum of potential tactics to prevent insect populations from causing unacceptable losses to ecosystem services. These tactics include cultural methods, pheromones, biological control, and limited use of insecticides (only as necessary and rotated among different modes of action). Using multiple tactics can effectively reduce pest population levels and prevent or delay genetic adaptation to any particular selection factor (Huffaker and Messenger 1976; Lowrance et al. 1984; Rabb et al. 1984; Kogan 1998; Reay-Jones et al. 2003; Kogan and Jepson 2007; Smith 2007; Gassman et al. 2008, 2009, 2011; Zalucki et al. 2009). *Preventive measures, rather than control, should be the first line of defense against potential pests.* Protection or incorporation of natural regulatory factors, such as host plant density and condition and native predator diversity (see Chapters 3 and 4), help to maintain insect populations below damage thresholds and thereby minimize the costs and uncertainty of control efforts. Remedial (control) tactics should be employed only as necessary when preventive tactics fail.

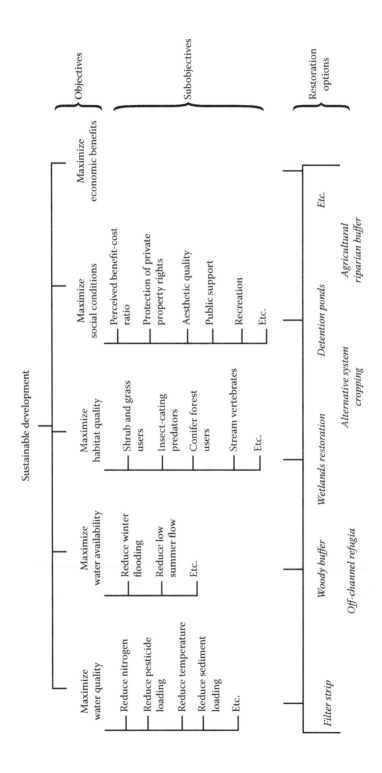

FIGURE 9.1
Hierarchy of objectives identified by multiple stakeholders in watershed management in western Oregon. Note direct attention to issues involving insects (e.g., reduce pesticide loading, insect-eating predators). Insects also affect stream vertebrates (salmon fisheries), conifer forest condition, plant cover in riparian buffers, and selection of alternative cropping systems. (From Lamy, F., J. Bolte, M. Santelmann, and C. Smith, *Journal of the American Water Resources Association*, 38, 517–529, 2002. With permission.)

9.5.1 Preventive Tactics

Preventive tactics include protection or augmentation of natural regulatory factors that ordinarily maintain target populations at small sizes (see Chapters 4 and 5). These tactics include maintaining or augmenting predator and parasite abundances and manipulating habitat or resource suitability and abundance.

9.5.1.1 Protecting or Augmenting Natural Enemies

Predation is a major factor in population regulation. Native predators and parasites should be favored over introduced biocontrol agents, although native predators may not always accept introduced, invasive prey (Adams et al. 2009). Introducing predators or parasites from a pest's region of origin entails numerous risks (Symondson et al. 2002; Louda et al. 2003; Delfosse 2005; van Lenteren et al. 2006; McCoy and Frank 2010; see Chapter 6). Despite refinement of quarantine and testing procedures, it is difficult to anticipate all consequences of species introductions, and any engineered solution may itself require future engineered solutions.

Practices that protect or augment the diversity and effectiveness of native predators and parasites avoid these problems. Agroforestry systems, in particular, provide habitat for insectivorous birds and other vertebrate, as well as invertebrate, predators. Noncrop features of the landscape may be necessary to sustain populations of these regulatory agents over areas larger than crop patches. Protection of refuges in field margins, hedgerows, or patches of native habitat in agricultural landscapes provides resources that sustain predators and parasites during periods when their primary hosts on crop plants may be absent (Landis et al. 2000, 2005; Pfiffner and Wyss 2004; Tscharntke et al. 2005, 2007; Wratten et al. 2007; Isaacs et al. 2009; Stephen and Berisford 2011). Such practices require cooperation among multiple landowners on a landscape scale.

9.5.1.2 Manipulating Habitat or Resource Suitability and Availability

A number of habitat and resource variables can be manipulated to reduce the likelihood of pest population growth. When possible, eliminating or preventing conditions that promote pest population growth can substantially reduce costs of control.

Incidence of insect-vectored diseases can be greatly reduced by proper sanitation and elimination of vector breeding habitats in and around human habitations. Elimination of artificial sources of standing water (such as discarded tires, garden pots or other containers, fountains, blocked gutters, and drainage ponds) will greatly reduce breeding sites for mosquitoes (Bartlett-Healy et al. 2012). Where fountains or ponds are desired or unavoidable, mosquito fish (e.g., *Gambusia* spp.) can be introduced to prevent mosquito breeding.

Properly maintained screens for windows and doors, sealed cracks around windows or doors, and good personal hygiene and sanitation in living quarters can prevent attraction, access, and reproduction by household nuisances such as cockroaches, ants, and termites and disease reservoirs (e.g., rodents) and vectors (e.g., fleas and lice). These practices substantially reduce the need for insecticides in the home.

Unfortunately, the human populations most vulnerable to outbreaks of insect-vectored diseases also are least able to prevent these (i.e., impoverished and refugee communities displaced by famine, natural disaster, or war). Currently, efforts to bring epidemic disease under control have suffered from insufficient funding, regional conflicts, and cultural impediments, such as suspicion of motives by the affected populations. The estimated US$2–3 billion per year to bring malaria under control would be a wise investment in improved health, quality of life, and economic development for much of the undeveloped world (Sachs and Malaney 2002; Snow et al. 2008). Governmental action also can be extremely effective. For example, Singapore imposes stiff fines on residents who violate regulations that require them to empty containers that can collect water and breed mosquitoes (Boo 2001). Such efforts ultimately contribute to global well-being to the extent that reducing the incidence of insect-vectored diseases anywhere reduces the number of infected individuals and the likelihood that disease will spread via immigration or international travel. Thus, investment in disease prevention worldwide protects all of us, as well as contributing to economic development and social stability in the poorest parts of the world.

In agricultural and forestry crops, *appropriate crop selection and diversification, host spacing, crop rotation between host and nonhost species, mixing host and nonhost species (including agro-forestry), and planting the most resistant crop varieties are traditional methods for reducing pest populations* (Riley 1878, 1885; Risch 1980, 1981; Schowalter and Turchin 1993; Abate et al. 2000; Smith 2005; Gliessman 2007; Kogan and Jepson 2007; Barbosa et al. 2009; Nealis et al. 2009). Selection of crops that do not require extensive irrigation can optimize plant condition and natural defenses and balance agricultural and urban needs for water. Mixing different crop species within fields (e.g., intercropping or agroforestry designs) increases diversity of hosts, nonhosts, and predators within crop patches. Diversity at the landscape level also can prevent pest problems. Many potential pests spread easily across landscapes representing continuous availability of (often unrelated) host species (Landis et al. 2000; 2005; Meagher and Nagoshi 2004; Jackson et al. 2008; Schowalter 2011). Some combinations of crop species are more conducive than others to pest problems. Population densities of soybean loopers, *Chrysodeixis includens*, are higher on soybean in landscapes dominated by cotton and soybean patches relative to landscapes with no cotton because soybean looper adults that feed on cotton flowers have higher rates of reproduction (Burleigh 1972). On the other hand, a landscape mosaic of host and nonhost crop patches, different forest age classes, and so on, can interrupt the spread of most pest species (Schowalter

2011). The proximity of native habitats within agricultural landscapes determines availability of refuges for pollinators and predators that maintain pollination and biological control functions (Riley 1878; Landis et al. 2000, 2005; Kremen et al. 2002, 2004; Klein et al. 2007; Wratten et al. 2007). Meehan et al. (2011) found that insecticide use increased with the proportion of landscape devoted to agricultural uses and decreased with the proportion of landscape in remnant natural or seminatural areas. Creating a landscape matrix that will limit spread of pest populations and facilitate movement of pollinators and predators will require cooperation among insect ecologists, land owners, and policymakers (Carriére et al. 2001; Zalucki et al. 2009).

Thinning dense pine forests has been shown to reduce the likelihood of outbreaks for many bark beetle species (Sartwell and Stevens 1975; Schowalter et al. 1981; Amman et al. 1988; Kolb et al. 1998; Schowalter 2012). Schowalter et al. (1981) recommended that average spacing between pine trees of at least 6 m would generally prevent development of bark beetle outbreaks and eliminate the need for control measures. Reay-Jones et al. (2003) estimated that planting resistant varieties of sugarcane could reduce crop losses to the Mexican rice borer, *Eoreuma loftini*, by 24%, and irrigation with 30 cm of water would reduce losses by 29%, substantially reducing the need for pest control. Mixing of different crops can reduce pest populations and crop losses substantially (Risch 1980, 1981; Zhou et al. 2009). Even mixing resistant and susceptible varieties of the same crop in the same field can provide significant benefit for pest management (Garrett and Mundt 1999).

As described in Chapter 4, development of most herbivorous insects is synchronized with growth and development of their host plants through mutual responses to temperature or photoperiod (Lawrence et al. 1997; Hunter and Elkinton 2000; Visser and Holleman 2001). Disruption of this synchrony between insect appearance and crop growth stage forces insects to feed on plants when they are less suitable for insect growth and reproduction. Advanced or delayed planting dates and cold water sprays on crops in spring can disrupt this synchrony (Miller 1983). These methods are feasible in some situations, especially small crop patch sizes.

Newer agricultural practices emphasize protection of soil moisture, fertility, and nutrient cycling processes that minimize crop stress and retain a diversity of predators and parasites to control herbivore populations (Landis et al. 2000, 2005; Altieri 2002; Denys and Tscharntke 2002; Pretty et al. 2003; Thies et al. 2003; Vincent et al. 2003; Tscharntke et al. 2005, 2007; Gliessman 2007; Kogan and Jepson 2007; Isaacs et al. 2009; Zhou et al. 2009). Examples include nontillage, intercrop or cover crop of nitrogen-fixing legumes, mulch or manure application, and conservation of noncrop patches and hedgerows within the agricultural landscape (Gliessman 2007). Healthy plants, even in agroecosystems, often can tolerate herbivory by replacing lost foliage or other tissues (Pedigo et al. 1986; Trumble et al. 1993). Another recent addition, trap cropping, involves planting a border strip of an attractive host variety around the crop field to concentrate pests in a small area, either to keep them out of the crop or to focus control measures more efficiently (Kogan and Turnipseed 1987; Hokkanen 1991). All of these practices can reduce the abundances of pest species. In addition, some of these practices provide benefits in terms of maintaining a diversity of native predators and pollinators within the agricultural landscape (Landis et al. 2000, 2005; Kremen et al. 2002; Thies et al. 2003; Tscharntke et al. 2005, 2007; Donald and Evans 2006; Hoehn et al. 2008) and contributing to food security in developing countries (Tscharntke et al. 2012).

Another promising new tool is the application of jasmonic acid (see Chapters 5 and 6) to induce increased production of chemical defenses by crop plants. Several studies indicate that application of jasmonic acid can elevate natural defenses and reduce pest damage (Thaler et al. 2001; Mészáros et al. 2012).

Clearly, these tactics will require trade-offs between maximizing production of particular crop species over large areas (which attracts and promotes pests) and costs of pest control above those currently paid for insecticides. Adoption of these tactics also will require cooperation among various landowners across landscapes (Carriére et al. 2001; Zalucki et al. 2009). However, the benefits have been demonstrated, especially for poor farmers who have had less access to expensive pesticide application equipment. Pretty et al. (2003) surveyed 208 farming projects (representing nine million farms covering 29 million hectares in 52 developing countries in Africa, Asia, and Latin America) in which farmers have adopted environmentally friendly practices and technologies. *Of 89 farming projects with reliable data, 93% showed an increase in food production per hectare resulting from increased water use efficiency, improved soil health and fertility, and minimal or zero pesticide use.* Clearly, a transition to more sustainable agricultural practices has substantial benefits for rural poor in developing countries.

9.5.2 Control Measures

Even with an emphasis on preventive measures, control will be warranted in some cases. Insecticides remain a component of crop protection and vector control but should be considered more often as a last resort, rather than as a first, for reasons described here and in earlier chapters. Although alternative control tactics may be equally, or nearly as, effective for many insects and more compatible with ecosystem services, some insects are capable of such rapid devastation, especially those that vector human or crop diseases, that ecological methods may not be sufficiently quick or effective to prevent unacceptable harm (Yang et al. 2006).

For example, mosquito vectors of malaria and yellow fever are responsible for more than one million deaths each year around the world. Improving human health and quality of life requires more immediate reduction in mosquito numbers, although methods for interrupting disease transmission and/or reducing the number of infected reservoirs (such as improved medical treatment or quarantine of patients) ultimately will provide more sustainable solutions with fewer effects on nontarget birds, fish, and other mosquito eating animals.

The Asian citrus psyllid, *Diaphorina citri*, vectors the bacterium, *Xanthomonas axonopodis*, that causes huanglongbing (citrus greening disease), an ultimately fatal disease that kills all commercial and native species of citrus (Yang et al. 2006). Failure to prevent spread of this disease would end citrus production (Yang et al. 2006). Other citrus pests have been controlled with biological control agents, but the prospect of losing all citrus production to greening disease has prompted employment of neonicotinoids that also kill biological control agents and pollinating honey bees (Claudianos et al. 2006; Halm et al. 2006; Yang et al. 2006). Unfortunately, sufficiently effective alternatives have not been discovered yet.

Insecticides have one advantage over other control tactics. They provide rapid reduction in abundance of susceptible pests and satisfy the needs of vulnerable human populations. When insecticide use is deemed necessary, chemicals with new modes of action should be used in place of older compounds and rotated with nonpesticidal tactics to combat resistant insects. New modes of action are necessary because insecticides with the same mode of action to which insects are resistant will trigger the same resistance response (see Box 6.1). Furthermore, population monitoring coupled with attention to action thresholds can improve the timing and efficacy of insecticide application, reducing the amount and number of applications required for adequate effectiveness.

Alternatives to synthetic insecticides include biopesticides, barrier treatments, biological control (see section 9.5.1.1) and pheromones or other attractants, and repellents. Biopesticides include a number of insecticidal plant or microbial products or analogues (including toxins from the neem tree, *Azadirachta indica*, neonicotinoids, and sponosyns; see Chapter 6) and IGRs (that interfere with insect developmental hormones), which often are more specific for target insects and might have fewer nontarget effects because they are naturally occurring materials (Miller 1996; Sparks 1996; see Chapter 6). Solutions of microbial pathogens, such as nuclear polyhedrosis virus (NPV), have the advantage of self-replication and can reach epizootic levels in dense insect populations with fewer nontarget effects (see case study introducing Chapter 1), although related nontarget species may suffer (Zangerl et al. 2001). Furthermore, despite concerns, transgenic plants incorporating microbial toxins have reduced the need for insecticide application and have fewer effects on nontarget species (Carriére et al. 2003; Cattaneo et al. 2006; Marvier et al. 2007).

Barrier treatments are effective in preventing access to hosts for some insects. For example, window and door screens and bed netting have a long history of use to prevent biting flies from gaining access to human or animal hosts. Barriers that reduce crowding among solitary locusts can prevent phase shift to the gregarious, plague phase and thereby prevent outbreaks at lower cost, compared with suppression of widespread plagues (Showler 2002). Barriers also can be used to stop and concentrate marching locusts where they can be killed without resort to chemicals (Riley 1878, 1883; Smith 1954). Duncan et al. (2009) and Sexton and Schowalter (1991) reported effective use of physical barriers to prevent flightless weevils from gaining access to resources in orchard tree crowns.

Pheromones have become valued tools because of their specificity in attracting target species (Tumlinson et al. 1969; Hemmann et al. 2008). Unfortunately, many insects are attracted only to specific blends of complex molecules, making identification and application of the proper pheromone difficult (Tumlinson et al. 1969; Hemmann et al. 2008). However, modern chemical analytical equipment has simplified the process for identifying the composition of chemical mixtures. Pheromones can be used in at least two different ways.

First, pheromones are particularly useful for detecting and eliminating newly introduced species while their populations are still small and restricted in area. Small populations are liable to disappear, even in the absence of interference, due to the difficulty of individuals finding mates. Consequently, isolated individuals in small populations depend on pheromones or other odors (particularly host odors) to locate each other. Too often, invasive species are ignored until they have become established and spread, when their damage is clearly recognized. At this point, their control is much more difficult and expensive. Early application of pheromones to detect, prevent mating, and reduce a small population below its extinction threshold would minimize future crop losses or disease transmission (Yamanaka and Liebhold 2009), as well as the future impacts of the insect or its control on ecosystem services. Baited traps placed in strategic locations attract, capture, and reduce the distribution and abundance of a target species (Boddum et al. 2009; Gries et al. 2009).

Second, pheromones can be impregnated in controlled release containers or plastic filaments and scattered in the environment to disrupt normal attraction of mates (Niwa et al. 1988; Alfaro et al. 2009; Vacas et al. 2009) or to repel insects from the treated site (Gillette et al. 2009). Baited traps also can be used to attract and trap target insects locally, reducing their abundance (Bray et al. 2009; Vargas et al. 2009). These applications are most effective for small or relatively isolated populations for which saturation of the populated area with the attractant is feasible and attraction of additional target insects from outside the populated area is limited, such as in orchards or on islands (Yamanaka and Liebhold 2009).

The most potent weapons in our arsenal should be reserved for situations in which other options have failed. Just as the most potent antibiotics are now reserved for cases involving resistant bacteria, the most effective insecticides should be saved for cases that require this option, in order to avoid insecticide resistance among target insects. The situations most demanding of fast-acting insecticides are those in which human lives or adequate food supply are at risk—especially in impoverished communities where insect-vectored diseases are most prevalent and stand in the way of economic development—in order to improve health, quality of life, and social stability. Accordingly, insecticides targeted for control of disease vectors should not be put at increased risk of resistance development by diverting them to widespread agricultural uses (Katima and Mng'anya 2009).

Although a priori identification of potentially invasive species remains problematic, in general, species that have wide tolerance ranges for temperature and moisture, colonize disturbed habitats rapidly, and have high reproductive and dispersal capabilities in their native habitats are most likely to become invasive in new habitats. Small, newly introduced populations of such species are most vulnerable to elimination because their small numbers interfere with finding each other for mating and reproduction (Yamanaka and Liebhold 2009). Pheromones targeted at mating disruption during this vulnerable period would have a high probability of preventing establishment. Modern pheromone libraries for various species and analytical equipment that can be used to identify pheromone blends of new invasives early can facilitate this effort (Durand et al. 2010).

9.5.3 Marketing Insects as Food

Another option (perhaps less palatable in some cultures) for controlling insects is to exploit them as food. Insects represent an efficient and nutritious food resource that supports a diversity of other species, including humans. Many fish and birds used as food by humans depend on insects for their production. Furthermore, because insects do not use as much of their assimilated energy for respiration as do homeothermic vertebrates (such as domestic livestock), they direct a greater proportion of consumed plant material into biomass. In a world starved for food, especially protein, insects seem to provide an obvious solution.

A variety of insects, including many important agricultural pests such as locusts, caterpillars, and beetles, have been valued food resources in many cultures (Mbata et al. 2002; Cerritos and Cano-Santana 2008; Ramos-Elorduy 2009; Yen 2009; see Chapter 2). Even if modern Americans cannot be persuaded to eat insects, insects could be collected from crops, packaged, and shipped to markets around the world where people would be grateful for such a valuable source of protein. In addition to becoming an international trade commodity, collected insects would reduce pest populations dramatically without insecticides, as done historically during locust plagues (Riley 1878, 1883). Cerritos and Cano-Santana (2008) reported a net benefit of more than US$3,000 per family from collecting grasshoppers for food instead of using insecticides, a substantial benefit for poor rural farmers.

9.6 Implementation of Integrated Pest Management Strategy

In practice, IPM requires a stepwise decision-making process, beginning with definition of management goals, desired future ecosystem conditions, or a sustainable level of ecosystem services (Figure 9.1). Defined management goals, desired future conditions, or sustainable levels of ecosystem services determine which available options are most appropriate in a given situation. Second, adequate estimation of insect population size and predicted time to reach the injury threshold establish the time frame for action (Schowalter et al. 1982). Third, available management options, including no action, should be evaluated on the basis of expected accomplishment of management goals, effectiveness within established time frame, economic and ecological costs and benefits, and environmental and social consequences, permitting ranking of options in terms of their net benefits relative to costs (Gatto et al. 2009). All management options have expected costs as well as benefits for various services and public interests that should be considered (see Chapter 8). Although insecticides often are viewed as the least expensive short-term option, increased future costs for controlling resistant target pests, as well as nontarget pests released from regulation by predators, and rebuilding damaged ecosystem services often are not considered in comparison with alternatives. In many cases, insect control is not the best option (Tables 9.1 and 9.2; see also introduction to Chapter 1). Obviously, if the pest problem is self-limiting, as typical in forest and grassland ecosystems with intact regulatory mechanisms (see Chapters 4 and 5), the costs of control can be avoided. However, even in managed (e.g., crop) ecosystems, control often is unnecessary if the insect population fails to reach the injury threshold, perhaps because the crop is capable of tolerating the damage or natural enemies truncate population growth (Pedigo et al. 1986; Ragsdale et al. 2011). Finally, management projects should be monitored and costs and benefits evaluated over time in order to improve future (adaptive) management.

Proper identification of target insects is crucial because inaccurate identification can lead to ineffective management strategies, or worse, given that even species within genera can differ in their adaptations to particular factors. Unfortunately, accurate identification is difficult for many ecologically and economically important insect groups. Some important families of insects have hundreds of species that can be distinguished only by a specialist, but retirements over the past decades combined with severe federal and state budget cuts have restricted the capacity of the remaining federal, state, and academic taxonomists to provide adequate identification services. For example, psyllids are important sap-sucking hemipterans that vector many serious crop diseases, but no one in the United States currently provides species-level identification for this group; European and Asian morphs of the gypsy moth are indistinguishable except by expensive DNA analysis but differ in their ability to disperse (European females do not fly whereas Asian females do) and, consequently, differ in the area that must be treated to prevent population growth and spread.

9.7 Integrated Pest Management: A Case Study

The value of the IPM approach can be illustrated by the history of attempts to control the cotton boll weevil, *Anthonomus grandis*, that defied 100 years of chemical control efforts in the southern United States but was finally eliminated as a cotton pest by applying the principles and tactics of IPM. The cotton boll weevil entered the United States from Mexico in 1892 (Hunter and Hinds 1904) and by 1922 had spread across the entire cotton growing region of the southern United States (Showler 2009). At that time, cotton was the primary crop in a largely agrarian region, but cotton production plummeted following the appearance of the weevil. For example, weevils destroyed the entire cotton crop of southwestern Alabama in 1913 (Smith 2007). The first year the boll weevil was found statewide in Alabama (1917), cotton production fell 70% to 500,000 bales, from a high of 1.7 million bales in 1914 (Smith 2007).

Early efforts to control the weevil included picking and burning infested cotton squares, drowning weevils in kerosene, and a variety of home remedies, none of which improved cotton yields. Destroying all cotton residue following harvest in the fall and volunteer cotton in early spring reduced weevil damage and was one of the earliest recommended control tactics (Howard 1896). Failed cotton crops caused land values to plunge, leading to bankruptcy and mass migration out of the region (Hardee and Harris 2003). However, cotton failure also led to diversification of crop production in the South and to a shift in cotton production to northern (and later western) regions where the weevil was unable to survive colder winters.

Passage by Congress of the Smith-Lever Act of 1914 created the Cooperative Extension Service, a state and federal cooperative that supported the hiring of entomologists to advise cotton growers. Calcium arsenate dust became available for weevil control in 1918. Initially, this was applied by hand as a dust, with no protection for applicators. Following World War I, aeronautical technology combined with veteran pilots led to aerial application methods that could cover larger areas with lower exposure to applicators (creating the term "crop duster").

Calcium arsenate application continued until the 1950s, when more effective chlorinated hydrocarbons became available after World War II. Initially, DDT was highly effective against the boll weevil. However, weevil populations in Louisiana were resistant to DDT by 1955 (Roussel and Clower 1957), and resistance was soon discovered in other areas. In addition, environmental research showed that these chemicals persisted in the environment and become more concentrated at higher trophic levels, leading to serious environmental concerns (Carson 1962).

During this period, organophospates such as methyl parathion and malathion were introduced as liquid insecticides. These compounds were highly toxic to weevils and applicators (as well as other vertebrates) but were relatively short-lived. During the 1970s, pyrethroids were introduced. These chemicals also were acutely toxic to other insects, especially bees, but had lower toxicity for vertebrates and lower persistence. Frequent application (12–16 per season) of these compounds was required for control. In addition, growers became more dependent on insecticides to control boll weevils and relied less on cultural and other nonchemical options.

Increasing concern about environmental costs and fear that boll weevils would become resistant to newer insecticides (as they had to chlorinated hydrocarbons) led to consideration of other options. Advances in understanding of weevil "diapause" (Brazzel and Newsom 1959) and pheromone communication (Tumlinson et al. 1969) suggested that targeting these key ecological attributes, using a combination of tactics, could eliminate this pest. Starting in the mid-1970s, mandated crop destruction soon after harvest deprived weevils of late season resources and overwintering sites; intensive crop monitoring with weevil pheromone bait trapped many weevils and indicated where weevils were most abundant, typically demes isolated by crop diversification. These more restricted sites were targeted for application of malathion, the most effective insecticide.

By the mid-1990s, no yield losses to boll weevil were reported in Alabama for the first time since 1910 (Smith 2007). Furthermore, insecticide use was reduced from an average of 10 to 14 applications per season to one to four (Smith 2007). The last boll weevil in Alabama was reported in 2003 (Smith 2007). The boll weevil was largely eliminated from most other states in the southern United States by 2010, although influx into Texas from Mexico requires continued monitoring and targeted control.

This century-long experience with the boll weevil demonstrated that adherence to the principles of integrated pest management can minimize pest damage, as well as undesirable effects on ecosystem services. Three points should be emphasized. First, the spread of the boll weevil initially was facilitated by extensive cultivation of a previously rare host in a new region. Second, *crop diversification in the South established a more stable economy and limited the spread of insects.* A monument to the boll weevil was dedicated in Enterprise, Alabama, in 1919 to recognize the weevil's contribution to crop diversification and an improved economy in Coffee County. Third, *the successful elimination of the boll weevil as a factor affecting cotton production was achieved ultimately by crop monitoring with pheromone traps to identify population sources combined with targeted use of multiple ecological tactics and limited use of insecticides* at critical stages in the weevil life history.

9.8 Urban Responsibilities

Clearly, land managers, commodity producers/harvesters, and policymakers who directly market ecosystem services have primary responsibility for implementing sustainable practices rather than maximizing short-term profits. However, for change to occur, the end users of these services must demand improvement and be willing to pay for the cost of changes in pest management practices or resource quality. Members of increasingly urban societies in developed and developing countries have a responsibility to encourage and support sustainable environmental management in the following ways.

First, *urban populations are increasingly the primary users of ecosystem services,* including agricultural productivity and water yield, and have a disproportional per capita effect on the sustainability of services. For example, the United States has 5% of the global population but uses 25% of the world's fossil fuel resources, much of this for extraction and transportation of ecosystem services to remote urban consumers (Worldwatch Institute 2012). Even efforts to protect ecosystems and their services in developed countries can increase global inequity if demand for resources is not reduced substantially. In the absence of reduced per capita demands, as well as incentives for ecosystem protection in undeveloped countries, protection of ecosystem services in developed countries will lead to increased resource export from undeveloped countries, contributing to further environmental degradation and a global loss of ecosystem services (Edwards and Laurance 2012; Koh and Lee 2012).

Much unsustainable agricultural and silvicultural use of insecticides is considered necessary to meet consumer demands for unblemished, inexpensive, and insect- and disease-free produce (Zalucki et al. 2009). Poorer farmers are more likely to use inexpensive insecticides than to risk their meager profit margin on sustainable, but more expensive, IPM practices. Farmers also are vulnerable to reduced prices for their crops if IPM practices are successful and increase yields, although prices are more stable for internationally traded commodities. *Insecticide use could be reduced dramatically if urban consumers were willing to pay higher prices for produce grown using sustainable pest management practices and/or to accept a greater degree of blemished, but otherwise edible, produce.*

Second, close cooperation between urban consumers and rural producers would benefit both. Organic produce has become more available to consumers who prefer food known to be free of synthetic insecticides, although biopesticides and other methods may be used to control weeds, insects, and pathogens. Farmers markets and fresh produce cooperatives are popular in many urban communities. Consumers with direct access to agricultural producers can support and encourage sustainable agricultural practices, ensuring a market for products grown using these practices and ensuring consumers food of known quality. Much imported food is produced using practices that are less sustainable than those required in the United States (e.g., employing insecticides such as DDT despite bans on its use in many countries) (Katima and Mng'anya 2009). *Many farmers would implement sustainable pest management practices if assured of stable markets for more expensive products.*

Third, urban consumers must become more aware of the needs and practices of rural agricultural communities and be willing to share the costs of sustainable ecosystem management practices (Matthews and Hammond 1999). Coalitions of consumers and commodity producers must insist that policymakers implement regulations that favor sustainable management practices and markets to support these practices. Unfortunately, many urban constituents lobby for policies that require sustainable practices but complain if prices increase to support the necessary changes in production practices.

Competition for water use is a particular problem in many areas (Soboll et al. 2011). Increasing water use by urban consumers reduces water available for crop production. In addition to directly reducing crop yields (Stambaugh et al. 2011), water limitation causes plant stress and increased vulnerability to herbivorous insects and pathogens (Mattson and Haack 1987; Reay-Jones et al. 2005). *Urban consumers can help reduce the need for agricultural pesticides by freeing more water for production of healthy crops.*

The world's increasingly urban population has additional responsibilities to ensure sustainable management of insects and ecosystem services. Public demands for pest-free housing typically drive decisions that favor the quickest and least expensive method to control insects, whether or not long-term goals of sustainability will be met. Insecticides typically meet these criteria, and the (currently) most effective option generally is selected. Widespread application of insecticides to control mosquitoes, ants, wasps, and cockroaches introduces a broad spectrum of toxins into residences and the urban environment.

Whereas professional pesticide applicators are licensed and required by law to follow all pesticide label directions for rates and location of application for particular pest species, homeowners are not similarly regulated. As described in this and preceding chapters, this practice quickly leads to insecticide resistance among target insects and undermines the effect of favored options. For example, within four years, bedbugs have become sufficiently resistant to pyrethroid insecticides that these are no longer used by pest control operators in the United States, despite the fact that there are few effective alternatives. This, in turn, requires increasing amounts of insecticides to achieve the same level of control or new insecticides with novel modes of action, an option that may not be available given the time necessary to adequately test and register new products. Increased amounts of insecticides pose hazards to human and pet health. *The urban populace can apply the principles of IPM by practicing preventive measures to avoid, or at least minimize, insect problems and the need for insecticides.*

Urban centers provide conducive conditions or habitats for insects that become pests. Such conditions include (1) widespread planting of exotic ornamental plant species that are most susceptible to herbivorous insects, (2) construction of ponds and fountains and neglected outdoor containers, tires, and debris that retain stagnant water and become breeding habitats for mosquitoes (Carlson et al. 2009), (3) accumulated organic building debris and mulch and leaking roofs or water lines that provide access and damp material to support termites, (4) open containers of human or pet food and unsanitary conditions in or near houses that provide habitat for filth flies, cockroaches, and other household pests, and (5) transport of infested luggage, furniture, or other household items that spread bedbugs, lice, and other human parasites.

Elimination of these conducive conditions would substantially reduce the abundance and spread of many insect pests and the need for insecticides. Pheromones or other attractants or repellents may be useful in controlling some household pests (Liang et al. 1998). Bt pellets can be used to prevent mosquito breeding in ponds and fountains (Russell and Kay 2008; Skovmand et al. 2009). However, mosquitoes also can become resistant to these bacteria (Singh and Prakash 2009). The recent identification of nonanal, produced by birds and humans, as a primary attractant for *Culex* mosquitoes (Syed and Leal 2009) offers the possibility of using or masking this compound to control mosquitoes around homes. Because mosquitoes and many other blood-seeking flies do not fly far from aquatic habitats, such targeted use to control mosquitoes in buffer zones around human habitations is practical (Mascari et al. 2007; Mascari and Foil 2009). Reduced insecticide use would prevent or delay insecticide resistance and improve the effectiveness of remaining insecticides, when used only as necessary. This may require policies and enforcement to minimize these conducive conditions on private properties.

In addition, *concerned urban homeowners can help to maintain ecosystem services and reduce pest problems through retention or restoration of patches of native ecosystems within urban landscapes.* Urban parks and other open space, roadside corridors, stormwater drainage systems, and backyard and rooftop gardens provide opportunities to maintain at least some ecosystem services within urban ecosystems. Butterflies, dragonflies, ladybird beetles, and tiger beetles are widely appreciated, offering opportunities to use them as examples for educational programs that promote the importance of insects for maintenance of ecosystem services and that encourage stewardship of biodiversity and natural resources. Urban landscapes and private gardens that provide native host plants and floral resources for larval and adult butterflies can partially offset the loss of natural habitats.

Biological control of mosquitoes in or near ponds, fountains, and other aquatic habitats (using Bt pellets, for example) can minimize negative effects on nontarget species, such as dragonflies, that provide natural regulation of mosquito populations. However, conservation of less popular biological control agents, such as spiders and wasps, is equally important and can reduce urban pest problems without resorting to chemical insecticides. Shrewsbury and Raupp (2006) tested the response of a common urban plant pest, the introduced azalea lace bug, *Stephanitis pyrioides*, to vegetation complexity in urban landscapes and concluded that predation, especially by spiders, was the major factor affecting lace bug abundance. Predator abundance was significantly higher and lace bug abundance significantly lower in complex urban landscapes, compared with simple landscapes. Complexity can be provided by homeowner landscaping, especially with patches of native vegetation, as well as by retention of native ecosystems as urban parks and other open spaces.

Urban ports and trade centers are primary avenues for introduction of invasive species. Although many transported species are unable to find suitable resources and to spread, many species can find suitable hosts among humans, pets, and exotic plant species in urban habitats and use these stepping-stones to spread across new continents, thereby threatening human health and ecosystem services (Suarez et al. 2001). Such species represent an enormous cost to ecosystem services, with at least 50,000 nonnative insects in the United States costing $120 billion annually in resource losses and control costs (Pimentel et al. 2005). In some cases, the threat to ecosystem services represented by invasive species is sufficient to justify harsh pesticides for control, at the expense of noninsectical options that could be used otherwise (Yang et al. 2006). Preventing introduction of such species will require more restrictive policies on imported materials. Most countries do not have sufficient inspection personnel or expertise to prevent introduction of contaminated material. For example, kiln drying of large raw timber products has demonstrated effectiveness in killing all insects and pathogens to maximum depth in wood shipments, whereas shipboard fumigation has not (Morrell 1995; Nzokou et al. 2008). Some fumigants are capable of sterilizing smaller wood products, provided circulation of fumigant in cargo containers is adequate (Barak et al. 2006). Nevertheless, shipboard fumigation currently meets requirements for timber imports into the United States.

9.9 Concluding Comments

Human influence on ecosystems and their services is global and pervasive (see Figure 6.5). Given recent changes in atmospheric chemistry, global warming, land use, invasive species, and roads through otherwise undisturbed ecosystems, it is unlikely that any ecosystems remain truly "natural." It is important to note that anthropogenic changes do not destroy ecosystems, but they do alter ecosystem conditions in ways that threaten the sustainability of the ecosystem services on which we depend. An estimated 60% of our global ecosystem services already are being degraded or being used unsustainably (Millennium Ecosystem Assessment 2005).

Currently, atmospheric CO_2 concentration is well above maximum values for the past 400,000 years, only about 40% of global forests remain in essentially undisturbed condition, and 38% of global agricultural lands have been degraded beyond their capacity for further production by unsustainable practices (Keeling et al. 1995; Cassel-Gintz and Petschel-Held 2000).

Furthermore, global warming is increasing the frequency of severe disturbances that affect species and ecosystem services (Gleason et al. 2008; Bender et al. 2010; Gutschick and BassiriRad 2010; Lubchenco and Karl 2012). Clearly, these changes threaten ecosystem capacity to support a growing human population. We can no longer afford to believe that we can control nature nor that we can solve our growing environmental problems with new manipulations. Engineered solutions to problems resulting from previous solutions is akin to the old lady in the nursery rhyme who swallowed a fly, then swallowed a spider to catch the fly, then swallowed a bird to catch the spider, and so on. That story does not end happily, and we also cannot predict the effects of our tampering with ecosystems well enough to avoid unintended consequences for critical services. As noted in Chapter 1, manipulating insects, resources, and ecosystems without sufficient knowledge of the consequences for long-term sustainability of ecosystem services is not truly "managing." We cannot undo changes already made that affect ecosystem capacity to maintain services at the global level. However, it is imperative that we implement new environmental policies and practices to protect the integrity of ecosystem processes and services that are essential to our survival on this planet.

Insects have responded to environmental changes in a variety of ways. Many are threatened by altered habitat or climatic conditions, and their disappearance eliminates their unique contributions to ecosystem processes and services. However, other members of this ancient group have demonstrated their capacity to thrive in a human-dominated environment. The swiftness of their adaptation to control tactics makes any single tool for long-term management of their populations very expensive and/or ultimately unsuccessful. Multiple tactics are necessary to prevent adaptation to any strong directional selection factor. Unfortunately, although integrated pest management represents a scientific approach to sustainable management, pest management decisions are based less often on real need or scientific recommendations than on public perceptions and policies that favor preemptive action to avoid potential pest problems. The result is rapid development of resistance to pesticides and induction of nontarget consequences that degrade ecosystem integrity and threaten sustainability of ecosystem services.

Improved pest management practices are fundamental to sustainability of ecosystem services. Human history demonstrates our capacity to degrade ecosystems on which we depend and our vulnerability to social unrest, population displacement, and conflict when food, water, and other necessary resources become inadequate. Insect-vectored diseases thrive in stressed, crowded human populations that are impoverished or displaced by famine or war. Although improved agricultural practices and vector control in poor countries can improve health, quality of life, and economic development, adoption of more sustainable practices in developed countries is in our own self-interest to the extent that changes contribute to social stability on a global scale.

Everyone is responsible for helping to drive environmental management policy toward more sustainable practices. Although resource managers and policymakers are ultimately responsible for implementing management practices that are more sustainable, an increasingly isolated urban consumer population has a responsibility to partner with commodity producers to encourage and guide more sustainable options and provide secure markets for sustainable products, even if this means accepting higher prices and lower quality for food and other services. Furthermore, public support for sustainable practices will improve environmental quality in ways that will contribute further to sustainability of ecosystem services.

Will this be easy? No. Some argue that the cost of making necessary adjustments in our environmental policies and use of ecosystem services will cripple our economy. On the other hand, not making these adjustments ultimately will bring famine, conflict, and disease outbreaks, as described throughout this book. It is imperative that we properly compare the long-term costs of losing ecosystem services with the long-term costs of ensuring their sustainability (Balmford et al. 2011). In some cases, expensive management policies have been deemed necessary to reduce nontarget impacts of pesticides on ecosystem services (Scholz et al. 2012). We also need to ensure that all stakeholders are able to participate in the planning of future directions and desired future conditions. This is the only way in which all parties can understand the needs and concerns that motivate other stakeholders in order to reach consensus on trade-offs that satisfy as many of these needs and concerns as possible, while giving priority to protection of ecosystem structures and processes that support services (Lamy et al. 2002; Maser 2010; Maser and Pollio 2011; Dickie et al. 2011; Bradford and D'Amato 2012).

Change occurs one resource manager and policymaker at a time. The changes recommended here require a commitment to continue through transition from reliance on insecticides to strategies that protect ecosystem integrity for greater long-term sustainability.

References

Abate, T., A. van Huis, and J. K. O. Ampofo. 2000. Pest management strategies in traditional agriculture: An African perspective. *Annual Review of Entomology* 45: 631–659.

Abt, E., J. V. Rodricks, J. I. Levy, L. Zeise, and T. A. Burke. 2010. Science and decisions: Advancing risk assessment. *Risk Analysis* 30: 1028–1036.

Acuña-Soto, R., D. W. Stahle, M. K. Cleaveland, and M. D. Therrell. 2002. Megadrought and megadeath in 16th century Mexico. *Emerging Infectious Diseases* 8: 360–362.

Adams, J. M., W. Fang, R. M. Callaway, D. Cipollini, E. Newell, and Transatlantic Acer Platanoides Invasion Network. 2009. A cross-continental test of enemy release hypothesis: Leaf herbivory on *Acer platanoides* (L.) is three times lower in North America than in its native Europe. *Biological Invasions* 11: 1005–1016.

Alfaro, C., V. Navarro-Llopis, and J. Primo. 2009. Optimization of pheromone dispenser density for managing the rice striped stem borer, *Chilo suppressalis* (Walker), by mating disruption. *Crop Protection* 28: 567–572.

Altieri, M. A. 2002. Agroecology: The science of natural resource management for poor farmers in marginal environments. *Agriculture, Ecosystems and Environment* 93: 1–24.

Amman, G. D., M. D. McGregor, R. F. Schmitz, and R. D. Oakes. 1988. Susceptibility of lodgepole pine to infestation by mountain pine beetles following partial cutting of stands. *Canadian Journal of Forest Research* 18: 688–695.

Archer, S. and D. A. Pyke. 1991. Plant-animal interactions affecting plant establishment and persistence on revegetated rangeland. *Journal of Range Management* 44: 558–565.

Arndt, D. S., M. O. Baringer, and M. R. Johnson, eds. 2010. State of the climate in 2009. *Bulletin of the American Meteorological Society* 91: S1–S224.

Ayala, A. J., B. Alberts, M. R. Berenbaum, B. Carvellas, M. T. Clegg, G. B. Dalrymple, R. M. Hazen, T. M. Horn, N. A. Moran, G. S. Omenn, et al. 2008. *Science, Evolution, and Creationism.* Washington, DC: National Academies Press.

Balmford, A., B. Fisher, R. E. Green, R. Naidoo, B. Strassburg, R. K. Turner, and A. S. L. Rodrigues. 2011. Bringing ecosystem services into the real world: An operational framework for assessing the economic consequences of losing wild nature. *Environmental and Resource Economics* 48: 161–175.

Barak, A. V., Y. Wang, G. Zhan, Y. Wu, L. Xu, and Q. Huang. 2006. Sulfuryl fluoride as a quarantine treatment for *Anoplophora glabripennis* (Coleoptera: Cerambycidae) in regulated wood packing material. *Journal of Economic Entomology* 99: 1628–1635.

Barbosa, P., J. Hines, I. Kaplan, H. Martinson, A. Szczepaniec, and Z. Szendrei. 2009. Associational resistance and associational susceptibility: Having right or wrong neighbors. *Annual Review of Ecology, Evolution and Systematics* 40: 1–20.

Bartlett-Healy, K., I. Unlu, P. Obenauer, T. Hughes, S. Healy, T. Crepeau, A. Farajollahi, B. Kesavaraju, D. Fonseca, G. Schoeler, et al. 2012. Larval mosquito habitat utilization and community dynamics of *Aedes albopictus* and *Aedes japonicus* (Diptera: Culicidae). *Journal of Medical Entomology* 49: 813–824.

Bender, M. A., T. R. Knutson, R. E. Tuleya, J. J. Sirutis, G. A. Vecchi, S. T. Garner, and I. M. Held. 2010. Modeled impact of anthropogenic warming on the frequency of intense Atlantic hurricanes. *Science* 327: 454–458.

Boddum, T., N. Skals, M. Wirén, R. Baur, S. Rauscher, and Y. Hillbur. 2009. Optimisation of the pheromone blend of the swede midge, *Contarinia nasturii*, for monitoring. *Pest Management Science* 65: 851–856.

Boo, C. S. 2001. Legislation for control of dengue in Singapore. *Dengue Bulletin* 25: 69–73.

Bora, S., I. Ceccacci, C. Delgado, and R. Townsend. 2010. *World Development Report 2011: Food Security and Conflict.* Agriculture and Rural Development Department, World Bank.

Bradbury, R. 1952. Sound of thunder. *Colliers Magazine* June 28: 20–21, 60–61.

Bradford, J. B. and A. W. D'Amato. 2012. Recognizing trade-offs in multi-objective land management. *Frontiers in Ecology and Management* 10: 210–216.

Bray, D. P., K. K. Brandi, R. P. Brazil, A. G. Oliveira, and J. G. C. Hamilton. 2009. Synthetic sex pheromone attracts the Leishmaniasis vector *Lutzomyia longipalpis* (Diptera: Psychodidae) to traps in the field. *Journal of Medical Entomology* 46: 428–434.

Bray, R. S. 1996. *Armies of Pestilence: The Impact of Disease on History*. New York: Barnes and Noble.

Brazzel, J. R. and L. D. Newsom. 1959. Diapause in *Anthomomus grandis* Boh. *Journal of Economic Entomology* 52: 603–611.

Briant, G., V. Gond, and S. G. W. Laurance. 2010. Habitat fragmentation and the desiccation of forest canopies: A case study from eastern Amazonia. *Biological Conservation* 143: 2763–2769.

Brouqui, P. 2011. Arthropod-borne diseases associated with political and social disorder. *Annual Review of Entomology* 56: 357–374.

Burleigh, J. G. 1972. Population dynamics and biotic controls of the soybean looper in Louisiana. *Environmental Entomology* 1: 290–294.

Carlson, J. C., L. A. Dyer, F. X. Omlin, and J. C. Beier. 2009. Diversity cascades and malaria vectors. *Journal of Medical Entomology* 46: 460–464.

Carpenter, S. R., H. A. Mooney, J. Agard, D. Capistrano, R. S. DeFries, S. Díaz, T. Dietz, A. K. Duraiappah, A. Oteng-Yeboah, H. M. Pereira, et al. 2009. Science for managing ecosystem services: Beyond the Millennium Ecosystem Assessment. *Proceedings of the National Academy of Sciences USA* 106: 1305–1312.

Carrière, Y., T. J. Dennehy, B. Pedersen, S. Haller, C. Ellers-Kirk, L. Antilla, Y-B. Liu, E. Willott, and B. E. Tabashnik. 2001. Large-scale management of insect resistance to transgenic cotton in Arizona: Can transgenic insecticidal crops be sustained? *Journal of Economic Entomology* 94: 315–325.

Carrière, Y., C. Ellers-Kirk, M. Sisterson, L. Antilla, M. Whitlow, T. J. Dennehy, and B. E. Tabashnik. 2003. Long-term regional suppression of pink bollworm by *Bacillus thuringiensis* cotton. *Proceedings of the National Academy of Sciences USA* 100: 1519–1523.

Carson, R. 1962. *Silent Spring*. New York: Houghton-Mifflin.

Carter, G. A. and A. K. Knapp. 2001. Leaf optical properties in higher plants: Linking spectral characteristics to stress and chlorophyll concentration. *American Journal of Botany* 88: 677–684.

Cassel-Gintz, M. and G. Petschel-Held. 2000. GIS-based assessment of the threat to world forests by patterns of non-sustainable civilisation nature interaction. *Journal of Environmental Management* 59: 279–298.

Cattaneo, M. G., C. Yafuso, C. Schmidt, C.-Y. Huang, M. Rahman, C. Olson, C. Ellers-Kirk, B. J. Orr, S. E. Marsh, L. Antilla, et al. 2006. Farm-scale evaluation of the impacts of transgenic cotton on biodiversity, pesticide use, and yield. *Proceedings of the National Academy of Sciences USA* 103: 7571–7576.

Cerritos, R. and Z. Cano-Santana. 2008. Harvesting grasshoppers *Sphenarium purpurascens* in Mexico for human consumption: A comparison with insecticidal control for managing pest outbreaks. *Crop Protection* 27: 473–480.

Christensen, N. L., Jr., S. V. Gregory, P. R. Hagenstein, T. A. Heberlein, J. Hendee, J. T. Olson, J. M. Peek, D. A. Perry, T. D. Schowalter, K. Sullivan, et al. 2000. *Environmental Issues in Pacific Northwest Forest Management*. Washington, DC: National Academy Press.

Clark, D. B. and D. A. Clark. 1985. Seedling dynamics of a tropical tree: Impacts of herbivory and meristem damage. *Ecology* 66: 1884–1892.

Claudianos, C., H. Ranson, R. M. Johnson, S. Biswas, M. A. Schuler, M. R. Berenbaum, R. Feyereisen, and J. G. Oakeshott. 2006. A deficit of detoxification enzymes: Pesticide sensitivity and environmental response in the honey bee. *Insect Molecular Biology* 15: 615–636.

Clausen, L. W. 1954. *Insect Fact and Folklore*. New York: MacMillan.

Cook, B. I., K. J. Anchukaitis, J. O. Kaplan, M. J. Puma, M. Kelley, and D. Gueyffier. 2012. Pre-Columbian deforestation as an amplifier of drought in Mesoamerica. *Geophysical Research Letters* 39: L16706.

Cooper, J. F. 1823. *The Pioneers*. New York: Charles Wiley.

Corbet, S. A. 1997. Role of pollinators in species preservation, conservation, ecosystem stability, and genetic diversity. In *Pollination: From Theory to Practise*, K. W. Richards, ed. 219–229. Proceedings of the 7th International Symposium on Pollination. Acta Horticulturae #437. Alberta, Canada: Lethbridge.

Delfosse, E. S. 2005. Risk and ethics in biological control. *Biological Control* 35: 319–329.

Denys, C. and T. Tscharntke. 2002. Plant-insect communities and predator-prey ratios in field margin strips, adjacent crop fields, and fallows. *Oecologia* 130: 315–324.

Diamond, J. 1999. *Guns, Germs, and Steel: The Fates of Human Societies*. New York: W.W. Norton.

Dickie, I. A., G. W. Yeates, M. G. St. John, B. A. Stevenson, J. T. Scott, M. C. Rillig, D. A. Peltzer, K. H. Orwin, M. U. F. Kirschbaum, J. E. Hunt, et al. 2011. Ecosystem service and biodiversity trade-offs in two woody successions. *Journal of Applied Ecology* 48: 926–934.

Donald, P. F. and A. D. Evans. 2006. Habitat connectivity and matrix restoration: The wider implications of agri-environmental schemes. *Journal of Applied Ecology* 43: 209–218.

Doran, P. T. and M. K. Zimmerman. 2009. Examining the scientific consensus on climate change. *Eos, Transactions of the American Geophysical Union* 90: 22–23.

Duncan, L. W., R. J. S Stuart, F. G. G Gmitter, and S.L. LaPointe. 2009. Use of landscape fabric to manage Diaprepes root weevil in citrus groves. *Florida Entomologist* 92: 74–79.

Durand, N., G. Carot-Sans, T. Chertemps, N. Montagné, E. Jacquin-Joly, S. Debernard, and M. Maïbèche-Coisne. 2010. A diversity of putative carboxylesterases are expressed in the antennae of the noctuid moth *Spodoptera littoralis*. *Insect Molecular Biology* 19: 87–97.

Edwards, D. P. and S. G. Laurance. 2012. Green labelling, sustainability, and the expansion of tropical agriculture: Critical issues for certification schemes. *Biological Conservation* 151: 60–64.

Ehrlich, P. and A. Ehrlich. 1981. *Extinction: The Causes and Consequences of the Disappearance of Species*. New York: Random House.

Ehrlich, P. R. and H. A. Mooney. 1983. Extinction, substitution, and ecosystem services. *BioScience* 33: 248–254.

Fedoroff, N. V., D. S. Battisti, R. N. Beachy, P. J. M. Cooper, D. A. Fischhoff, C. N. Hodges, V. C. Knauf, D. Lobell, B. J. Mazur, D. Molden, et al. 2010. Radically rethinking agriculture for the 21st century. *Science* 327: 833–834.

Felland, C. M., H. N. Pitre, R. G. Luttrell, and J. L. Hamer. 1990. Resistance to pyrethroid insecticides in soybean looper (Lepidoptera: Noctuidae) in Mississippi. *Journal of Economic Entomology* 83: 35–40.

Foley, J. A., M. T. Coe, M. Scheffer, and G. Wang. 2003a. Regime shifts in the Sahara and Sahel: Interactions between ecological and climatic systems in northern Africa. *Ecosystems* 6: 524–539.

Foley, J. A., M. H. Costa, C. Delire, N. Ramankutty, and P. Snyder. 2003b. Green surprise? How terrestrial ecosystems could affect earth's climate. *Frontiers in Ecology and the Environment* 1: 38–44.

Garrett, K. A. and C. C. Mundt. 1999. Epidemiology in mixed host populations. *Phytopathology* 89: 984–990.

Gassman, A. J., S. P. Stock, M. Sisterson, Y. Carriére, and B. E. Tabashnik. 2008. Synergism between entomopathogenic nematodes and Bt crops: Integrating biological control and resistance management. *Journal of Applied Ecology* 45: 957–966.

Gassman, A. J., J. A. Fabrick, M. Sisterson, E. R. Hannon, S. P. Stock, Y. Carriére, and B. E. Tabashnik. 2009. Effects of pink bollworm resistance to Bacillus thuringiensis on phenoloxidase activity and susceptibility to entomopathogenic nematodes. *Journal of Economic Entomology* 102: 1224–1232.

Gassmann, A. J., J. L. Petzold-Maxwell, R. S. Keweshan, and M. W. Dunbar. 2011. Field-evolved resistance to Bt maize by western corn rootworm. *PLoS ONE* 6(7): e22629.

Gatto, P., A. Zocca, A. Battisti, M. J. Barrento, M. Branco, and M. R. Paiva. 2009. Economic assessment of managing processionary moth in pine forests: A case study in Portugal. *Journal of Environmental Management* 90: 683–691.

Gillette, N. D., N. Erbilgin, J. N. Webster, L. Pederson, S. R. Mori, J. D. Stein, D. R. Owen, K. M. Bischel, and D. L. Wood. 2009. Aerially applied verbenone-releasing laminated flakes protect *Pinus contorta* stands from attack by *Dendroctonus ponderosae* in California and Idaho. *Forest Ecology and Management* 257: 1405–1412.

Geier, P. W. 1966. Management of insect pests. *Annual Review of Entomology* 11: 471–490.

Gleason, K. L., J. H. Lawrimore, D. H. Levinson, T. R. Karl, and D. J. Karoly. 2008. A revised U.S. climate extremes index. *Journal of Climate* 21: 2124–2137.

Gliessman, S. R. 2007. *Agroecology: The Ecology of Sustainable Food Systems*, 2nd ed. Boca Raton, FL: CRC Press.

Goodman, C. A. and A. J. Mills. 1999. The evidence base on the cost-effectiveness of malaria control measures in Africa. *Health Policy and Planning* 14: 301–312.

Gray, N. J. and L. M. Campbell. 2008. Science, science policy, and marine protected areas. *Conservation Biology* 23: 460–468.

Gries, R., P. W. Schaefer, T. Gotoh, S. Takács, and G. Gries. 2009. Spacing of traps baited with species-specific *Lymantria* pheromones to prevent interference by antagonistic components. *Canadian Entomologist* 141: 145–152.

Gutschick, V. P. and H. BassiriRad. 2010. Biological extreme events: A research framework. *Eos, Transactions of the American Geophysical Union* 91: 85–86.

Halm, M.-P., A. Rortais, G. Arnold, J. N. Taséi, and S. Rault. 2006. New risk assessment approach for systemic insecticides: The case of honey bees and imidacloprid (Gaucho). *Environmental Science and Technology* 40: 2448–2454.

Hardee, D. D. and F. A. Harris. 2003. Eradicating the boll weevil (Coleoptera: Curculionidae): A clash between a highly successful insect, good scientific achievement, and differing agricultural philosophies. *American Entomologist* 49: 82–97.

Hemmann, D. J., J. D. Allison, and K. F. Haynes. 2008. Trade-off between sensitivity and specificity in the cabbage looper response to sex pheromone. *Journal of Chemical Ecology* 34: 1476–1486.

Hoehn, P., T. Tscharntke, J. M. Tylianakis, and I. Steffan-Dewenter. 2008. Functional group diversity of bee pollinators increases crop yield. *Proceedings of the Royal Society B* 275: 2283–2291.

Hokkanen, H. M. T. 1991. Trap cropping in pest management. *Annual Review of Entomology* 36: 119–138.

Howard, L. O. 1896. *The Mexican Cotton Boll Weevil*. USDA Bureau of Entomology Circular 14. Washington, DC: Government Printing Office.

Hsiang, S. M., K. C. Meng, and M. A. Cane. 2011. Civil conflicts are associated with the global climate. *Nature* 476: 438–441.

Huffaker, C. B. and P. S. Messenger, eds. 1976. *Theory and Practice of Biological Control.* New York: Academic.

Hunter, A. F. and J. S. Elkinton. 2000. Effects of synchrony with host plant on populations of a spring-feeding lepidopteran. *Ecology* 81: 1248–1261.

Hunter, W. D. and W. E. Hinds. 1904. *The Mexican Cotton Boll Weevil.* USDA Division of Entomology Bulletin 45. Washington, DC: Government Printing Office.

Isaacs, R., J. Tuell, A. Fiedler, M. Gardiner, and D. Landis. 2009. Maximizing arthropod-mediated ecosystem services in agricultural landscapes: The role of native plants. *Frontiers in Ecology and the Environment* 7: 196–203.

Jackson, R. E., J. R. Bradley, J. Van Duyn, B. R. Leonard, K. C. Allen, R. Luttrell, J. Ruberson, J. Adamczyk, J. Gore, D. D. Hardee, et al. 2008. Regional assessment of *Helicoverpa zea* populations on cotton and non-cotton crop hosts. *Entomologia Experimentalis et Applicata* 126: 89–106.

Janssen, R. H. H., M. B. J. Meinders, E.H. van Nes, and M. Scheffer. 2008. Microscale vegetation-soil feedback boosts hysteresis in a regional vegetation-climate system. *Global Change Biology* 14: 1104–1112.

Juang, J.-Y., G. G. Katul, A. Porporato, P. C. Stoy, M. S. Sequeira, M. Detto, H.-S. Kim, and R. Oren. 2007. Eco-hydrological controls on summertime convective rainfall triggers. *Global Change Biology* 13: 887–896.

Katima, J. H. Y. and S. Mng'anya. 2009. African NGOs outline commitment to malaria control without DDT. *Pesticide News* 84: 5.

Keeling, C. D., T. P. Whorf, M. Wahlen, and J. van der Pilcht. 1995. Interannual extremes in the rate of rise of atmospheric carbon dioxide since 1980. *Science* 375: 666–670.

Klein, A.-M, B. E. Vaissière, J. H. Cane, I. Steffan-Dewenter, S. A. Cunningham, C. Kremen, and T. Tscharntke. 2007. Importance of pollinators in changing landscapes for world crops. *Proceedings of the Royal Society B* 274: 303–313.

Knight, T. M., J. A. Steets, J. A. Vamosi, S. J. Mazer, M. Burd, D. R. Campbell, M. R. Dudash, M. O. Johnston, R. J. Mitchell, and T.-L. Ashman. 2005. Pollen limitation of plant reproduction: Pattern and process. *Annual Review of Ecology, Evolution and Systematics* 36: 467–497.

Kogan, M. 1998. Integrated pest management: Historical perspectives and contemporary developments. *Annual Review of Entomology* 43: 243–270.

Kogan, M. and P. Jepson. 2007. Ecology, sustainable development and IPM: The human factor. In *Perspectives in Ecological Theory and Integrated Pest Management*, M. Kogan and P. Jepson, eds. 1–44. Cambridge, UK: Cambridge University Press.

Kogan, M. and S. G. Turnipseed. 1987. Ecology and management of soybean arthropods. *Annual Review of Entomology* 32: 507–538.

Koh, L. P. and T. M. Lee. 2012. Sensible consumerism for environmental sustainability. *Biological Conservation* 151: 3–6.

Kolb, T. E., K. M. Holmberg, M. R. Wagner, and J. E. Stone. 1998. Regulation of ponderosa pine foliar physiology and insect resistance mechanisms by basal area treatments. *Tree Physiology* 18: 375–381.

Kremen, C., N. M. Williams, R. L. Bugg, J. P. Fay, and R. W. Thorp. 2004. The area requirements of an ecosystem service: Crop pollination by native bee communities in California. *Ecology Letters* 7: 1109–1119.

Kremen, C., N. M. Williams, and R. W. Thorp. 2002. Crop pollination from native bees at risk from agricultural intensification. *Proceedings of the National Academy of Sciences USA* 99: 16812–16816.

Lamy, F., J. Bolte, M. Santelmann, and C. Smith. 2002. Development and evaluation of multiple-objective decision-making methods for watershed management planning. *Journal of the American Water Resources Association* 38: 517–529.

Landis, D. A., F. D. Menalled, A. C. Costamagna, and T. K. Wilkinson. 2005. Manipulating plant resources to enhance beneficial arthropods in agricultural landscapes. *Weed Science* 53: 902–908.

Landis, D. A., S. D. Wratten, and G. M. Gurr. 2000. Habitat management to conserve natural enemies of arthropod pests in agriculture. *Annual Review of Entomology* 45: 175–201.

Lawrence, R. K., W. J. Mattson, and R. A. Haack. 1997. White spruce and the spruce budworm: Defining the phenological window of susceptibility. *Canadian Entomologist* 129: 291–318.

Lentz, D. L. and B. Hockaday. 2009. Tikal timbers and temples: Ancient Maya forestry and the end of time. *Journal of Archaeology* 36: 1342–1353.

Leopold, A. 1949. *Sand County Almanac*. Oxford: Oxford University Press.

Liang, D., A. Zhang, R. J. Kopanic, Jr., W. L. Roelofs, and C. Schal. 1998. Field and laboratory evaluation of the female sex pheromone for detection, monitoring, and management of brownbanded cockroaches (Dictyoptera: Blattelidae). *Journal of Economic Entomology* 91: 480–485.

Lorenz, E. N. 1993. *The Essence of Chaos*. Seattle: University of Washington Press.

Louda, S. M., R. W. Pemberton, M. T. Johnson, and P. A. Follett. 2003. Non-target effects—the Achilles' heel of biocontrol? Retrospective analyses to assess risk associated with biocontrol introductions. *Annual Review of Entomology* 48: 365–396.

Lowrance, R., B. R. Stinner, and G. J. House, eds. 1984. *Agricultural Ecosystems: Unifying Concepts*. New York: Wiley.

Lubchenco, J. and T. R. Karl. 2012. Predicting and managing extreme weather events. *Physics Today* 65(3): 31–37.

Marvier, M., C. McCreedy, J. Regetz, and P. Kareiva. 2007. A meta-analysis of effects of Bt cotton and maize on non-target invertebrates. *Science* 316: 1475–1477.

Mascari, T. M. and L. D. Foil. 2009. Evaluation of rhodamine B as an orally delivered biomarker for rodents and a feed-through transtadial biomarker for phlebotamine sand flies (Diptera: Psychodidae). *Journal of Medical Entomology* 46: 1131–1137.

Mascari, T. M., M. A. Mitchell, E. D. Rowton, and L. D. Foil. 2007. Laboratory evaluation of novaluron as a feed-through for control of immature sand flies (Diptera: Psychodidae). *Journal of Medical Entomology* 44: 714–717.

Maser, C. 2005. *Our Forest Legacy: Today's Decisions, Tomorrow's Consequences*. Washington, DC: Maisonneuve Press.

Maser, C. 2010. *Social-Environmental Planning: The Design Interface Between Everyforest and Everycity*. Boca Raton, FL: CRC Press/Taylor & Francis.

Maser, C. and C. A. Pollio. 2011. *Resolving Environmental Conflicts*, 2nd ed. Boca Raton, FL: CRC Press/Taylor & Francis.

Matthews, E. and A. Hammond. 1999. *Critical Consumption Trends and Implications: Degrading Earth's Resources*. Washington, DC: World Resources Institute.

Mattson, W. J. and R. A. Haack. 1987. The role of drought in outbreaks of plant-eating insects. *BioScience* 37: 110–118.

Mbata, K. J., E. N. Chidumayo, and C. M. Lwatula. 2002. Traditional regulation of edible caterpillar exploitation in the Kopa area of Mpika district in northern Zambia. *Journal of Insect Conservation* 6: 115–130.

McCoy, E. D. and J. H. Frank. 2010. How should the risk associated with the introduction of biological control agents be estimated? *Agricultural and Forest Entomology* 12: 1–8.

Meagher, R. L. and R. N. Nagoshi. 2004. Population dynamics and occurrence of *Spodoptera frugiperda* host strains in southern Florida. *Ecological Entomology* 29: 614–620.

Meehan, T. D., B. P. Werling, D. A. Landis, and C. Gratton. 2011. Agricultural landscape simplification and insecticide use in the Midwestern United States. *Proceedings of the National Academy of Sciences USA* 108: 11500–11505.

Mészáros, A., J. M. Beuzelin, M. J. Stout, P. L. Bommireddy, M. R. Riggio, and B. R. Leonard. 2012. Jasmonic acid-induced resistance to the fall armyworm, *Spodoptera frugiperda*, in conventional and transgenic cottons expressing *Bacillus thuringiensis* insecticidal proteins. *Entomologia Experimentalis et Applicata* 140: 226–237.

Millennium Ecosystem Assessment. 2005. *Ecosystems and Human Well-Being: Biodiversity Synthesis*. Washington, DC: World Resources Institute.

Miller, G. E. 1983. Evaluation of the effectiveness of cold-water misting of trees in seed orchards for control of Douglas-fir cone gall midge (Diptera: Cecidomyiidae). *Journal of Economic Entomology* 76: 916–919.

Miller, J. D., E. C. Scott, and S. Okamoto. 2006. Public acceptance of evolution. *Science* 313: 765–766.

Miller, T. A. 1996. Resistance to pesticides: Mechanisms, development, and management. In *Cotton Insects and Mites: Characterization and Management*, E. G. King, J. R. Phillips, and R. J. Coleman, eds. 323–378. Memphis, TN: The Cotton Foundation.

Mooney, H. A. 2010. The ecosystem-service chain and the biological diversity crisis. *Philosophical Transactions of the Royal Society B* 365: 31–39.

Morrell, J. J. 1995. Importation of unprocessed logs into North America: A review of pest mitigation procedures and their efficacy. *Forest Products Journal* 45: 41–50.

Müller, M. and H. Job. 2009. Managing natural disturbance in protected areas: Tourists' attitude towards the bark beetle in a German national park. *Biological Conservation* 142: 375–383.

Myers, M. 1999. The world's forests and their ecosystem services. In *Nature's Services: Societal Dependence on Natural Ecosystems*, G. C. Daily, ed. 215–235. Washington, DC: Island Press.

Naeem, S. 1998. Species redundancy and ecosystem reliability. *Conservation Biology* 12: 39–45.

Nansen, C., T. Macedo, R. Swanson, and D. K. Weaver. 2009. Use of spatial structure analysis hyperspectral data cubes for detection of insect-induced stress in wheat plants. *International Journal of Remote Sensing* 30: 2447–2464.

Nansen, C., A. J. Sidumo, and S. Capareda. 2010. Variogram analysis of hyperspectral data to characterize the impact of biotic and abiotic stress of maize plants and to estimate biofuel potential. *Applied Spectroscopy* 64: 627–636.

Nealis, V. G., M. K. Noseworthy, R. Turnquist, and V. R. Waring. 2009. Balancing risks of disturbance from mountain pine beetle and western spruce budworm. *Canadian Journal of Forest Research* 39: 839–848.

Niwa, C. G., G. E. Daterman, C. Sartwell, and L. L. Sower. 1988. Control of *Rhyacionia zozana* (Lepidoptera: Tortricidae) by mating disruption with synthetic sex pheromone. *Environmental Entomology* 17: 593–595.

Norman, E. M., K. Rice, and S. Cochran. 1992. Reproductive biology of *Asimina parviflora* (Annonaceae). *Bulletin of the Torrey Botanical Club* 119: 1–5.

Nzokou, P., S. Tourtellot, and D. P. Kamdem. 2008. Kiln and microwave heat treatment of logs infested by the emerald ash borer (*Agrilus planipennis* Fairmaire) (Coleoptera: Buprestidae). *Forest Products Journal* 58: 68–72.

Pedigo, L. P., S. H. Hutchins, and L. G. Higley. 1986. Economic injury levels in theory and practice. *Annual Review of Entomology* 31: 341–368.

Perry, D. A., R. Oren, and S. C. Hart 2008. *Forest Ecosystems*, 2nd ed. Baltimore, MD: Johns Hopkins University Press.

Pfiffner, L. and E. Wyss. 2004. Use of sown wildflower strips to enhance natural enemies of agricultural pests. In *Ecological Engineering for Pest Management: Advances in Habitat Manipulation for Arthropods*, G. M. Gurr, S. D. Wratten, and M. A. Altieri, eds, 165–186. Canberra, Australia: CSIRO Publishing.

Pimentel D., R. Zuniga, and D. Morrison. 2005. Update on the environmental and economic costs associated with alien-invasive species in the United States. *Ecological Economics* 52: 273–288.

Pretty, J. N., J. I. L. Morison, and R. E. Hine. 2003. Reducing food poverty by increasing agricultural sustainability in developing countries. *Agriculture, Ecosystems and Environment* 95: 217–234.

Rabb, R. L., G. K. DeFoliart, and G. G. Kennedy. 1984. An ecological approach to managing insect populations. In *Ecological Entomology*, C. B. Huffaker and R. L. Rabb, eds. 697–728. New York: John Wiley & Sons.

Ragsdale, D. W., D. A. Landis, J. Brodeur, G. E. Heimpel, and N. Desneux. 2011. Ecology and management of the soybean aphid in North America. *Annual Review of Entomology* 56: 375–399.

Ramos-Elorduy, J. 2009. Anthro-entomophagy: Cultures, evolution, and sustainability. *Entomological Research* 39: 271–288.

Reay-Jones, R. P. F., A. T. Showler, T. E. Reagan, B. L. Legendre, M. O. Way, and E. B. Moser. 2005. Integrated tactics for managing the Mexican rice borer (Lepidoptera: Crambidae) in sugarcane. *Environmental Entomology* 34: 1558–1565.

Reay-Jones, F. P. F., M. O. Way, M. Sétamou, B. L. Legendre, and T. E. Reagan. 2003. Resistance to the Mexican rice borer (Lepidoptera: Crambidae) among Louisiana and Texas sugarcane cultivars. *Journal of Economic Entomology* 96: 1929–1934.

Ricketts, T. H., J. Regetz, I. Steffan-Dewenter, S. A. Cunningham, C. Kremen, A. Bogdanski, B. Gemmill-Herren, S. S. Greenleaf, A. M. Klein, M. M. Mayfield, et al. 2008. Landscape effects on crop pollinator services: Are there general patterns? *Ecology Letters* 11: 499–515.

Riley, C. V. 1878. *First Annual Report of the United States Entomological Commission for the Year 1877 Relating to the Rocky Mountain Locust and the Best Methods of Preventing Its Injuries and of Guarding Against Its Invasions, in Pursuance of an Appropriation Made by Congress for This Purpose*. Washington, DC: U.S. Department of Agriculture.

Riley, C. V. 1883. *Third Report of the United States Entomological Commission, Relating to the Rocky Mountain Locust, the Western Cricket, the Army-Worm, Canker Worms, and the Hessian Fly, Together with Descriptions of Larvae of Injurious Forest Insects,*

Studies on the Embryological Development of the Locust and of Other Insects, and on the Systematic Position of the Orthoptera in Relation to Other Orders of Insects. Washington, DC: U.S. Department of Agriculture.

Riley, C. V. 1885. *Fourth Report of the United States Entomological Commission, Being a Revised Edition of Bulletin No. 3, and the Final Report on the Cotton Worm, Together with a Chapter on the Boll Worm.* Washington, DC: U.S. Department of Agriculture.

Risch, S. 1980. The population dynamics of several herbivorous beetles in a tropical agroecosystem: The effect of intercropping corn, beans and squash in Costa Rica. *Journal of Applied Ecology* 17: 593–612.

Risch, S. J. 1981. Insect herbivore abundance in tropical monocultures and polycultures: An experimental test of two hypotheses. *Ecology* 62: 1325–1340.

Roussel, J. S. and D. F. Clower 1957. Resistance to the chlorinated hydrocarbon insecticides in the boll weevil. *Journal of Economic Entomology* 50: 463–468.

Russell, T. L. and B. H. Kay. 2008. Biologically based insecticides for the control of immature Australian mosquitoes: A review. *Australian Journal of Entomology* 47: 232–242.

Sachs, J. and P. Malaney. 2002. The economic an social burden of malaria. *Nature* 415: 680–685.

Sartwell, C. and R. E. Stevens. 1975. Mountain pine beetle in ponderosa pine: Prospects for silvicultural control in second-growth stands. *Journal of Forestry* 73: 136–140.

Schlenker, W. and M. J. Roberts. 2009. Nonlinear temperature effects indicate severe damages to U.S. crop yields under climate change. *Proceedings of the National Academy of Sciences USA* 106: 15594–15598.

Scholz, N. L., E. Fleishman, L. Brown, I. Werner, M. L. Johnson, M. L. Brooks, C. L. Mitchelmore, and D. Schlenk. 2012. A perspective on modern pesticides, pelagic fish declines, and unknown ecological resilience in highly managed ecosystems. *BioScience* 62: 428–434.

Schowalter, T. D. 2011. *Insect Ecology: An Ecosystem Approach*, 3rd ed. San Diego: Elsevier/Academic.

Schowalter, T. D. 2012. Ecology and management of bark beetles (Coleoptera: Curculionidae: Scolytinae) in southern pine forests. *Journal of Integrated Pest Management* 3(2): A1–A7.

Schowalter, T. D., R. N. Coulson, R. H. Turnbow, and W. S. Fargo. 1982. Accuracy and precision of procedures for estimating populations of the southern pine beetle (Coleoptera: Scolytidae) by using host tree correlates. *Journal of Economic Entomology* 75: 1009–1016.

Schowalter, T. D., D. N. Pope, R. N. Coulson, and W. S. Fargo. 1981. Patterns of southern pine beetle (*Dendroctonus frontalis* Zimm.) infestation enlargement. *Forest Science* 27: 837–849.

Schowalter, T. D. and P. Turchin. 1993. Southern pine beetle infestation development: Interaction between pine and hardwood basal areas. *Forest Science* 39: 201–210.

Sexton, J. M. and T. D. Schowalter. 1991. Physical barriers to reduce *Lepesoma lecontei* (Coleoptera: Curculionidae) damage to conelets in a Douglas-fir seed orchard in western Oregon. *Journal of Economic Entomology* 84: 212–214.

Showler, A. T. 2002. A summary of control strategies for the desert locust, *Schistocerca gregaria* (Forskål). *Agriculture, Ecosystems and Environment* 90: 97–103.

Showler, A. T. 2009. Roles of host plants in boll weevil range expansion beyond tropical Mesoamerica. *American Entomologist* 55: 234–242.

Shrewsbury, P. M. and M. J. Raupp. 2006. Do top-down or bottom-up forces determine *Stephanitis pyrioides* abundance in urban landscapes? *Ecological Applications* 16: 262–272.

Singh, G. and S. Prakash. 2009. Efficacy of *Bacillus sphaericus* against larvae of malaria and filarial vectors: An analysis of early resistance detection. *Parasitology Research* 104: 763–766.

Skovmand, O., T. D. A. Ouedraogo, E. Sanogo, H. Samuelsen, L. P. Toé, and T. Baldet. 2009. Impact of slow-release *Bacillus sphaericus* granules on mosquito populations followed in a tropical urban environment. *Journal of Medical Entomology* 46: 67–76.

Smith, C. M. 2005. *Plant Resistance to Arthropods: Molecular and Conventional Approaches.* Dordrecht, The Netherlands: Springer.

Smith, R. C. 1954. An analysis of 100 years of grasshopper populations in Kansas (1854 to 1954). *Transactions of the Kansas Academy of Science* 57: 397–433.

Smith, R. H. 2007. *History of the Boll Weevil in Alabama.* Alabama Agricultural Experiment Station Bulletin 670. Auburn, AL: Auburn University.

Snow, R. W., C. A. Guerra, J. J. Mutheu, and S. I. Hay. 2008. International funding for malaria control in relation to populations at risk of stable *Plasmodium falciparum* transmission. *PLoS Medicine* 5(7): e142.

Soboll, A., M. Elbers, R. Barthel, J. Schmude, A. Ernst, and R. Ziller. 2011. Integrated regional modelling and scenario development to evaluate future water demand under global change conditions. *Mitigation and Adaptation Strategies for Global Change* 16: 477–498.

Sparks, T. C. 1996. Toxicology of insecticides and acaricides. In *Cotton Insects and Mites: Characterization and Management*, E. G. King, J. R. Phillips, and R. J. Coleman, eds. 283–322. Memphis, TN: The Cotton Foundation.

Stambaugh, M. C., R. P. Guyette, E. R. McMurry, E. R. Cook, D. M. Meko, and A. R. Lupo. 2011. Drought duration and frequency in the U.S. Corn Belt during the last millennium (AD 992–2004). *Agricultural and Forest Meteorology* 151: 154–162.

Stephen, F. M. and C. W. Berisford. 2011. Biological control of southern pine beetle. In *Southern Pine Beetle. II. U.S. Forest Service Southern Research Station General Technical Report SRS-140*, R. N. Coulson and K. D. Klepzig, eds. 415–427. Asheville, NC: U.S. Forest Service Southern Research Station.

Stern, V. M., R. F. Smith, R. van den Bosch, and K. S. Hagen. 1959. The integration of chemical and biological control of the spotted alfalfa aphid. Part 1. The integrated control concept. *Hilgardia* 29: 81–101.

Suarez, A. V., D. A. Holway, and T. J. Case. 2001. Patterns of spread in biological invasions dominated by long-distance jump dispersal: Insights from Argentine ants. *Proceedings of the National Academy of Sciences USA* 98: 1095–1100.

Syed, Z. and W. S. Leal. 2009. Acute olfactory response of Culex mosquitoes to a human- and bird-derived attractant. *Proceedings of the National Academy of Sciences USA* 106: 18803–18808.

Symondson, W. O. C., K. D. Sunderland, and M. H. Greenstone. 2002. Can generalist predators be effective biocontrol agents? *Annual Review of Entomology* 47: 561–594.

Thaler, J. S., M. J. Stout, R. Karban, and S. S. Duffey. 2001. Jasmonate-mediated induced plant resistance affects a community of herbivores. *Ecological Entomology* 26: 312–324.

Therrell, M. D., D. W. Stahle, and R. Acuña-Soto. 2004. Aztec drought and the "curse of one rabbit." *Bulletin of the American Meteorological Society* 85: 1263–1272.

Thies, C., I. Steffan-Dewenter, and T. Tscharntke. 2003. Effects of landscape context on herbivory and parasitism at different spatial scales. *Oikos* 101: 18–25.

Thompson, D. C. and K. T. Gardner. 1996. Importance of grasshopper defoliation period on southwestern blue grama-dominated rangeland. *Journal of Range Management* 49: 494–498.

Tilman, D. and J. A. Downing. 1994. Biodiversity and stability in grasslands. *Nature* 367: 363–365.

Torell, L. A., J. H. Davis, E. W. Huddleston, and D. C. Thompson. 1989. Economic injury levels for interseasonal control of rangeland insects. *Journal of Economic Entomology* 82: 1289–1294.

Trumble, J. T., D. M. Kolodny-Hirsch, and I. P. Ting. 1993. Plant compensation for arthropod herbivory. *Annual Review of Entomology* 38: 93–119.

Tscharntke, T., R. Bommarco, Y. Clough, T. O. Crist, T. Kleijn, T. A. Rand, J. M. Tylianakis, S. van Nouhoys, and S. Vidal. 2007. Conservation biological control and enemy diversity on a landscape scale. *Biological Control* 43: 294–309.

Tscharntke, T., Y. Clough, T. C. Wanger, L. Jackson, I Motzke, I. Perfecto, J. Vandermeer, and A. Whitbread. 2012. Global food security, biodiversity conservation, and the future of agricultural intensification. *Biological Conservation* 151: 53–59.

Tscharntke, T., A. M. Klein, A. Kruess, I. Steffan-Dewenter, and C. Thies. 2005. Landscape perspectives on agricultural intensification and biodiversity—ecosystem service management. *Ecology Letters* 8: 857–874.

Tumlinson, J. H., D. D. Hardee, R. C. Gueldner, A. C. Thompson, P. A. Hedin, and J. P. Minyard. 1969. Sex pheromones produced by male boll weevils: Isolation, identification, and synthesis. *Science* 166: 1010–1012.

Vacas, S., C. Alfaro, V. Navarro-Llopis, M. Zarzo, and J. Primo. 2009. Study on the optimal pheromone release rate for attraction of *Chilo suppressalis* (Lepidoptera: Pyralidae). *Journal of Economic Entomology* 102: 1094–1100.

van Lenteren, J. C., J. Bale, F. Bigler, H. M. T. Hokkanen, and A. J. M. Loomans. 2006. Assessing risks of releasing exotic biological control agents of arthropod pests. *Annual Review of Entomology* 51: 609–634.

Vargas, R. I., J. C. Piñero, R. F. L. Mau, J. D. Stark, M. Hertlein, A. Mafra-Neto, R. Coler, and A. Getchell. 2009. Attraction and mortality of oriental fruit flies to SPLAT-MAT-methyl eugenol with spinosad. *Entomologia Experimentalis et Applicata* 131: 286–293.

Vincent, C., G. Hallman, B. Panneton, and F. Fleurat-Lessard. 2003. Management of agricultural insects with physical control methods. *Annual Review of Entomology* 48: 261–281.

Visser, M. E. and L. J. M. Holleman. 2001. Warmer springs disrupt the synchrony of oak and winter moth phenology. *Proceedings of the Royal Society* B 268: 289–294.

Worldwatch Institute. 2012. The state of consumption today. http://www.world watch.org/node/810.

Wratten, S. D., D. F. Hochuli, G. M. Gurr, J. Tylianakis, and S. L. Scarratt. 2007. Conservation, biodiversity, and integrate pest management. In *Perspectives in Ecological Theory and Integrated Pest Management*, M. Kogan and P. Jepson, eds. 223–245. Cambridge, UK: Cambridge University Press.

Xue, Y., K. N. Liou, and A. Kashahara. 1990. Investigation of biophysical feedback on the African climate using a two-dimensional model. *Journal of Climate* 3: 337–352.

Yamanaka, T. and A. M. Liebhold. 2009. Spatially implicit approaches to understand the manipulation of mating success for insect invasion management. *Population Ecology* 51: 427–444.

Yang, Y., M. Huang, G. A. C. Beattie, Y. Xia, G. Ourang, and J. Xiong. 2006. Distribution, biology, ecology, and control of the psyllid *Diaphorina citri* Kuwayama, a major pest of citrus: A status report for China. *International Journal of Pest Management* 52: 343–352.

Yen, A. L. 2009. Entomophagy and insect conservation: Some thoughts for digestion. *Journal of Insect Conservation* 13: 667–670.

Zalucki, M. P., D. Adamson, and M. J. Furlong. 2009. The future of IPM: Whither or wither? *Australian Journal of Entomology* 48: 85–96.

Zangerl, A. R., D. McKenna, C. L. Wraight, M. Carroll, P. Ficarello, R. Warner, and M. R. Berenbaum. 2001. Effects of exposure to event 176 *Bacillus thuringiensis* corn pollen on monarch and black swallowtail caterpillars under field conditions. *Proceedings of the National Academy of Sciences USA* 98: 11908–11912.

Zhang, D. D., P. Brecke, H. F. Lee, Y. Q. He, and J. Zhang. 2007. Global climate change, war, and population decline in recent human history. *Proceedings of the National Academy of Sciences USA* 104: 19214–19219.

Zheng, X. and E. A. B. Eltahir. 1998. The role of vegetation in the dynamics of West African monsoons. *Journal of Climate* 11: 2078–2096.

Zhou, H.-B., J.-L. Chen, D.-F. Cheng, Y. Liu, and J.-R. Sun. 2009. Effects of wheat-pea intercropping on the population dynamics of *Sitobion avenae* (Homoptera: Aphididae) and it main natural enemies. *Acta Entomologica Sinica* 52: 775–782.

Index

A

Abiotic changes, direct effects of, 67
  air and water quality, 69–70
  precipitation extremes, 68–69
  temperature extremes, 67–68
  wind speed and water flow, 69
Acidic precipitation, 195
Actias luna, 215
Acyrthosiphum pisum, 153
Adaptive attributes, 53–55
Adelges tsugae, 58, 200
Aedes aegypti, 39, 41
Aedes albopictus, 119–120
Aeneolamia albofasciata, 66
Aflatoxins, 74
Agelastica alni, 69
Agrilus planipennis, 200
Alabama argillacea, 97
Alfalfa aphid, 153
Alkaloids, 76
Allee effect, 117
Alnus rubra, 246
Altered site of action resistance, 210–211
Anhydrobiosis, 68
Anopheles darlingi, 255, 256
Anopheles gambiae, 40, 194
Anopheles stephensi, 212
Anoplophora glabripennis, 200
Anthocharis cardamines, 59
Anthonomus grandis, 43, 106, 198
Anthropogenic changes and
    management, 189
  alternative control options, 214–216
  ecosystems, effects of humans on, 190
    disturbances, 190–193
    fragmentation and conversion,
      201–205
    global change, 193–196
    insect problems, induction of,
      205–208
    pollution and invasive species,
      196–201
  insecticide effects, 208–214

Anthropogenic extermination of
    species, 173
Anthropogenic versus natural
    landscapes, 192–193, 192f
Ants. *See specific types*
Aphaenogaster barbigula, 250
Aphaenogaster cockerelli, 158
Aphidius ervi, 153
Aphids, control of, 301
Aphis glycines, 196, 301
Apis mellifera, 199, 283
Aquatic ecosystems, 148–149
Aquatic insects, 238
Argentine ant, 200
Army ant, 71, 121
Armyworm, 152
Arthropod communities, 169
Asclepias humistrata, 73
Asian citrus psyllid, 313
Asian longhorned beetle, 200
Asphondylia borrichiae, 116
Atrax spp., 39
Atta laevigata, 71, 199f
Atta vollenweideri, 170, 250
Avermectins, 209

B

Bacillus thuringiensis, 83, 210, 213, 214
Bacteriovores, 235
Baculovirus spp., 1
Ballooning, 57
Balsam fir sawfly, 78
Bark beetles, 59, 67, 104, 169
  control of, 299
Barrier treatments, 305, 313–314
Bedbugs, 36
Beech bark disease fungus, 200
Beech scale insect, 200
Beehives, 44
Beekeeping in trees, 25
Beeswax, 26, 27, 30–31
Beetles. *See specific types*

Printed and bound by CPI Group (UK) Ltd, Croydon, CR0 4YY

01/11/2024

01782625-0006